Springer Monographs in Mathematics

For further volumes:
http://www.springer.com/series/3733

Leonid P. Lebedev • Iosif I. Vorovich
Michael J. Cloud

Functional Analysis in Mechanics

Second Edition

Leonid P. Lebedev
Department of Mathematics
National University of Colombia
Bogotá, Colombia

Iosif I. Vorovich†
Department of Mathematics & Mechanics
Rostov State University
Rostov-on-Don, Russia

Michael J. Cloud
Department of Electrical Engineering
Lawrence Technological University
Southfield, MI, USA

ISSN 1439-7382
ISBN 978-1-4899-9756-2 ISBN 978-1-4614-5868-5 (eBook)
DOI 10.1007/978-1-4614-5868-5
Springer New York Heidelberg Dordrecht London

Mathematics Subject Classification (2010): 4601, 4602, 7401, 7402, 74B05, 74K25, 35JXX, 35A01, 35A02

Printed on acid-free paper

Springer is part of Springer Science+Business Media (www.springer.com)

Preface

In Russia, a university Mechanics department will typically exist within a "Mathematical Faculty." Such a department is not an engineering department in the western sense, but is something intermediate between a mathematics department and an engineering department. It will offer courses on calculus, linear algebra, analysis, differential geometry, differential equations, and so on, along with extensive courses on analytical mechanics, the strength of materials, continuum mechanics, elasticity, fluid mechanics, and more specialized subjects.

When the first author of this book was a student of the second author, functional analysis was not in the curriculum for mechanicists. In 1971, Professor Vorovich offered a short course on functional analysis to a broad audience consisting of mathematicians and mechanicists, students and professors. It included a simple and minimal introduction to the theory of Banach and Hilbert spaces that opened the door to understanding (with some difficulty on the part of the non-mathematicians) certain interesting applications in mechanics. The mathematicians were surprised at how abstract theorems could be applied to mechanics and, moreover, that these theorems could actually be rooted in mechanics. It was emphasized that strain energy is not only a physical notion, but a measure by which a norm and inner product can be imposed on the set of deformations or velocities of a body. This idea was developed by Vorovich in his doctoral dissertation on the nonlinear theory of elastic shallow shells in 1957, but was imbedded in just a few long examples. Many subsequent publications of the idea were made in *Doklady AN USSR*, the central scientific journal of the USSR, where the results were presented without proof.

Later, Professor Vorovich's lectures were extended and became the basis for a regular course at Rostov State University and many other institutions across the USSR. The course contents, including the applications considered, continued to evolve, but the present book preserves the main ideas of the original course. As the course was just one semester in length, it contained only a minimal subset of the abstract theory that enabled students to understand the applications. The first edition of this book contained some abstract material that was not presented in the course. This second edition includes more extended coverage of the classical, abstract portions of functional analysis, as well as additional mechanics problems. Taken together, the first three chapters now constitute a regular text on applied functional analysis. This potential use of the book is supported by a significantly extended set of exercises with hints and solutions.

The Introduction (pages 1–7) is an unnumbered chapter with single-digit internal numbering of its equations, examples, and theorems. Chapters 1 through 4 employ a three-digit scheme where the first two numbers are the chapter and section numbers (hence Remark 1.3.2 is the second labeled remark in Section 1.3). The Introduction closes with some additional remarks on notation.

Acknowledgments. We would like to thank Senior Editor Achi Dosanjh and Associate Editor Donna Chernyk for assistance with this second edition. We are also grateful to Elena Vorovich and to our wives, Natasha Lebedeva and Beth Lannon-Cloud, for their understanding and support.

Bogota, Colombia *Leonid Lebedev*

Okemos, Michigan, USA *Michael Cloud*

August 2012

Contents

Introduction

Long ago it was traditional to apply mathematics only to mechanics and physics. Now it difficult to find an area of knowledge in which mathematics is not used as a tool to create new models and to simulate them. This is due mainly to the fantastic ability of computers to process models having thousands of parameters.

Fortunately, mathematics tends to produce methods of great generality. Functional analysis, in particular, allows us to approach different mathematical facts and methods from a unified point of view. Let us consider some examples.

Example 1. A system of linear algebraic equations

$$x_i = \sum_{j=1}^{n} a_{ij}x_j + c_i \qquad (i = 1, \ldots, n) \tag{1}$$

can be solved by an iterative scheme of the form

$$\begin{aligned} x_i^{(0)} &= c_i\,, \\ x_i^{(k+1)} &= \sum_{j=1}^{n} a_{ij}x_j^{(k)} + c_i \qquad (i = 1, \ldots, n, \quad k = 0, 1, 2, \ldots)\,. \end{aligned}$$

To establish convergence, let us consider the difference

$$x_i^{(k+1)} - x_i^{(k)} = \sum_{j=1}^{n} a_{ij}[x_j^{(k)} - x_j^{(k-1)}]\,.$$

We have

$$\max_{1\le i\le n} |x_i^{(k+1)} - x_i^{(k)}| \le \max_{1\le i\le n} \sum_{j=1}^{n} |a_{ij}|\,|x_j^{(k)} - x_j^{(k-1)}| \le q \cdot \max_{1\le j\le n} |x_j^{(k)} - x_j^{(k-1)}|$$

where

$$q = \max_{1 \le i \le n} \sum_{j=1}^{n} |a_{ij}| .$$

If $q < 1$, then convergence is ensured and a solution to (1) is $(z_1, \ldots, z_n)$ where

$$z_i = \lim_{k \to \infty} x_i^{(k)} \qquad (i = 1, \ldots, n) .$$

Now consider a system of integral equations

$$x_i(t) = \sum_{j=1}^{n} \int_0^1 a_{ij}(t, s) x_j(s)\, ds + c_i(t) \qquad (i = 1, \ldots, n) \tag{2}$$

where $c_i(t)$ and $a_{ij}(t, s)$ are given continuous functions on the interval $[0, 1]$ and the square $[0, 1] \times [0, 1]$, respectively. The iterative scheme

$$\begin{aligned} x_i^{(0)}(t) &= c_i(t), \\ x_i^{(k+1)}(t) &= \sum_{j=1}^{n} \int_0^1 a_{ij}(t, s) x_j^{(k)}(s)\, ds + c_i(t) \qquad (i = 1, \ldots, n) \end{aligned}$$

produces functions $x_i^{(k)}(t)$ that satisfy

$$x_i^{(k+1)}(t) - x_i^{(k)}(t) = \sum_{j=1}^{n} \int_0^1 a_{ij}(t, s)[x_j^{(k)}(s) - x_j^{(k-1)}(s)]\, ds .$$

We have

$$|x_i^{(k+1)}(t) - x_i^{(k)}(t)| \le \sum_{j=1}^{n} \int_0^1 |a_{ij}(t, s)|\, |x_j^{(k)}(s) - x_j^{(k-1)}(s)|\, ds$$

so that

$$\max_{\substack{1 \le i \le n \\ 0 \le t \le 1}} |x_i^{(k+1)}(t) - x_i^{(k)}(t)| \le q \cdot \max_{\substack{1 \le j \le n \\ 0 \le s \le 1}} |x_j^{(k)}(s) - x_j^{(k-1)}(s)|$$

where

$$q = \max_{\substack{1 \le i \le n \\ 0 \le t \le 1}} \sum_{j=1}^{n} \int_0^1 |a_{ij}(t, s)|\, ds .$$

If $q < 1$, the component sequences $\{x_i^{(k)}(t)\}$ $(i = 1, \ldots, n)$ are uniformly convergent on $[0, 1]$. Hence a solution to (2) is $(z_1(t), \ldots, z_n(t))$ where

$$z_i(t) = \lim_{k \to \infty} x_i^{(k)}(t) \qquad (i = 1, \ldots, n) .$$

The obvious similarity between the treatments of (1) and (2) suggests that a general approach might cover these and other cases of interest. Later we will present the generalization, known as Banach's contraction mapping principle.

Example 2. In what follows, we deal mainly with spaces of infinite dimension. For example, the wave equation

$$\frac{\partial^2 u}{\partial t^2} = \frac{\partial^2 u}{\partial x^2} \tag{3}$$

describes the vibrations $u = u(x,t)$ of a stretched string. Let the ends of the string be fixed:

$$u(0,t) = u(1,t) = 0 .$$

It is natural to seek a solution with finite potential and kinetic energies, i.e., with

$$\int_0^1 \left(\frac{\partial u}{\partial x}\right)^2 dx < \infty , \qquad \int_0^1 \left(\frac{\partial u}{\partial t}\right)^2 dx < \infty .$$

We could seek a solution in the form of a Fourier series

$$u(x,t) = \sum_{k,m} A_{km} \sin \pi kx \sin \pi mt .$$

This solution is evidently described by a countable set of numbers A_{km}, which can be regarded as the components of a vector having infinitely many components. The set of such "vectors" clearly constitutes a space that is infinite dimensional.

One difficulty in dealing with an infinite dimensional space is that the Bolzano–Weierstrass principle (that any bounded infinite sequence contains a convergent subsequence) breaks down. For example, we cannot select a convergent subsequence from the bounded sequence of functions $\{\sin kx\}$.

Example 3. In contemporary mathematical physics, generalized solutions are typical. Without going into too much detail, we may briefly consider the problem of a beam with clamped ends bending under a load $q(x)$. A corresponding boundary value problem is

$$(B(x)y''(x))'' - q(x) = 0 , \qquad y(0) = y'(0) = y(l) = y'(l) = 0 , \tag{4}$$

where $B(x)$ and l are the stiffness and length, respectively, of the beam. This formulation supposes $y = y(x)$ to possess derivatives up to fourth order.

The same boundary value problem can be posed differently through the use of variational principles. It can be shown that the total potential energy functional of the beam, defined by

$$I(y) = \frac{1}{2} \int_0^l [B(y'')^2 - 2q(x)y]\, dx ,$$

assumes a minimum value at an equilibrium state of the beam (here all the functions $y(x)$ under consideration must satisfy the boundary conditions stated in (4)). The first variation of $I(y)$,

$$\delta I(y, \varphi) = \int_0^l [B(x)y''(x)\varphi''(x) - q(x)\varphi(x)]\, dx ,$$

vanishes for any sufficiently smooth function $\varphi(x)$ satisfying the boundary conditions

$$\varphi(0) = \varphi'(0) = \varphi(l) = \varphi'(l) = 0 \tag{5}$$

if $y(x)$ satisfies (4); that is,

$$\delta I(y, \varphi) = 0 . \tag{6}$$

A function $y(x)$ is called a generalized solution to the problem (4) if equation (6) holds for any sufficiently smooth function $\varphi(x)$ satisfying the conditions (5). So a generalized solution, which can have no more than two derivatives on the interval $[0, l]$, satisfies the equilibrium equation in the sense of Lagrange's variational principle. For a moving system, we can introduce generalized solutions using Hamilton's variational principle. All we require of such a solution is that we can calculate its potential and kinetic energy.

Since the smoothness restrictions for generalized solutions are milder than those for classical solutions, the above approach extends the circle of problems we may investigate and solve numerically. In particular, problems with non-smooth loads occur in applications. The generalized approach also arises naturally when we study convergence of the finite element method (FEM)— a powerful tool in mathematical physics. In the FEM, a generalized problem setup is called a *weak setup*. We prefer the term *energy setup*, as this type of setup can be obtained from the well-known energetic variational principles of mechanics.

At this point we hope the reader has begun to picture functional analysis as a powerful tool in applications. It emerged as a generalization of certain aspects of classical linear algebra and mathematical analysis to infinite dimensional cases, incorporating portions of the calculus of variations and topology. With an aim toward applications in the various mathematical sciences, it also inherited ideas from physics and, in particular, mechanics. In Chapter 1 we will present a more systematic study of its fundamentals.

We would like to add a point for the practitioner who is new to functional analysis and doubtful of its utility. When an engineer deals with some mechanical structure, he carries a mental image of how the structure deforms or behaves dynamically under a given range of loads. Rather than regarding the solution to his boundary value problem as a collection of pointwise calculated values, he concerns himself with how the solution *as a whole object* changes under certain changes in the loads. This is very similar to how one considers a mechanical problem by the methods of functional analysis. In this approach, a load is defined pointwise but is considered as a whole object — an element F of a functional space. Next, a unique characteristic of the structure's response to this load is identified; suppose it is the displacement field u corresponding to the load F. While u is a function (or vector function) depending on the space coordinates and possibly on time, it can also be regarded as a whole object — another element of some functional space. The relation between F and u can be written quite simply as

$$u = u(F) , \tag{7}$$

provided it is remembered that this is not a pointwise dependence: here, u in total depends on F as the entire set of loads acting on the structure. This viewpoint on the correspondence is more or less similar to the way an engineer imagines structural behavior. On one hand, (7) is a convenient representation for the solution of the problem. On the other hand, it should be understood that this representation, even for problems of linear mechanics, is much more complicated than those encountered in the finite dimensional problems of linear algebra. So it is useful to know which properties of a problem in linear algebra can be carried over for use with the continuous problems of linear mechanics and, just as importantly, which ones cannot. Such questions, pertaining to general relations of the type (7), may be pursued through the tools of functional analysis.

Let us close this introduction by presenting three important theorems from classical analysis. They are all named after Karl Weierstrass (1815–1897) and find frequent application in the book. Recall that in $\mathbb{R}^k$, the term *compact set* refers to a closed and bounded set.

Theorem 1. Let $f(\mathbf{x})$ be a function continuous on a compact set $\Omega \subset \mathbb{R}^k$. Then $f(\mathbf{x})$ is bounded on Ω and attains its supremum and infimum on Ω. That is,

1. there exists a constant c such that $|f(\mathbf{x})| \le c$ for all $\mathbf{x} \in \Omega$, and
2. there exist points $\mathbf{x}_* \in \Omega$ and $\mathbf{x}^* \in \Omega$ such that

$$f(\mathbf{x}_*) = \inf_{\mathbf{x}\in\Omega} f(\mathbf{x}) , \qquad f(\mathbf{x}^*) = \sup_{\mathbf{x}\in\Omega} f(\mathbf{x}) .$$

Hence the values $\max_{\mathbf{x}\in\Omega} f(\mathbf{x})$ and $\min_{\mathbf{x}\in\Omega} f(\mathbf{x})$ exist.

Theorem 2. Suppose a sequence $\{f_n(\mathbf{x})\}$ of functions continuous on a compact set $\Omega \subset \mathbb{R}^k$ converges uniformly; that is, for any $\varepsilon > 0$ there is an integer $N = N(\varepsilon)$ such that

$$|f_{n+m}(\mathbf{x}) - f_n(\mathbf{x})| < \varepsilon$$

for any $n > N$, $m > 0$, and $\mathbf{x} \in \Omega$. Then the limit function

$$f(\mathbf{x}) = \lim_{n\to\infty} f_n(\mathbf{x})$$

exists and is continuous on Ω.

Theorem 3. Let $f(\mathbf{x})$ be a function continuous on a compact set $\Omega \subset \mathbb{R}^k$. For any $\varepsilon > 0$ there is a polynomial $P_n(\mathbf{x})$ of the nth degree such that

$$|f(\mathbf{x}) - P_n(\mathbf{x})| < \varepsilon$$

for any $\mathbf{x} \in \Omega$.

Some familiarity with the main inequalities of analysis is assumed. These are summarized in an appendix beginning on p. 265.

Some Remarks on Notation

In this book we employ standard set and logic symbols such as

$\{x\colon P\}$	set of all x having property P
$\overline{S}$	closure of a set S
$\in$	set membership
$\subseteq, \subset$	subset relation, proper subset relation
$\cup$	union
$\cap$	intersection
$A \setminus B$	set difference
$\mathbb{R}$	set of real numbers
$\mathbb{Q}$	set of rational numbers
$\Longrightarrow$	logical implication
$\Longleftrightarrow$	logical equivalence (if and only if)

A proof, remark, or problem is punctuated with a right-justified empty square: □

We denote sequences using braces. The notation $\{x_n\} \subset S$ means that $\{x_n\}$ is a sequence with $x_n \in S$ for each n. When the range of indices must be specified more precisely we use notation such as $\{x_n\}_{n=0}^{\infty}$ or x_n $(n = 0, 1, 2, \ldots)$.

Subsequences play an important role in our development. They are denoted using compound subscripts, with a couple of distinct conventions employed where convenient.

1. Given a sequence $\{x_n\}$, we may choose an element x_{i_1}, a second element x_{i_2} (with $i_2 > i_1$), and so forth, thereby selecting a subsequence denoted by $\{x_{i_k}\}$:

$$\{x_{i_k}\} \equiv x_{i_1}\,,\ x_{i_2}\,,\ x_{i_3}\,,\ \ldots.$$

   Note that $i_k \geq k$ for each k. In such cases, for simplicity, we often renumber the resulting subsequence immediately and rename it as $\{x_n\}$.

2. In other cases, however, we begin with a sequence $\{x_n\}$ and "sift" it repeatedly with the aim of constructing a *diagonal sequence*. In such cases the notation $\{x_{i_1}\}$ will be used for the subsequence that results from the first sifting,

$$\{x_{i_1}\} = x_{1_1}\,,\ x_{2_1}\,,\ x_{3_1}\,,\ \ldots,$$

   the notation $\{x_{i_2}\}$ will be used for the subsequence that results from the second sifting,

$$\{x_{i_2}\} = x_{1_2}\,,\ x_{2_2}\,,\ x_{3_2}\,,\ \ldots,$$

   and so forth. The nth sifting step $(n = 1, 2, 3, \ldots)$ is done in such a way that the resulting subsequence satisfies a certain property, $\mathcal{P}_n$ say. Because the diagonal sequence

$$\{x_{n_n}\} = x_{1_1}\,,\ x_{2_2}\,,\ x_{3_3}\,,\ \ldots$$

can be considered as a subsequence of any of the sifted subsequences, it satisfies all of the properties $\mathcal{P}_n$ for $n = 1, 2, 3, \ldots$.

In contrast to authors who use $C(\Omega)$ to denote the set of functions continuous on a domain Ω, we reserve the symbol $C(\Omega)$ for the *space* of functions continuous on Ω with the sup metric. As Ω is a closed and bounded set throughout most of our discussion, the max metric is usually appropriate. The symbol $\overline{\Omega}$, for the closure of Ω, appears in just a few spots where Ω is permitted to be open for comparison with the treatments of certain topics in other books. On those occasions when we impose another type of metric on the set of continuous functions on Ω, we refrain from using the symbol $C(\Omega)$ for the resulting metric space. Even in such cases, however, we may still use statements such as $f \in C(\Omega)$ to indicate that f is continuous on Ω (while making no reference to the usual max metric on the space $C(\Omega)$).

The norm symbol $\|\cdot\|$ is formally introduced on p. 39, but is used a few times prior to that on the assumption that the reader has encountered it in more elementary courses. The same holds for the terms *operator* and *functional*, introduced formally on p. 72.

We avoid the operator notation

$$f\colon D(f) \subseteq X \to Y\,,$$

preferring to say that f *acts from* X *to* Y (or, if $Y = X$, that f *acts in* X). By this we do not necessarily mean that f is defined on all of X.

When space permits, we write out partial derivatives as

$$\frac{\partial u}{\partial x}\,,\quad \frac{\partial^2 u}{\partial x^2}\,,\quad \text{and so on.}$$

Otherwise we make use of the more compact subscript notation u_x, u_{xx}, etc.

The *Kronecker delta* symbol

$$\delta_{mn} = \begin{cases} 1\,, & m = n\,, \\ 0\,, & m \neq n\,, \end{cases}$$

is used when convenient.

Chapter 1
Metric, Banach, and Hilbert Spaces

1.1 Preliminaries

Consider a set of particles $P_1, \ldots, P_n$. To locate these particles in the space $\mathbb{R}^3$, we need a reference system. Let the Cartesian coordinates of particle P_i be (ξ_i, η_i, ζ_i). Identifying (ξ_1, η_1, ζ_1) with the triple (x_1, x_2, x_3), (ξ_2, η_2, ζ_2) with (x_4, x_5, x_6), and so on, we obtain a vector $\mathbf{x}$ of the Euclidean space $\mathbb{R}^{3n}$ with coordinates $(x_1, x_2, \ldots, x_{3n})$. Then $\mathbf{x}$ determines the positions of all particles in the set.

To distinguish different configurations $\mathbf{x}$ and $\mathbf{y}$ of the system, we can introduce a distance from $\mathbf{x}$ to $\mathbf{y}$:

$$d_E(\mathbf{x}, \mathbf{y}) = \left[\sum_{i=1}^{3n} (x_i - y_i)^2 \right]^{1/2} .$$

This is the *Euclidean distance* (or *metric*) of $\mathbb{R}^{3n}$. Alternatively, we could characterize the distance from $\mathbf{x}$ to $\mathbf{y}$ using the function

$$d_S(\mathbf{x}, \mathbf{y}) = \max_{1 \le i \le 3n} |x_i - y_i| .$$

It is easily seen that d_E and d_S each satisfy the following properties, known as the *metric axioms*:

D1. $d(\mathbf{x}, \mathbf{y}) \ge 0$;

D2. $d(\mathbf{x}, \mathbf{y}) = 0$ if and only if $\mathbf{x} = \mathbf{y}$;

D3. $d(\mathbf{x}, \mathbf{y}) = d(\mathbf{y}, \mathbf{x})$;

D4. $d(\mathbf{x}, \mathbf{y}) \le d(\mathbf{x}, \mathbf{z}) + d(\mathbf{z}, \mathbf{y})$ for any third vector $\mathbf{z}$.

Any real-valued function $d(\mathbf{x}, \mathbf{y})$ defined for all $\mathbf{x}, \mathbf{y} \in \mathbb{R}^{3n}$ is called a *metric* on $\mathbb{R}^{3n}$ if it satisfies the properties D1–D4. Property D1 is the *axiom of positiveness*, property D3 is the *axiom of symmetry*, and property D4 is the *triangle inequality*.

Problem 1.1.1. Let a real-valued function $d(\mathbf{x}, \mathbf{y})$ be defined for all $\mathbf{x}, \mathbf{y} \in \mathbb{R}^n$. Show that if d satisfies D2, D3, and D4, then it also satisfies D1. Confirm that this does not depend on the nature of the elements $\mathbf{x}$ and $\mathbf{y}$.[1] □

Remark 1.1.1. It follows from Problem 1.1.1 that the metric axioms can be restricted to just D2, D3, and D4. □

The metrics d_E and d_S are *equivalent* relative to sequence convergence in $\mathbb{R}^{3n}$, since there exist positive constants m_1 and m_2 independent of $\mathbf{x}$ and $\mathbf{y}$ such that

$$0 < m_1 \leq \frac{d_E(\mathbf{x}, \mathbf{y})}{d_S(\mathbf{x}, \mathbf{y})} \leq m_2 < \infty \tag{1.1.1}$$

whenever $\mathbf{x}, \mathbf{y} \in \mathbb{R}^{3n}$ and $\mathbf{x} \neq \mathbf{y}$. So

$$\lim_{k\to\infty} d_E(\mathbf{x}_k, \mathbf{x}) = 0 \iff \lim_{k\to\infty} d_S(\mathbf{x}_k, \mathbf{x}) = 0 .$$

Remark 1.1.2. In what follows, we shall use notation such as "m_i" for those constants whose exact values are not important. An example occurred in (1.1.1). □

Inequality (1.1.1) shows that, in a certain way, $d_E(\mathbf{x}, \mathbf{y})$ and $d_S(\mathbf{x}, \mathbf{y})$ have the same standing as metrics on $\mathbb{R}^{3n}$. We can introduce other functions on $\mathbb{R}^{3n}$ satisfying axioms D1–D4, e.g.,

$$d_p(\mathbf{x}, \mathbf{y}) = \left(\sum_{i=1}^{3n} |x_i - y_i|^p\right)^{1/p} \qquad (p = \text{constant} \geq 1)$$

and

$$d_k(\mathbf{x}, \mathbf{y}) = \left(\sum_{i=1}^{3n} k_i\,|x_i - y_i|^2\right)^{1/2} \qquad (k_i > 0)\ .$$

Problem 1.1.2. Show that any two of the metrics introduced above are equivalent on $\mathbb{R}^n$. Note that two metrics $d_1(\mathbf{x}, \mathbf{y})$ and $d_2(\mathbf{x}, \mathbf{y})$ on $\mathbb{R}^n$ are equivalent if there exist m_1 and m_2 such that

$$0 < m_1 \leq \frac{d_1(\mathbf{x}, \mathbf{y})}{d_2(\mathbf{x}, \mathbf{y})} \leq m_2 < \infty \tag{1.1.2}$$

for any $\mathbf{x}, \mathbf{y} \in \mathbb{R}^n$ with $\mathbf{x} \neq \mathbf{y}$. □

In general, the equivalence of a pair of metrics is defined so that any sequence convergent relative to one metric is also convergent relative to the other one. The existence of constants $m_k > 0$ satisfying (1.1.2) is sufficient but not necessary; the reader may wish to find a pair of equivalent metrics on $\mathbb{R}^n$ for which (1.1.2) does not hold.

The notion of metric generalizes the notion of distance in $\mathbb{R}^3$. It can be applied to particle locations, velocities, accelerations, and masses in order to distinguish between different states of a given system or between different systems of particles.

[1] Hints for many of the problems are contained in an appendix beginning on p. 269.

The same holds for any mechanical system described by a finite number of parameters (such as forces or temperatures).

Let us extend the idea of distance to continuum problems. Take a string, with fixed ends, of length π. For a loaded string, we can use the Fourier expansion

$$u(s) = \sum_{k=1}^{\infty} x_k \sin ks \tag{1.1.3}$$

to describe the resulting deflection $u(s)$. Any state of the string can be identified with a vector $\mathbf{x}$ having infinitely many coordinates $x_1, x_2, \ldots$. The dimension of the space S of all such vectors is obviously not finite.

We can modify the metric of $\mathbb{R}^n$ to determine the distance from $\mathbf{x}$ to $\mathbf{y}$ in S. The necessary changes are evident; we can use

$$d(\mathbf{x}, \mathbf{y}) = \Big(\sum_{i=1}^{\infty} |x_i - y_i|^p \Big)^{1/p} \qquad (p \geq 1)$$

or

$$d(\mathbf{x}, \mathbf{y}) = \sup_i |x_i - y_i| \ .$$

The resulting distances satisfy D1–D4 and are metrics. The analogy between $\mathbb{R}^n$ and S is not perfect, however. Consider the distance from $\mathbf{0} = (0, 0, 0, \ldots)$ to $\mathbf{x}_0 = (1, 1/2, 1/3, \ldots)$ using the metrics

$$d_1(\mathbf{x}, \mathbf{y}) = \sum_{i=1}^{\infty} |x_i - y_i| \quad \text{and} \quad d_2(\mathbf{x}, \mathbf{y}) = \sup_i |x_i - y_i| \ .$$

Since

$$d_1(\mathbf{x}_0, \mathbf{0}) = \sum_{i=1}^{\infty} 1/i \quad \text{and} \quad d_2(\mathbf{x}_0, \mathbf{0}) = 1$$

and the series diverges, we do not have

$$d_1(\mathbf{x}_0, \mathbf{0}) / d_2(\mathbf{x}_0, \mathbf{0}) \leq m_2 < \infty \ .$$

So an inequality of the form (1.1.2) is not satisfied. Moreover, as convergence in one metric does not necessitate convergence in the other, the two metrics are not equivalent. This example also shows why a metric must be *well-defined*, i.e., uniquely defined and finite for any pair of elements $\mathbf{x}, \mathbf{y}$. In the case of S above, we can only introduce a metric on some subset of S over which the metric is well-defined. This yields various metric spaces of infinite dimensional vectors. For general sets, this idea is realized in the following

Definition 1.1.1. Let X be a nonempty set of elements (often called *points*). A function $d(x, y)$, uniquely defined and finite for all $x, y \in X$, is a *metric* on X if it satisfies the metric axioms D1–D4. We refer to the pair (X, d) as a *metric space*.

Although a metric space consists of a set X and a metric d, certain standard metric spaces (for which the metric is understood) can safely be denoted by the underlying set X alone.

Remark 1.1.3. In the following pages, we shall not distinguish between metric spaces based on the same set of elements if their metrics are equivalent. If d_1 and d_2 are equivalent metrics, we shall not distinguish between (X, d_1) and (X, d_2). However, if d_1 and d_2 are not equivalent, then (X, d_1) and (X, d_2) are different metric spaces. We also saw that metric spaces could be formed from the elements of S (the set of infinite dimensional vectors) only by restriction to certain subsets of S. So S is not a metric space, but merely a (linear) set of infinite dimensional vectors whose linear subsets (subspaces), together with their metrics, can constitute various metric spaces of vectors having infinitely many coordinates. □

In the definition of metric space the nature of the elements of the space is unimportant. The elements could be abstract objects, even ordinary objects such as chairs or tables — it is merely necessary that we can introduce, for each pair of elements of the set, a function satisfying the metric axioms. In mathematical physics, metric spaces of functions are often employed. These are the spaces to which solutions of certain equations or the parameters of a problem must belong. During the rigorous investigation of such problems, restrictions are always imposed on the properties of the solutions sought. This is due not only to a desire for rigor and formalism; a mathematical problem can have several solutions, certain parts of which may contradict our ideas about the nature of the process described by the problem. Additional restrictions based on the physical nature of the problem allow us to select physically reasonable solutions. One way to impose such restrictions is to require that the solution belong to a metric space. Thus the choice of space in which one seeks a solution can be crucial for addressing a realistic problem. Depending on this choice, we may or may not have existence and uniqueness of a solution, etc. Metric spaces in mathematical physics are usually linear and infinite dimensional.

We now list some metric spaces of infinite dimensional vectors

$$\mathbf{x} = (x_1, x_2, \ldots) .$$

A vector with infinitely many components may be regarded as a sequence $\mathbf{x} = \{x_i\}$, and the following spaces are often called *sequence spaces*.

1. *The metric space m.* The space m is the set of all bounded sequences $\mathbf{x}$. For $\mathbf{x} = (x_1, x_2, \ldots)$ and $\mathbf{y} = (y_1, y_2, \ldots)$, the metric is given by

$$d(\mathbf{x}, \mathbf{y}) = \sup_i |x_i - y_i| . \tag{1.1.4}$$

2. *The metric space ℓ^p.* The space ℓ^p $(p \geq 1)$ consists of the set of all sequences

$$\mathbf{x} = (x_1, x_2, \ldots) \quad \text{for which} \quad \sum_{i=1}^{\infty} |x_i|^p < \infty ,$$

along with the metric

$$d_p(\mathbf{x}, \mathbf{y}) = \left(\sum_{i=1}^{\infty} |x_i - y_i|^p \right)^{1/p} . \tag{1.1.5}$$

3. *The metric space c.* The space c is the linear subspace of m that consists of all convergent sequences; the metric is (1.1.4).

4. *The metric space c_0.* The space c_0 is the subspace of c consisting of all sequences converging to 0; again, the metric is (1.1.4).

The metrics of these spaces were introduced by analogy with metrics on $\mathbb{R}^n$. Similarly we can introduce (with no change in notation) spaces of infinite dimensional vectors with complex components. Of course, any proposed metric space must be checked for compliance with Definition 1.1.1. For the spaces m, c, c_0, and ℓ^1 this task is almost trivial. For ℓ^p with $p > 1$ (and even for $\mathbb{R}^n$ with the corresponding metrics) the task is not so simple and will be carried out later.

We now consider another class of metrics: the energy metrics.

5. *The energy space for a string.* The potential energy of a string is proportional to

$$\int_0^{\pi} \left(\frac{\partial u}{\partial s} \right)^2 ds = \pi \sum_{k=1}^{\infty} k^2 x_k^2 ,$$

x_k being defined by the representation (1.1.3). We can compare two states

$$u(s) = \sum_{k=1}^{\infty} x_k \sin ks \quad \text{and} \quad v(s) = \sum_{k=1}^{\infty} y_k \sin ks$$

of the string via the metric

$$d(u, v) \equiv d(\mathbf{x}, \mathbf{y}) = \left[\sum_{k=1}^{\infty} k^2 (x_k - y_k)^2 \right]^{1/2} . \tag{1.1.6}$$

The energy space for the string is the set of all sequences of Fourier coefficients x_k such that

$$\sum_{k=1}^{\infty} k^2 x_k^2 < \infty .$$

The metric is given by (1.1.6).

Problem 1.1.3. Show that (1.1.4) and (1.1.6) are indeed metrics on their respective sets. *Preliminary hint*: For (1.1.4) the task is simple. Now consider (1.1.6). For each element of the energy space for a string, defined by Fourier coefficients x_k, we introduce another infinite vector $\mathbf{z}$ with components $z_k = kx_k$. By the definition of the energy metric, it induces the metric of ℓ^2 over the set of all such $\mathbf{z}$. Because ℓ^2 is a metric space, so is the energy space of Fourier coefficients for the string. □

Energy spaces are advantageous when applied to mechanics problems, as we shall see later.

6. *The metric space of straight lines.* The notion of metric space is abstract, and the elements of the underlying set need not be vectors. Consider the set M of all straight lines in the plane which do not pass through the coordinate origin. A straight line L is given by the equation

$$x \cos \alpha + y \sin \alpha - p = 0 .$$

Let us show that

$$d(L_1, L_2) = \left[(p_1 - p_2)^2 + 4 \sin^2 \frac{\alpha_1 - \alpha_2}{2} \right]^{1/2}$$

is a metric on M. Axioms D1 and D3 are obviously satisfied. Consider D2. Certainly $d(L_1, L_2) = 0$ whenever $L_1 = L_2$. Conversely, $d(L_1, L_2) = 0$ implies both

$$p_1 = p_2 \quad \text{and} \quad \sin(\alpha_1 - \alpha_2)/2 = 0 ;$$

the latter condition gives

$$\alpha_1 - \alpha_2 = 2\pi n \qquad (n = 0, \pm 1, \pm 2, \ldots)$$

hence $L_1 = L_2$. Finally, consider D4. Since

$$4 \sin^2 \frac{\alpha_1 - \alpha_2}{2} = (\sin \alpha_1 - \sin \alpha_2)^2 + (\cos \alpha_1 - \cos \alpha_2)^2$$

we have

$$d(L_1, L_2) = \left[(p_1 - p_2)^2 + (\sin \alpha_1 - \sin \alpha_2)^2 + (\cos \alpha_1 - \cos \alpha_2)^2 \right]^{1/2} .$$

Let $(p_i, \sin \alpha_i, \cos \alpha_i)$ be the coordinates of a point A_i in three-dimensional Euclidean space. Noting that $d(L_i, L_j)$ equals the Euclidean distance from A_i to A_j in $\mathbb{R}^3$, we see that D4 is also satisfied. However, the metric $d(L_1, L_2)$ lacks an immediate geometrical interpretation.

To prove that ℓ^p with $p \geq 1$ is a metric space, we require the inequalities of the following section.

1.2 Hölder's Inequality and Minkowski's Inequality

We start with *Hölder's inequality* for sums,

$$\left| \sum_{k=1}^{n} x_k y_k \right| \leq \left(\sum_{k=1}^{n} |x_k|^p \right)^{1/p} \left(\sum_{k=1}^{n} |y_k|^q \right)^{1/q} , \tag{1.2.1}$$

where $p > 1$ and

$$1/p + 1/q = 1 . \tag{1.2.2}$$

The constants p and q satisfying (1.2.2) are called *conjugate exponents*. Introducing the notation

$$\|\mathbf{x}\|_p = \left(\sum_{k=1}^{n} |x_k|^p \right)^{1/p}$$

(later we verify that this expression is a norm[2] of $\mathbf{x}$ on $\mathbb{R}^n$) we rewrite (1.2.1) as

$$\left| \sum_{k=1}^{n} x_k y_k \right| \le \|\mathbf{x}\|_p \, \|\mathbf{y}\|_q \; .$$

We begin to prove Hölder's inequality.

Proof. Because

$$\left| \sum_{k=1}^{n} x_k y_k \right| \le \sum_{k=1}^{n} |x_k y_k| \, ,$$

Hölder's inequality follows from the relation

$$\sum_{k=1}^{n} |x_k y_k| \le \|\mathbf{x}\|_p \, \|\mathbf{y}\|_q \; . \tag{1.2.3}$$

If either $\mathbf{x}$ or $\mathbf{y}$ is zero, then (1.2.3) is trivial. So we suppose they are both nonzero and prove the equivalent result

$$\sum_{k=1}^{n} \frac{|x_k|}{\|\mathbf{x}\|_p} \frac{|y_k|}{\|\mathbf{y}\|_q} \le 1 \; .$$

Now we need an elementary inequality for positive numbers a and b:

$$a^{1/p} b^{1/q} \le \frac{a}{p} + \frac{b}{q} \quad \text{where} \quad \frac{1}{p} + \frac{1}{q} = 1 \, , \tag{1.2.4}$$

which we prove later. Let us take

$$a = \frac{|x_k|^p}{\|\mathbf{x}\|_p^p} \, , \qquad b = \frac{|y_k|^q}{\|\mathbf{y}\|_q^q} \; .$$

We get

$$\frac{|x_k|}{\|\mathbf{x}\|_p} \cdot \frac{|y_k|}{\|\mathbf{y}\|_q} \le \frac{|x_k|^p}{p\,\|\mathbf{x}\|_p^p} + \frac{|y_k|^q}{q\,\|\mathbf{y}\|_q^q} \qquad (k = 1, \ldots, n) \; .$$

Summing these over k, we have

---

[2] As stated in the Introduction, the norm symbol $\|\cdot\|$ is formally introduced on p. 39. It is used here on the assumption that the reader has seen it in more elementary courses.

$$\sum_{k=1}^{n} \frac{|x_k|\,|y_k|}{\|\mathbf{x}\|_p \|\mathbf{y}\|_q} \leq \frac{\sum_{k=1}^{n} |x_k|^p}{p \sum_{k=1}^{n} |x_k|^p} + \frac{\sum_{k=1}^{n} |y_k|^q}{q \sum_{k=1}^{n} |y_k|^q} = \frac{1}{p} + \frac{1}{q} = 1\,.$$

To complete the proof, we show (1.2.4). Taking the derivative of the function

$$f(t) = \frac{t}{p} + \frac{1}{q} - t^{1/p}$$

we see that the minimum point of f for $t \geq 0$ is $t = 1$. Moreover, $f(1) = 0$, so for $t \geq 0$ we have

$$t^{1/p} \leq \frac{t}{p} + \frac{1}{q}\,.$$

Putting $t = a/b$, we get

$$\frac{a^{1/p}}{b^{1/p}} \leq \frac{a}{pb} + \frac{1}{q}\,.$$

Multiplying this by b and noting that $1/q = 1 - 1/p$, we get (1.2.4). □

Provided the series

$$\sum_{k=1}^{\infty} |x_k|^p \quad \text{and} \quad \sum_{k=1}^{\infty} |y_k|^q$$

both converge, the limit passage as $n \to \infty$ in (1.2.3) shows that the series

$$\sum_{k=1}^{\infty} x_k y_k$$

converges absolutely. This yields Hölder's inequality for series:

$$\left| \sum_{k=1}^{\infty} x_k y_k \right| \leq \left(\sum_{k=1}^{\infty} |x_k|^p \right)^{1/p} \left(\sum_{k=1}^{\infty} |y_k|^q \right)^{1/q}. \tag{1.2.5}$$

Now we prove *Minkowski's inequality*

$$\left(\sum_{k=1}^{n} |x_k + y_k|^p \right)^{1/p} \leq \left(\sum_{k=1}^{n} |x_k|^p \right)^{1/p} + \left(\sum_{k=1}^{n} |y_k|^p \right)^{1/p} \tag{1.2.6}$$

for $p \geq 1$, which can be rewritten as

$$\|\mathbf{x} + \mathbf{y}\|_p \leq \|\mathbf{x}\|_p + \|\mathbf{y}\|_p\,.$$

This constitutes the triangle inequality for the norm $\|\cdot\|_p$ in $\mathbb{R}^n$. For $p = 1$ the result is trivial so we consider it for $p > 1$. Multiplying both sides of (1.2.6) by the quantity

$$\left(\sum_{k=1}^{n} |x_k + y_k|^p\right)^{1/q},$$

we get an equivalent inequality

$$\sum_{k=1}^{n} |x_k + y_k|^p \le \left[\left(\sum_{k=1}^{n} |x_k|^p\right)^{1/p} + \left(\sum_{k=1}^{n} |y_k|^p\right)^{1/p}\right]\left(\sum_{k=1}^{n} |x_k + y_k|^p\right)^{1/q}, \tag{1.2.7}$$

which we prove. We start with a simple inequality

$$\begin{aligned}\sum_{k=1}^{n} |x_k + y_k|^p &= \sum_{k=1}^{n} |x_k + y_k|\,|x_k + y_k|^{p-1} \\ &\le \sum_{k=1}^{n} |x_k|\,|x_k + y_k|^{p-1} + \sum_{k=1}^{n} |y_k|\,|x_k + y_k|^{p-1}\,. \end{aligned} \tag{1.2.8}$$

Using (1.2.1), we get

$$\sum_{k=1}^{n} |x_k|\,|x_k + y_k|^{p-1} \le \left(\sum_{k=1}^{n} |x_k|^p\right)^{1/p}\left(\sum_{k=1}^{n} |x_k + y_k|^{q(p-1)}\right)^{1/q}.$$

But $q = p/(p-1)$ and we have

$$\sum_{k=1}^{n} |x_k|\,|x_k + y_k|^{p-1} \le \left(\sum_{k=1}^{n} |x_k|^p\right)^{1/p}\left(\sum_{k=1}^{n} |x_k + y_k|^{p}\right)^{1/q}.$$

Similarly,

$$\sum_{k=1}^{n} |y_k|\,|x_k + y_k|^{p-1} \le \left(\sum_{k=1}^{n} |y_k|^p\right)^{1/p}\left(\sum_{k=1}^{n} |x_k + y_k|^{p}\right)^{1/q}.$$

Substituting these into (1.2.8) we get (1.2.7) and hence (1.2.6). If the series

$$\sum_{k=1}^{\infty} |x_k|^p \quad \text{and} \quad \sum_{k=1}^{\infty} |y_k|^p$$

converge, then the limit as $n \to \infty$ produces Minkowski's inequality for series:

$$\left(\sum_{k=1}^{\infty} |x_k + y_k|^p\right)^{1/p} \le \left(\sum_{k=1}^{\infty} |x_k|^p\right)^{1/p} + \left(\sum_{k=1}^{\infty} |y_k|^p\right)^{1/p}. \tag{1.2.9}$$

Now we can prove

Theorem 1.2.1. For $p \ge 1$, the space ℓ^p, consisting of the vectors $\mathbf{x}$ such that $||\mathbf{x}||_p < \infty$, is a metric space.

Proof. The case $p = 1$ is trivial so let $p > 1$. Clearly $||\mathbf{x}||_p$, which is the distance from $\mathbf{x}$ to the zero vector, is finite by hypothesis. Verification of axioms D1–D3 is trivial. If $\mathbf{x} = (x_1, x_2, \ldots)$, $\mathbf{y} = (y_1, y_2, \ldots)$, and $\mathbf{z} = (z_1, z_2, \ldots)$, then by Minkowski's inequality

$$d_p(\mathbf{x}, \mathbf{y}) \equiv \left(\sum_{k=1}^{\infty} |x_k - y_k|^p\right)^{1/p} = \left(\sum_{k=1}^{\infty} |(x_k - z_k) + (z_k - y_k)|^p\right)^{1/p}$$
$$\leq \left(\sum_{k=1}^{\infty} |x_k - z_k|^p\right)^{1/p} + \left(\sum_{k=1}^{\infty} |z_k - y_k|^p\right)^{1/p} = d_p(\mathbf{x}, \mathbf{z}) + d_p(\mathbf{z}, \mathbf{y})$$

and D4 is satisfied. □

Problem 1.2.1. Let $h_1, h_2, \ldots$ be positive numbers. Show that the space of infinite sequences $\mathbf{x} = (x_1, x_2, \ldots)$ satisfying

$$\|\mathbf{x}\|_{w,p} = \left(\sum_{k=1}^{\infty} h_k |x_k|^p\right)^{1/p} < \infty ,$$

with the weighted metric $d_{w,p}(\mathbf{x}, \mathbf{y}) = \|\mathbf{x} - \mathbf{y}\|_{w,p}$, is a metric space. □

For the spaces of infinite dimensional vectors considered above, a vector $\mathbf{x}$ belongs to a space if and only if $d(\mathbf{x}, \mathbf{0}) < \infty$. So we should establish this inequality. The task is more complicated for function spaces. Another note concerns terminology. Although for $\mathbf{x} = (x_1, x_2, \ldots)$ we use the terms "sequence" and "infinite dimensional vector" interchangeably, implicitly assuming the x_k to be the components of the vector $\mathbf{x}$, we will never carry out the change of basis that is so central to ordinary linear algebra. This is because the issue of basis is not as simple in infinite dimensional spaces as it is in $\mathbb{R}^n$.

1.3 Metric Spaces of Functions

Functions of two or three variables serve to describe the behavior or change in state of a body in space. Displacements, velocities, loads, and temperatures are all functions of position. The notion of metric space is an appropriate tool for distinguishing between states of a body. In continuum mechanics we deal mostly with real-valued continuous or differentiable functions.

Let Ω be a closed and bounded (i.e., compact) domain in $\mathbb{R}^n$. A natural measure of the deviation between two continuous functions $f(\mathbf{x})$ and $g(\mathbf{x})$, $\mathbf{x} \in \Omega$, is the *max metric* given by

$$d(f, g) = \max_{\mathbf{x} \in \Omega} |f(\mathbf{x}) - g(\mathbf{x})| . \tag{1.3.1}$$

Clearly $d(f, g)$ satisfies axioms D1–D3 on p. 9. Let us verify D4. Since the function $|f(\mathbf{x}) - g(\mathbf{x})|$ is continuous on Ω, there is a point $\mathbf{x}_0 \in \Omega$ such that

$$d(f, g) = \max_{\mathbf{x} \in \Omega} |f(\mathbf{x}) - g(\mathbf{x})| = |f(\mathbf{x}_0) - g(\mathbf{x}_0)|$$

(recall Theorem 1 on p. 5). For any function $h(\mathbf{x})$ which is continuous on Ω, we get

$$d(f,g) = |f(\mathbf{x}_0) - g(\mathbf{x}_0)| \le |f(\mathbf{x}_0) - h(\mathbf{x}_0)| + |h(\mathbf{x}_0) - g(\mathbf{x}_0)| \le d(f,h) + d(h,g)\,.$$

Thus $d(f,g)$ in (1.3.1) is a metric.

Definition 1.3.1. Let Ω be a closed and bounded domain in $\mathbb{R}^n$. We denote by $C(\Omega)$ the metric space consisting of the set of all continuous functions on Ω supplied with the metric (1.3.1).

Remark 1.3.1. When $\Omega = [a,b]$, a finite interval along the real line, we will denote $C(\Omega)$ by $C(a,b)$ instead of $C([a,b])$ or $C[a,b]$. □

We can also introduce the metric space $C(V)$ of the bounded functions that are continuous on an open set $V \subseteq \mathbb{R}^n$. Because Weierstrass' theorem is not valid on V, however, we use

$$d(f,g) = \sup_{\mathbf{x}\in V} |f(\mathbf{x}) - g(\mathbf{x})|\,. \tag{1.3.2}$$

The reader may easily verify the metric axioms in this case. The definition of the metric space of continuous functions can also be extended to the case in which the functions are continuous with respect to an abstract argument belonging to some topological space V. The metric is defined by (1.3.2).

To account for the derivatives of functions, we use a metric such as

$$d(f,g) = \sum_{|\alpha|\le k} \max_{\mathbf{x}\in\Omega} |D^\alpha f(\mathbf{x}) - D^\alpha g(\mathbf{x})| \tag{1.3.3}$$

where

$$D^\alpha f = \frac{\partial^{|\alpha|} f}{\partial x_1^{\alpha_1} \cdots \partial x_n^{\alpha_n}}\,, \qquad |\alpha| = \alpha_1 + \cdots + \alpha_n\,, \tag{1.3.4}$$

and α is regarded as an n-tuple $\alpha = (\alpha_1, \ldots, \alpha_n)$. The notation introduced in (1.3.4) is often called *multi-index notation*.

Problem 1.3.1. Verify the metric axioms for the space $C^{(k)}(\Omega)$ of all continuous functions on a closed and bounded domain Ω whose derivatives up to order k are continuous on Ω, with the metric (1.3.3). □

Problem 1.3.2. Interpret the notation (1.3.4) for the case $n = 3$ and $\alpha = (1,0,2)$. □

On the set of all continuous functions on Ω, let us consider another metric:

$$d(f,g) = \left(\int_\Omega |f(\mathbf{x}) - g(\mathbf{x})|^p \, d\Omega \right)^{1/p} \qquad (p \ge 1) \tag{1.3.5}$$

where Ω is a Jordan measurable compact domain in $\mathbb{R}^n$. Jordan measurability, which ensures that the Riemann integral exists for any function under consideration, guarantees that (1.3.5) is well-defined for any functions f, g that are continuous on Ω. In applications we use domains occupied by physical bodies of comparatively simple shape. We will always assume such domains are Jordan measurable. We denote the set of continuous functions with metric (1.3.5) by $\tilde{L}^p(\Omega)$. The notation $L^p(\Omega)$ will be reserved for a wider set of functions, with the same metric, to be introduced later. For our purposes, the space $L^p(\Omega)$ is used to characterize the classes of loads for mechanical problems and the properties of generalized solutions to those problems. In what follows, we tacitly assume that Ω is a Jordan measurable domain.

Regarding (1.3.5), the only nontrivial metric axiom to be verified is D4; its validity follows from Minkowski's inequality for integrals

$$\left(\int_\Omega |f_1(\mathbf{x}) + f_2(\mathbf{x})|^p \, d\Omega\right)^{1/p} \le \left(\int_\Omega |f_1(\mathbf{x})|^p \, d\Omega\right)^{1/p} + \left(\int_\Omega |f_2(\mathbf{x})|^p \, d\Omega\right)^{1/p} \tag{1.3.6}$$

which holds for any $p \ge 1$. With

$$f_1(\mathbf{x}) = f(\mathbf{x}) - h(\mathbf{x}) \,, \qquad f_2(\mathbf{x}) = h(\mathbf{x}) - g(\mathbf{x}) \,,$$

inequality (1.3.6) becomes

$$d(f, g) \le d(f, h) + d(h, g) \,,$$

showing that the set of continuous functions on Ω is also a metric space under the metric expression (1.3.5).

Although

$$\left(\int_\Omega |f(\mathbf{x}) - g(\mathbf{x})|^p \, d\Omega\right)^{1/p} \le (\text{mes}\, \Omega)^{1/p} \max_{\mathbf{x}\in\Omega} |f(\mathbf{x}) - g(\mathbf{x})|$$

where

$$\text{mes}\, \Omega = \int_\Omega 1 \, d\Omega$$

is the *measure* of Ω, we cannot find a constant m such that the inequality

$$\max_{\mathbf{x}\in\Omega} |f(\mathbf{x}) - g(\mathbf{x})| \le m \left(\int_\Omega |f(\mathbf{x}) - g(\mathbf{x})|^p d\Omega\right)^{1/p}$$

holds for all continuous functions $f(\mathbf{x})$ and $g(\mathbf{x})$. Hence the metrics (1.3.1) and (1.3.5) are not equivalent on the set of continuous functions.

Remark 1.3.2. For $0 < p < 1$, the expression $d(f, g)$ in (1.3.5) is not a metric. The inequality sign reverses in Minkowski's inequality for this range of p. □

Minkowski's inequality (1.3.6) follows from Hölder's inequality for integrals:

$$\left|\int_\Omega f(\mathbf{x}) g(\mathbf{x}) \, d\Omega\right| \le \left(\int_\Omega |f(\mathbf{x})|^p \, d\Omega\right)^{1/p} \left(\int_\Omega |g(\mathbf{x})|^q \, d\Omega\right)^{1/q} \tag{1.3.7}$$

where $1/p + 1/q = 1$ and $p > 1$. These useful inequalities are established below.

Problem 1.3.3. Show that the function

$$d(f, g) = \int_0^1 |f'(x) - g'(x)|\, dx$$

is not a metric on the set of functions continuous on $[0, 1]$. □

Hölder's Inequality for Integrals. The proof of Hölder's inequality (1.3.7) is based on Young's inequality (1.2.4). Since (1.3.7) clearly holds if f or g is zero, we suppose f and g are not zero and that the integrals on the right-hand side do not vanish. Setting

$$a = \frac{|f(\mathbf{x})|^p}{\int_\Omega |f(\mathbf{x})|^p\, d\Omega}, \qquad b = \frac{|g(\mathbf{x})|^q}{\int_\Omega |g(\mathbf{x})|^q\, d\Omega},$$

in (1.2.4), we get

$$\frac{|f(\mathbf{x})|}{\left(\int_\Omega |f(\mathbf{x})|^p\, d\Omega\right)^{1/p}} \cdot \frac{|g(\mathbf{x})|}{\left(\int_\Omega |g(\mathbf{x})|^q\, d\Omega\right)^{1/q}} \le \frac{|f(\mathbf{x})|^p}{p\int_\Omega |f(\mathbf{x})|^p\, d\Omega} + \frac{|g(\mathbf{x})|^q}{q\int_\Omega |g(\mathbf{x})|^q\, d\Omega}.$$

Integration over Ω yields

$$\frac{\int_\Omega |f(\mathbf{x})\, g(\mathbf{x})|\, d\Omega}{\left(\int_\Omega |f(\mathbf{x})|^p\, d\Omega\right)^{1/p}\left(\int_\Omega |g(\mathbf{x})|^q\, d\Omega\right)^{1/q}} \le \frac{\int_\Omega |f(\mathbf{x})|^p\, d\Omega}{p\int_\Omega |f(\mathbf{x})|^p\, d\Omega} + \frac{\int_\Omega |g(\mathbf{x})|^q\, d\Omega}{q\int_\Omega |g(\mathbf{x})|^q\, d\Omega}$$

$$= \frac{1}{p} + \frac{1}{q} = 1.$$

In view of the triangle inequality

$$\left|\int_\Omega f(\mathbf{x})\, g(\mathbf{x})\, d\Omega\right| \le \int_\Omega |f(\mathbf{x})\, g(\mathbf{x})|\, d\Omega,$$

we obtain (1.3.7). Note that we did not require continuity of f and g, but only convergence of the integrals

$$\int_\Omega |f(\mathbf{x})|^p\, d\Omega \quad \text{and} \quad \int_\Omega |g(\mathbf{x})|^q\, d\Omega.$$

Minkowski's Inequality. Here the proof is also analogous to that for sequences. Let us estimate the integral

$$\int_\Omega |f(\mathbf{x}) + g(\mathbf{x})|^p\, d\Omega \qquad (p > 1).$$

We have

$$\int_\Omega |f(\mathbf{x}) + g(\mathbf{x})|^p \, d\Omega = \int_\Omega |f(\mathbf{x}) + g(\mathbf{x})| \, |f(\mathbf{x}) + g(\mathbf{x})|^{p-1} \, d\Omega$$

$$\leq \int_\Omega |f(\mathbf{x}) + g(\mathbf{x})|^{p-1} |f(\mathbf{x})| \, d\Omega + \int_\Omega |f(\mathbf{x}) + g(\mathbf{x})|^{p-1} |g(\mathbf{x})| \, d\Omega$$

$$\leq \left(\int_\Omega |f(\mathbf{x}) + g(\mathbf{x})|^{q(p-1)} \, d\Omega \right)^{1/q} \left(\int_\Omega |f(\mathbf{x})|^p \, d\Omega \right)^{1/p}$$

$$+ \left(\int_\Omega |f(\mathbf{x}) + g(\mathbf{x})|^{q(p-1)} \, d\Omega \right)^{1/q} \left(\int_\Omega |g(\mathbf{x})|^p \, d\Omega \right)^{1/p}.$$

Since $q(p-1) = p$, we can divide through by the quantity

$$\left(\int_\Omega |f(\mathbf{x}) + g(\mathbf{x})|^p \, d\Omega \right)^{1/q}$$

to get (1.3.6) with $f_1 = f$ and $f_2 = g$. Again, only convergence of the integrals of $|f(\mathbf{x})|^p$ and $|g(\mathbf{x})|^q$ over Ω is assumed.

1.4 Some Relations for the Metrics in $\tilde{L}^p(\Omega)$ and ℓ^p

We have shown that $\tilde{L}^p(\Omega)$ is a metric space. Later we show that the distance between $f \in \tilde{L}^p(\Omega)$ and the zero function, denoted by

$$\|f\|_{L^p} = \left(\int_\Omega |f(\mathbf{x})|^p \, d\Omega \right)^{1/p},$$

plays an important role in the theory of integrable functions. In Sect. 1.11 we call it the norm on the space $L^p(\Omega)$, but for now it is just notation.

If Ω is bounded, Hölder's inequality yields

$$\int_\Omega |f(\mathbf{x})| \, d\Omega = \int_\Omega |f(\mathbf{x})| \cdot 1 \, d\Omega \leq \left(\int_\Omega |f(\mathbf{x})|^p \, d\Omega \right)^{1/p} \left(\int_\Omega 1^q \, d\Omega \right)^{1/q}$$

$$= (\text{mes } \Omega)^{1/q} \left(\int_\Omega |f(\mathbf{x})|^p \, d\Omega \right)^{1/p}. \tag{1.4.1}$$

This also holds for discontinuous functions for which the integral $\int_\Omega |f(\mathbf{x})|^p \, d\Omega$ converges, implying that such functions are absolutely integrable.

If $1 \leq r < p$, then (1.4.1) yields an inequality for the elements of the spaces $\tilde{L}^p(\Omega)$ and $L^p(\Omega)$:

$$\|f\|_{L^r} \leq m_{r,p} \, \|f\|_{L^p} \,, \tag{1.4.2}$$

with some constant $m_{r,p}$ for all $f \in \tilde{L}^p(\Omega)$. Indeed, Hölder's inequality with exponents $p_1 = p/r$ and $1/q_1 = 1 - 1/p_1$ yields

$$\int_\Omega |f(\mathbf{x})|^r \cdot 1\, d\Omega \le (\text{mes}\,\Omega)^{1/q_1} \left(\int_\Omega |f(\mathbf{x})|^p \, d\Omega \right)^{1/p_1}$$

which we can raise to the power $1/r$. Note that (1.4.1) and (1.4.2) hold only when mes Ω is finite.

It is instructive to compare (1.4.2) with a somewhat analogous inequality for sequences. For $p \ge 1$ we introduce the norm of $\mathbf{x} \in \ell^p$ as the distance between the sequence $\mathbf{x} = \{x_k\}$ and the zero sequence:

$$\|\mathbf{x}\|_{\ell^p} = \left(\sum_{k=1}^{\infty} |x_k|^p \right)^{1/p}.$$

Suppose $1 \le r < p$ and $\mathbf{x} = \{x_k\} \in \ell^r$. We will show that

$$\|\mathbf{x}\|_{\ell^p} \le \|\mathbf{x}\|_{\ell^r} \tag{1.4.3}$$

and so $\mathbf{x} \in \ell^p$ as well. Note that, relative to the inequality sign, the exponents r and p are not situated as they are in (1.4.2). Because (1.4.3) holds trivially if $\mathbf{x} = \mathbf{0}$, we take $\|\mathbf{x}\|_{\ell^r} \ne 0$. For a number a such that $0 \le a \le 1$, we have $a^p \le a^r$. Clearly

$$0 \le \frac{|x_k|}{\left(\sum_{j=1}^{\infty} |x_j|^r \right)^{1/r}} \le 1$$

so

$$\left[\frac{|x_k|}{\left(\sum_{j=1}^{\infty} |x_j|^r \right)^{1/r}} \right]^p \le \left[\frac{|x_k|}{\left(\sum_{j=1}^{\infty} |x_j|^r \right)^{1/r}} \right]^r \qquad (k = 1, 2, \ldots).$$

Summing these over k, raising to the $1/p$ power, and rearranging slightly, we get

$$\frac{\left(\sum_{j=1}^{\infty} |x_j|^p \right)^{1/p}}{\left(\sum_{j=1}^{\infty} |x_j|^r \right)^{1/r}} \le \frac{\left(\sum_{j=1}^{\infty} |x_j|^r \right)^{1/p}}{\left(\sum_{j=1}^{\infty} |x_j|^r \right)^{1/p}} = 1,$$

from which (1.4.3) follows. Because the ℓ^p norm is majorized by the ℓ^r norm, the elements of ℓ^r constitute part of the space ℓ^p if $1 \le r < p$.

1.5 Metrics in Energy Spaces

We have already introduced the energy space for a string. Let us consider other examples. In what follows, we shall employ only dimensionless variables, parameters, and functions of state of a body.

Bending of a Beam. In the Introduction we considered the problem of bending a clamped beam, governed by equation (4). The beam has strain energy

$$\mathcal{E}_1(y) = \frac{1}{2}\int_0^l B(x)(y'')^2\,dx\,.$$

The stiffness $B(x)$ of the beam is a positive function that is continuous on $[0, l]$ except, possibly, for a few jump discontinuities.

On the set S consisting of all functions $y(x) \in C^{(2)}(0, l)$ satisfying

$$y(0) = y'(0) = y(l) = y'(l) = 0\,, \tag{1.5.1}$$

let us consider

$$d(y_1, y_2) = \Big(2\mathcal{E}_1(y_1 - y_2)\Big)^{1/2} = \left(\int_0^l B(x)[y_1''(x) - y_2''(x)]^2\,dx\right)^{1/2}$$

as a metric. Axioms D1 and D3 obviously hold. Satisfaction of D4 follows from the fact that $\mathcal{E}_1(y)$ is quadratic in y. To verify D2, we need only show that $d(y, z) = 0$ implies $y(x) = z(x)$. But $d(y, z) = 0$ implies $[y(x) - z(x)]'' = 0$, hence $y(x) - z(x) = a_1 x + a_2$ where a_1, a_2 are constants. Imposing (1.5.1), we arrive at $a_1 = a_2 = 0$. So $d(y_1, y_2)$ is a metric on S.

Elastic Membrane. The strain energy of a membrane occupying a domain $\Omega \subset \mathbb{R}^2$ is proportional to the quantity

$$\mathcal{E}_2(u) = \iint_\Omega \left[\left(\frac{\partial u}{\partial x}\right)^2 + \left(\frac{\partial u}{\partial y}\right)^2\right] dx\,dy\,.$$

So we can try

$$d(u, v) = (\mathcal{E}_2(u - v))^{1/2} \tag{1.5.2}$$

as a metric on the functions $u = u(x, y)$ that describe the transverse displacement of the membrane. We first consider the case in which the edge of the membrane is clamped, i.e.,

$$u\big|_{\partial\Omega} = 0 \tag{1.5.3}$$

where $\partial\Omega$ is the boundary of Ω. Expression (1.5.2) is a metric on the subset of $C^{(1)}(\Omega)$ consisting of the functions that satisfy (1.5.3). Axioms D1 and D3 hold obviously; D2 holds by (1.5.3), and D4 holds by the quadratic nature of $\mathcal{E}_2(u)$ and Minkowski's inequality. This space is appropriate for investigating the *Dirichlet problem* for Poisson's equation,

$$\Delta u = -f\,, \qquad u\big|_{\partial\Omega} = 0\,,$$

which governs the clamped membrane under a load $f = f(x, y)$.

A *Neumann problem*, on the other hand, may incorporate a homogeneous boundary condition of the form

$$\left.\frac{\partial u}{\partial n}\right|_{\partial\Omega} = 0\,. \tag{1.5.4}$$

According to the calculus of variations, a minimizer of the functional[3]

$$J(u) = \frac{1}{2}\iint_\Omega \left[\left(\frac{\partial u}{\partial x}\right)^2 + \left(\frac{\partial u}{\partial y}\right)^2 - 2fu\right] dx\,dy$$

is a solution to this problem. This Neumann problem can be cast in variational form as in the following exercise.

Problem 1.5.1. Given $f(x, y) \in C(\Omega)$, find a minimizer $u(x, y)$ of $J(u)$ such that $u(x, y) \in C^{(1)}(\Omega)$. □

The boundary condition (1.5.4) appears as a natural boundary condition in Problem 1.5.1; that is, it arises as one of the equations to be satisfied by a minimizer u of the functional $J(u)$. We need not formulate it in advance. That is why we do not require boundary conditions on functions constituting the energy space for the Neumann problem. If we take (1.5.2) as a metric for this energy space, we see that metric axiom D2 is not satisfied: from $d(u, v) = 0$ it follows only that $u(x, y) - v(x, y) =$ constant. There are two ways in which we can use the energy metric in this problem. One is to define a space whose elements are equivalence classes of functions, two functions belonging to the same class (and hence identified with each other) if their difference is a constant on Ω. This approach takes into account the stress of the membrane, but not its displacements as a "rigid" whole. Another approach, avoiding *rigid displacements*[4] of the form $u(x, y) = c$, imposes an additional integral-type restriction on all functions of the space, e.g.,

$$\iint_\Omega u(x, y)\,dx\,dy = 0\,.$$

Both approaches allow (1.5.2) as a metric on an energy space for a Neumann problem. We shall consider this in more detail later. Solvability of Neumann's problem will require an additional balance condition

$$\iint_\Omega f(x, y)\,dx\,dy = 0$$

---

[3] The term *functional* is defined on p. 72. We use it occasionally prior to that on the assumption that the reader has encountered it in earlier courses. Roughly, a *real-* or *complex-valued functional* is a mapping from a linear space X to the real or complex numbers. In an application, it will often be represented by an integral dependent on certain functions that are regarded as variables.

[4] A *rigid displacement* is defined as a displacement that does not change the strain energy. In particular, if the body is not deformed, the strain energy for this displacement is zero. The term *rigid motion* is often used instead, even in statics problems.

on the applied forces. Note that we can use the same energy space and a similar approach when considering the nonhomogeneous Neumann boundary condition

$$\left.\frac{\partial u}{\partial n}\right|_{\partial\Omega} = \psi .$$

In this case solvability will require

$$\iint_{\Omega} f(x,y)\,dx\,dy + \int_{\partial\Omega} \psi(s)\,ds = 0 ,$$

meaning that the resultant of all normal forces acting on the membrane must vanish, and the functional $J(u)$ must include an additional term $\int_{\partial\Omega} \psi u\,ds$.

A Plate. A linearly elastic plate has strain energy

$$\mathcal{E}_3(w) = \iint_{\Omega} \frac{D}{2}\left\{(\Delta w)^2 + 2(1-\nu)\left[\left(\frac{\partial^2 w}{\partial x \partial y}\right)^2 - \frac{\partial^2 w}{\partial x^2}\frac{\partial^2 w}{\partial y^2}\right]\right\} dx\,dy \tag{1.5.5}$$

where D is the bending stiffness of the plate, ν is Poisson's ratio ($0 < \nu < 1/2$), and $w(x,y)$ is the normal displacement of the midsurface, which is denoted by Ω in the xy-plane. If the edge of the plate is clamped we get

$$\left.w\right|_{\partial\Omega} = \left.\frac{\partial w}{\partial n}\right|_{\partial\Omega} = 0 . \tag{1.5.6}$$

Since $D > 0$, the plate energy takes nonnegative values for all smooth w. If $\mathcal{E}_3(w) = 0$, then all second derivatives of w vanish so that $w = a + bx + cy$ and, from (1.5.6), $w = 0$. Therefore D2 is satisfied by the distance function

$$d(w_1, w_2) = (2\mathcal{E}_3(w_1 - w_2))^{1/2} .$$

The other metric axioms are easily checked; in particular, satisfaction of D4 follows from the quadratic form of the energy functional and Minkowski's inequality. So $d(w_1, w_2)$ is a metric on the subset of $C^{(2)}(\Omega)$ consisting of all functions satisfying (1.5.6). This is the energy space for the plate.

If the edge of the plate is free from geometrical fixation (such as clamping), the situation is similar to the Neumann problem of membrane theory: we must eliminate "rigid" displacements of the plate. We consider this in detail later.

Problem 1.5.2. Verify that $\mathcal{E}_3(w) = 0$ implies $w = \alpha + \beta x + \gamma y$, where α, β, γ are constants. □

Linear Elasticity. The strain energy functional for an elastic body occupying a bounded domain $\Omega \subset \mathbb{R}^3$ is

$$\mathcal{E}_4(\mathbf{u}) = \frac{1}{2}\iiint_{\Omega} c^{ijkl}\,\epsilon_{kl}\,\epsilon_{ij}\,d\Omega \tag{1.5.7}$$

where the c^{ijkl} are components of the tensor of elastic moduli. In the homogeneous, isotropic case, these components are determined by Young's elastic constant E and Poisson's ratio ν. In Cartesian coordinates (x_1, x_2, x_3), the strain tensor with components (ϵ_{ij}) is defined by

$$\epsilon_{ij} \equiv \epsilon_{ij}(\mathbf{u}) = \frac{1}{2}\left(\frac{\partial u_i}{\partial x_j} + \frac{\partial u_j}{\partial x_i}\right), \qquad \mathbf{u} = (u_1, u_2, u_3)\,.$$

Note that the repeated index convention for tensors is in effect here.

From elasticity theory, the elastic moduli satisfy the following conditions:

(a) The tensor is symmetric:

$$c^{ijkl} = c^{klij} = c^{jikl}\,. \tag{1.5.8}$$

(b) The tensor is positive definite; for any symmetric tensor (ϵ_{ij}) with $\epsilon_{ij} = \epsilon_{ji}$, the inequality

$$c^{ijkl}\,\epsilon_{kl}\,\epsilon_{ij} \ge c_0 \sum_{i,j=1}^{3} \epsilon_{ij}^2 \tag{1.5.9}$$

holds with a positive constant c_0 that does not depend on (ϵ_{ij}).

Let us introduce a metric

$$d(\mathbf{u}, \mathbf{v}) = (2\mathcal{E}_4(\mathbf{u} - \mathbf{v}))^{1/2} \tag{1.5.10}$$

on the set of continuously differentiable vector functions $\mathbf{u}(\mathbf{x})$ representing displacements of points in the body. When $d(\mathbf{u}, \mathbf{v}) = 0$, condition (1.5.9) implies $\epsilon_{ij}(\mathbf{u} - \mathbf{v}) = 0$ for all $i, j = 1, 2, 3$. As is known from the theory of elasticity,

$$\mathbf{u}(\mathbf{x}) - \mathbf{v}(\mathbf{x}) = \mathbf{a} + \mathbf{x} \times \mathbf{b}$$

where $\mathbf{a}$ and $\mathbf{b}$ are constant vectors and the right-hand side is the general form of a rigid displacement for an elastic body. If we restrict the set of vector functions by the boundary condition

$$\mathbf{u}\big|_{\partial\Omega} = \mathbf{0} \tag{1.5.11}$$

(i.e., clamp the boundary), we get $\mathbf{u}(\mathbf{x}) - \mathbf{v}(\mathbf{x}) = \mathbf{0}$. The other metric axioms hold as well. Thus we can impose the metric (1.5.10) on the set of all continuously differentiable vector functions $\mathbf{u}(\mathbf{x})$ satisfying (1.5.11) and obtain an energy space for the elastic body.

Later we shall consider other boundary conditions that appear in the boundary value problems of linear elasticity. For detailed explanations, the interested reader can consult [36], [24], [8].

We have not introduced a special notation for the energy spaces discussed thus far, since they are not the spaces we shall actually use. They form only the basis of the actual energy spaces, and to finish our preparation we require the notions of Lebesgue integral and generalized derivative.

1.6 Sets in a Metric Space

By analogy with Euclidean space, we may introduce a few concepts.

Definition 1.6.1. In a metric space X, the set of points $x \in X$ such that

$$d(x_0, x) < r$$

is the *open ball* of radius r about x_0. It will usually be denoted by $B(x_0, r)$. The set of points $x \in X$ such that $d(x_0, x) \le r$ is the *closed ball* of radius r about x_0. It will usually be denoted by $\overline{B}(x_0, r)$.

These definitions coincide with the notions of ball in elementary geometry. However, even in Euclidean space, the use of non-Euclidean metrics can yield balls having unconventional shapes. In $\mathbb{R}^3$ with the metric $d(\mathbf{x}, \mathbf{y}) = \sup_i |x_i - y_i|$, the ball $B(\mathbf{0}, 1)$ is a cube of edge length 2.

Definition 1.6.2. A subset S of a metric space X is *open* if for each $x \in S$ there exists r such that the open ball $B(x, r)$ is contained in S.

Problem 1.6.1. Show that an open ball is an open set. □

In a metric space we can introduce the figures (e.g., ellipses) whose definitions require only a notion of distance. In a concrete metric space, we can introduce some sets using special properties of their elements. For example, in c (see p. 13) a cube C may be defined by

$$C = \{\mathbf{x} = (x_1, x_2, \ldots) \in c \colon |x_k - x_{k0}| \le a \ \text{ for all } k\}$$

where $\mathbf{x}_0 = (x_{10}, x_{20}, \ldots)$ is a fixed point of c. Although C is a "cube" by analogy with cubes in $\mathbb{R}^3$, it is also a ball in the metric $d(\mathbf{x}, \mathbf{y}) = \sup_i |x_i - y_i|$ by Definition 1.6.1.

So far we have not used the notion of linear space and, where possible in this chapter, we shall not exploit it. But the algebraic nature of a linear space X allows us to consider the line defined by

$$tx + (1 - t)y \qquad (x, y \in X) \tag{1.6.1}$$

where $t \in (-\infty, \infty)$ is a parameter. If we restrict $t \in [0, 1]$, then (1.6.1) yields a *segment* in X. When necessary, we shall also use the notions of plane, subspace, etc.

Definition 1.6.3. A set S in X is *convex* if together with each pair of its points it contains the segment connecting those points. That is, S is convex if

$$tx + (1 - t)y \in S \qquad (0 \le t \le 1)$$

for every pair of elements $x, y \in S$.

Definition 1.6.4. A set in a metric space X is *bounded* if there is a ball of finite radius that contains all the elements of the set.

Problem 1.6.2. (a) Show that the union of two bounded sets is a bounded set. (b) Show that the open and closed balls $B(0, r)$ and $\overline{B}(0, r)$ are convex sets. (c) Show that if A and B are convex sets, then so is the intersection $A \cap B$. □

1.7 Convergence in a Metric Space

While geometric figures in metric spaces can be interesting, we are more interested in properties that extend the usual ideas of calculus to function spaces. In large part the required definitions mimic those from calculus. Let us introduce the notion of sequence convergence in a metric space.

In a metric space X, an infinite sequence $\{x_i\}$ has *limit* x if, for every $\varepsilon > 0$, there exists a number $N = N(\varepsilon)$ such that whenever $i > N$ we have $d(x_i, x) < \varepsilon$. In other words, $x_i \in B(x, \varepsilon)$ for all $i > N$. We write

$$x = \lim_{i\to\infty} x_i \quad \text{or} \quad x_i \to x \text{ as } i \to \infty .$$

We also say that $\{x_i\}$ *converges to* x or is *convergent*. This notion generalizes the notion of convergence from calculus, and possesses similar properties:

1. There exists no more than one limit of a convergent sequence. Indeed suppose to the contrary that $\{x_i\}$ has two distinct limits x^* and x^{**}. Then $d(x^*, x^{**}) = a > 0$, say. Take $\varepsilon = a/3$; by definition, there exists N such that for all $i \geq N$ we have $d(x_i, x^*) \leq a/3$ and $d(x_i, x^{**}) \leq a/3$. But

   $$a = d(x^*, x^{**}) \leq d(x^*, x_i) + d(x_i, x^{**}) \leq a/3 + a/3 = 2a/3 ,$$

   a contradiction.
2. A sequence which is convergent in a metric space is bounded.

Problem 1.7.1. Prove the second property. □

The ease of obtaining these and similar results might lead us to try to generalize other classical results like the Bolzano–Weierstrass theorem. However, as mentioned before, not all results extend to spaces of infinite dimension.

A sequence $\{x_i\}$ is called a *Cauchy sequence*[5] if for each $\varepsilon > 0$ there exists $N = N(\varepsilon)$ such that $d(x_n, x_m) < \varepsilon$ whenever $m, n \geq N$. The next problem shows that this is not in general equivalent to the notion of convergence.

Problem 1.7.2. Construct a sequence of functions continuous on the interval $[0, 1]$ such that the sequence converges to a discontinuous function in the space $\tilde{L}(0, 1)$ where

[5] In the Russian literature, the term *fundamental sequence* is used instead of *Cauchy sequence*.

$$d(f,g) = \int_0^1 |f(x) - g(x)|\, dx \tag{1.7.1}$$

specifies the metric. □

Problem 1.7.3. (a) Prove that any subsequence of a convergent sequence is convergent. (b) Show that every Cauchy sequence is bounded. □

1.8 Completeness

Definition 1.8.1. A metric space is *complete* if every Cauchy sequence in the space has a limit in the space; otherwise, it is *incomplete*.

The space $\mathbb{R}$ of real numbers under the metric $d(x, y) = |x-y|$ provides an example of a complete metric space. Its subset $\mathbb{Q}$ of rational numbers yields an incomplete space; there exist Cauchy sequences of rational numbers whose limits are irrational.

Another example of a complete metric space is $C(\Omega)$ for compact $\Omega \subset \mathbb{R}^n$. This follows from Theorem 2 on p. 5. (The reader should verify that a Cauchy sequence in $C(\Omega)$ is uniformly convergent.)

Problem 1.7.2 shows that, depending on the metric imposed, the same set can yield a complete or incomplete metric space. The space of continuous functions on a compact set Ω with the metric (1.7.1) is incomplete.

Definition 1.8.2. An element x of a metric space X is an *accumulation point* of a set S if any ball centered at x contains a point of S different from x. We say that S is a *closed set in* X if it contains all of its accumulation points.

It is clear that for x to be an accumulation point of S it suffices that there exists a countable sequence of open balls centered at x, with radii $\varepsilon_n \to 0$, each containing a point of S distinct from x.

Definition 1.8.3. The *closure* $\overline{S}$ of a set S is the union of S with the set of all its accumulation points.

Note that $S \subseteq \overline{S}$, while $\overline{S} = S$ if and only if S is closed.

Remark 1.8.1. In general, it is false that $\overline{B}(x_0, r) = \overline{B(x_0, r)}$ in a metric space. That is, the closure of an open ball is not always the same as the corresponding closed ball. In a normed space they coincide. □

Problem 1.8.1. Prove the following propositions. (a) A closed ball is a closed set. (b) A set is closed if and only if it contains the limits of all its convergent sequences. (c) A set is closed if and only if its complement is open. (d) $\overline{S}$ is the smallest closed set containing S. (e) We have $x \in \overline{S}$ if and only if for any $\varepsilon > 0$ there exists $x' \in S$ that $\rho(x', x) < \varepsilon$. (f) If $x \in \overline{S}$ then $x_n \to x$ for some $\{x_n\} \subset S$. (g) The union of a finite number of closed sets is also a closed set. (h) The intersection of any family of closed sets is a closed set. □

If X is a complete metric space, the definition of an accumulation point x in S states that there is a Cauchy sequence belonging to S, whose elements are different from x, for which x is a limit element. Conversely, if we have a Cauchy sequence belonging to S in a complete metric space then there is a limit point in X. If S is closed, there are only two possibilities for this point:

1. it is an accumulation point of S;
2. it is an isolated point belonging to S.

These facts permit another form of Definition 1.8.2 for a complete metric space.

Definition 1.8.2′. A set S in a complete metric space X is closed if any Cauchy sequence whose elements are in S has a limit belonging to S.

The next theorem is evident.

Theorem 1.8.1. A subset S of a complete metric space X, supplied with the metric of X, is a complete metric space if and only if S is closed in X.

Definition 1.8.4. A set A is *dense* in a metric space X if for every $x \in X$ any ball of nonzero radius about x contains an element of A.

Weierstrass' theorem (Theorem 3 on p. 5) states that the set of all polynomials is dense in $C(\Omega)$, where Ω is any compact set in $\mathbb{R}^n$.

Completeness is important as numerous limit passages are needed to justify numerical methods, existence theorems, etc. The energy spaces of continuously differentiable functions introduced above are all incomplete. Because these spaces are so convenient in mechanics, we require the material of the next section.

Problem 1.8.2. Prove the *principle of nested balls.* Suppose a complete metric space X contains a sequence of closed balls $\{\overline{B}(x_n, r_n)\}_{n=1}^{\infty}$ such that $\overline{B}(x_{n+1}, r_{n+1}) \subseteq \overline{B}(x_n, r_n)$ for each n, and such that the radii $r_n \to 0$. There is a unique point $x \in X$ such that $x \in \bigcap_{n=1}^{\infty} \overline{B}(x_n, r_n)$. □

1.9 Completion Theorem

Definition 1.9.1. A one-to-one correspondence between two metric spaces (M_1, d_1) and (M_2, d_2) is a *one-to-one isometry* if the correspondence between the elements of these spaces preserves the distances between the elements; that is, if a pair of elements $x, y \in M_1$ corresponds to a pair $u, v \in M_2$, then $d_1(x, y) = d_2(u, v)$.

The next result is the *completion theorem.*

Theorem 1.9.1 (completion). For a metric space M, there is a one-to-one isometry between M and a set $\tilde{M}$ which is dense in a complete metric space M^*. We call M^* the *completion* of M. If M is a linear space, the isometry preserves algebraic operations.

Remark 1.9.1. The elements of $\tilde{M}$ differ in nature from those of the "base set" M. However, in what follows we identify them and refer to the elements of $\tilde{M}$ as if they belonged to M. □

Before we can prove the completion theorem, we need

Definition 1.9.2. Two sequences $\{x_n\}$ and $\{y_n\}$ in M are *equivalent* if $d(x_n, y_n) \to 0$ as $n \to \infty$.

Proof (of Theorem 1.9.1). The proof is constructive. First we introduce the set M^*. Then we verify that it has the properties indicated.

Let $\{x_n\}$ be a Cauchy sequence in M. Collect all Cauchy sequences in M that are equivalent to $\{x_n\}$ and call the collection an equivalence class X. Any Cauchy sequence from X is called a *representative* of X. To any $x \in M$ there corresponds the equivalence class which contains the stationary sequence $(x, x, x, \ldots)$; it is called a *stationary equivalence class*. Denote all equivalence classes X by M^* and all stationary equivalence classes by $\tilde{M}$. Introducing on M^* the metric given by

$$d(X, Y) = \lim_{n\to\infty} d(x_n, y_n) \tag{1.9.1}$$

where $\{x_n\}$ and $\{y_n\}$ are representatives of the equivalence classes X and Y respectively, we obtain the needed spaces M^* and $\tilde{M}$, and the correspondence.

We must show that (a) expression (1.9.1) is actually a metric, i.e., it does not depend on the choice of representatives and satisfies the metric axioms, (b) the space M^* is complete, and (c) the set $\tilde{M}$ is dense in M^*.

(a) Validity of (1.9.1). Let us establish that the limit $d(X, Y)$ exists and is independent of the choice of representative sequences. Because $d(x, y)$ satisfies metric axiom D4, we have

$$d(x_n, y_n) \le d(x_n, x_m) + d(x_m, y_m) + d(y_m, y_n)$$

so that

$$d(x_n, y_n) - d(x_m, y_m) \le d(x_n, x_m) + d(y_m, y_n) \,.$$

Interchanging m and n we have

$$d(x_m, y_m) - d(x_n, y_n) \le d(x_m, x_n) + d(y_n, y_m) \,,$$

hence

$$|d(x_n, y_n) - d(x_m, y_m)| \le d(x_n, x_m) + d(y_n, y_m) \to 0 \text{ as } n, m \to \infty$$

because $\{x_n\}$ and $\{y_n\}$ are Cauchy sequences. So $\{d(x_n, y_n)\}$ is a Cauchy sequence in $\mathbb{R}$ and the limit in (1.9.1) exists. Similarly, the reader can verify that this limit does not depend on the choice of representatives of X and Y.

Problem 1.9.1. Provide this verification. □

We now verify the metric axioms for (1.9.1).

D1: We have

$$d(X, Y) = \lim_{n\to\infty} d(x_n, y_n) \geq 0 .$$

D2: If $X = Y$ then $d(X, Y) = 0$. Conversely, if $d(X, Y) = 0$ then X and Y contain the same set of equivalent Cauchy sequences.

D3: We have

$$d(X, Y) = \lim_{n\to\infty} d(x_n, y_n) = \lim_{n\to\infty} d(y_n, x_n) = d(Y, X) .$$

D4: For $x_n, y_n, z_n \in M$,

$$d(x_n, y_n) \leq d(x_n, z_n) + d(z_n, y_n) .$$

Passage to the limit gives

$$d(X, Y) \leq d(X, Z) + d(Z, Y)$$

for the equivalence classes X, Y, Z containing $\{x_n\}$, $\{y_n\}$, $\{z_n\}$, respectively.

(*b*) *Completeness of* M^*. Let $\{X^i\}$ be a Cauchy sequence in M^*. We shall show that $\{X^i\}$ tends to a limit X as $i \to \infty$. From each class X^i we first choose a Cauchy sequence $\{x_j^{(i)}\}$ and from this an element denoted x_i such that

$$d(x_i, x_j^{(i)}) < \frac{1}{i} \quad \text{for all } j > i ,$$

which is possible since $\{x_j^{(i)}\}$ is a Cauchy sequence. Let us show that $\{x_i\}$ is a Cauchy sequence. Denote by X_i the equivalence class containing the stationary sequence $(x_i, x_i, \ldots)$ and observe that

$$d(X_i, X^i) = \lim_{j\to\infty} d(x_i, x_j^{(i)}) \leq \lim_{j\to\infty} \frac{1}{i} = \frac{1}{i} \quad \text{for all } i .$$

Then

$$\begin{aligned} d(x_i, x_j) = d(X_i, X_j) &\leq d(X_i, X^i) + d(X^i, X^j) + d(X^j, X_j) \\ &\leq \frac{1}{i} + d(X^i, X^j) + \frac{1}{j} \to 0 \quad \text{as } i, j \to \infty . \end{aligned}$$

Now let us denote by X the equivalence class containing the Cauchy sequence $\{x_i\}$. We shall show that $X^i \to X$ as $i \to \infty$. We have

$$\begin{aligned} d(X^i, X) \leq d(X^i, X_i) + d(X_i, X) &\leq \frac{1}{i} + d(X_i, X) \\ &= \frac{1}{i} + \lim_{j\to\infty} d(x_i, x_j) \to 0 \quad \text{as } i \to \infty \end{aligned}$$

since $\{x_i\}$ is a Cauchy sequence. This completes the proof of (b).

(c) *Density of $\tilde{M}$ in M^**. Let X be an equivalence class containing a representative sequence $\{x_n\}$. Denoting by X_n the stationary class for the stationary sequence $(x_n, x_n, \ldots)$, we have

$$d(X_n, X) = \lim_{m\to\infty} d(x_n, x_m) \to 0 \quad \text{as } n \to \infty$$

since $\{x_n\}$ is a Cauchy sequence.

Finally, the equality $d(X, Y) = d(x, y)$ if X and Y are stationary classes corresponding to x and y, respectively, gives the one-to-one isometry between M and $\tilde{M}$. Preservation of algebraic operations in M is obvious, and this completes the proof of Theorem 1.9.1. □

It is worth noting what happens if M is complete. It is clear that we can determine a one-to-one correspondence between any equivalence class and the unique element which is the limit of a representative sequence of this class. Thus we can identify a complete metric space with its completion. However, the natures of the elements of the space and its completion are different: the elements of the completion are equivalence classes.

Because Theorem 1.9.1 is of great importance, let us review its main points:

1. M^* is a complete metric space whose elements are classes of equivalent Cauchy sequences composed of elements taken from M.
2. M is isometrically identified with $\tilde{M}$, which is the set of all stationary equivalence classes.
3. $\tilde{M}$ is dense in M^*.

We can sometimes establish a property of a limit of a representative sequence of X that does not depend on the particular choice of representative. In that case we shall say that the class X possesses this property. This is typical for Sobolev and energy spaces; the formulation of such properties is the basis of the so-called imbedding theorems.

The following sections entail applications of Theorem 1.9.1.

1.10 Lebesgue Integrals and the Space $L^p(\Omega)$

By arguments similar to those in Sect. 1.8, the set of all functions continuous on a compact domain $\Omega \subset \mathbb{R}^n$ with metric

$$d(f(\mathbf{x}), g(\mathbf{x})) = \left(\int_\Omega |f(\mathbf{x}) - g(\mathbf{x})|^p \, d\Omega \right)^{1/p} \qquad (p \geq 1) \qquad (1.10.1)$$

is an incomplete metric space.

Let us apply Theorem 1.9.1 to this case. The completion is denoted by $L^p(\Omega)$. (When $p = 1$ we usually omit the superscript and write $L(\Omega)$ instead of $L^1(\Omega)$.) An element of $L^p(\Omega)$ is an equivalence class of Cauchy sequences of functions continuous on Ω. Note that in the present context $\{f_n(\mathbf{x})\}$ is a Cauchy sequence if

$$\int_\Omega |f_n(\mathbf{x}) - f_m(\mathbf{x})|^p \, d\Omega \to 0 \quad \text{as } n, m \to \infty$$

and two sequences $\{f_n(\mathbf{x})\}$ and $\{g_n(\mathbf{x})\}$ are equivalent if

$$\int_\Omega |f_n(\mathbf{x}) - g_n(\mathbf{x})|^p \, d\Omega \to 0 \quad \text{as } n \to \infty \, .$$

Remark 1.10.1. In the classical theory of functions of a real variable, it is shown that for any equivalence class in $L^p(\Omega)$ there is a function (or, more precisely, a class of equivalent functions) which is a limit, in a certain sense, of a representative sequence of the class; for this function, the so-called Lebesgue integral is introduced. Our constructions of $L^p(\Omega)$ and the Lebesgue integral are equivalent to those of the classical theory. In view of this, we sometimes refer to an equivalence class of $L^p(\Omega)$ as a "function." □

Remark 1.10.2. By Weierstrass' theorem, a function continuous on Ω can be approximated by a polynomial in the metric of $C(\Omega)$, and hence in that of $L^p(\Omega)$. An interpretation is that any equivalence class of $L^p(\Omega)$ contains a Cauchy sequence whose elements are infinitely differentiable functions (moreover, polynomials), hence we may obtain $L^p(\Omega)$ on the basis of just this subset of $C(\Omega)$. □

Remark 1.10.3. Because Riemann integration appears in (1.10.1), we must exclude some "exotic" domains Ω that are allowed in the classical theory of Lebesgue integration. It is possible to extend the present approach to achieve the same degree of generality, but our applications do not require this. The reader may bridge this gap if desired. Furthermore, Ω need not be bounded in order to construct the theory. □

The Lebesgue Integral. An element of $L^p(\Omega)$ (an equivalence class) is denoted by $F(\mathbf{x})$. To construct the Lebesgue integral, we use the Riemann integral. We first consider how to define the integral

$$\int_\Omega |F(\mathbf{x})|^p \, d\Omega \, , \qquad F(\mathbf{x}) \in L^p(\Omega) \, .$$

We take a representative Cauchy sequence $\{f_n(\mathbf{x})\}$ from $F(\mathbf{x})$ and consider the sequence $\{K_n\}$ given by

$$K_n = \left(\int_\Omega |f_n(\mathbf{x})|^p \, d\Omega \right)^{1/p} .$$

This is a Cauchy sequence of numbers. Indeed,

$$\begin{aligned} |K_n - K_m| &= \left| \left(\int_\Omega |f_n(\mathbf{x})|^p \, d\Omega \right)^{1/p} - \left(\int_\Omega |f_m(\mathbf{x})|^p \, d\Omega \right)^{1/p} \right| \\ &\le \left(\int_\Omega |f_n(\mathbf{x}) - f_m(\mathbf{x})|^p \, d\Omega \right)^{1/p} \to 0 \quad \text{as } m, n \to \infty \, , \end{aligned}$$

as a consequence of the inequality

$$\left(\int_\Omega |f(\mathbf{x}) - g(\mathbf{x}) + g(\mathbf{x})|^p \, d\Omega\right)^{1/p} \le \left(\int_\Omega |f(\mathbf{x}) - g(\mathbf{x})|^p \, d\Omega\right)^{1/p} + \left(\int_\Omega |g(\mathbf{x})|^p \, d\Omega\right)^{1/p}$$

(which follows from Minkowski's inequality) and a similar result obtained by swapping f and g. So there exists K such that

$$K = \lim_{n\to\infty} K_n = \lim_{n\to\infty} \left(\int_\Omega |f_n(\mathbf{x})|^p \, d\Omega\right)^{1/p}.$$

Problem 1.10.1. Show that K is independent of the choice of representative sequence. □

The number K^p is called the Lebesgue integral of $|F(\mathbf{x})|^p$:

$$K^p \equiv \int_\Omega |F(\mathbf{x})|^p \, d\Omega = \lim_{n\to\infty} \int_\Omega |f_n(\mathbf{x})|^p \, d\Omega .$$

Let $F(\mathbf{x}) \in L^p(\Omega)$ where Ω is a compact domain. We show that $F(\mathbf{x}) \in L^r(\Omega)$ whenever $1 \le r \le p$. By Hölder's inequality,

$$\left|\int_\Omega 1 \cdot |f(\mathbf{x})|^r \, d\Omega\right| \le \left(\int_\Omega 1^q \, d\Omega\right)^{1/q} \left(\int_\Omega (|f(\mathbf{x})|^r)^{p/r} \, d\Omega\right)^{r/p}$$
$$= (\operatorname{mes} \Omega)^{1/q} \left(\int_\Omega |f(\mathbf{x})|^p \, d\Omega\right)^{r/p}$$

if $1/q + r/p = 1$. For any r such that $1 \le r \le p$, it follows that

$$\left(\int_\Omega |f_n(\mathbf{x}) - f_m(\mathbf{x})|^r \, d\Omega\right)^{1/r} \le (\operatorname{mes} \Omega)^{1/r - 1/p} \left(\int_\Omega |f_n(\mathbf{x}) - f_m(\mathbf{x})|^p \, d\Omega\right)^{1/p}.$$

Hence a sequence of functions which is a Cauchy sequence in the metric (1.10.1) of $L^p(\Omega)$ is also a Cauchy sequence in the metric of $L^r(\Omega)$ whenever $1 \le r < p$. In similar fashion we can show that any two sequences equivalent in $L^p(\Omega)$ are also equivalent in $L^r(\Omega)$. Therefore any element of $L^p(\Omega)$ also belongs to $L^r(\Omega)$ if $1 \le r < p$, and we can say that $L^p(\Omega)$ is a subset of $L^r(\Omega)$. Thus we can determine an integral

$$\int_\Omega |F(\mathbf{x})|^r \, d\Omega$$

for $1 \le r < p$. Moreover, passage to the limit shows that

$$\left(\int_\Omega |F(\mathbf{x})|^r \, d\Omega\right)^{1/r} \le (\operatorname{mes} \Omega)^{1/r - 1/p} \left(\int_\Omega |F(\mathbf{x})|^p \, d\Omega\right)^{1/p}. \tag{1.10.2}$$

Now we can introduce the Lebesgue integral for an element $F(\mathbf{x}) \in L^p(\Omega)$ where $p \ge 1$. Take a representative sequence $\{f_n(\mathbf{x})\}$ of the class $F(\mathbf{x})$. That the sequence of numbers

$$\left\{ \int_\Omega f_n(\mathbf{x})\, d\Omega \right\}$$

is a Cauchy sequence follows from the inequality

$$\left| \int_\Omega f(\mathbf{x})\, d\Omega \right| \le \int_\Omega |f(\mathbf{x})|\, d\Omega\,.$$

Problem 1.10.2. Provide the details. □

So the quantity

$$\int_\Omega F(\mathbf{x})\, d\Omega = \lim_{n\to\infty} \int_\Omega f_n(\mathbf{x})\, d\Omega$$

is uniquely determined for $F(\mathbf{x})$ and is called the *Lebesgue integral* of $F(\mathbf{x}) \in L^p(\Omega)$ over Ω. Note that for the Lebesgue integral we have

$$\left| \int_\Omega F(\mathbf{x})\, d\Omega \right| \le (\text{mes}\,\Omega)^{1/q} \left(\int_\Omega |F(\mathbf{x})|^p\, d\Omega \right)^{1/p} \tag{1.10.3}$$

if $1/q + 1/p = 1$.

Problem 1.10.3. Prove (1.10.3). □

In mechanics, the work of external forces often takes the form

$$\int_\Omega F(\mathbf{x}) G(\mathbf{x})\, d\Omega\,.$$

Let us determine this integral when

$$F(\mathbf{x}) \in L^p(\Omega)\,, \qquad G(\mathbf{x}) \in L^q(\Omega)\,, \qquad \frac{1}{p} + \frac{1}{q} = 1\,.$$

Consider

$$I_n = \int_\Omega f_n(\mathbf{x}) g_n(\mathbf{x})\, d\Omega$$

where $\{f_n(\mathbf{x})\}$ and $\{g_n(\mathbf{x})\}$ are representative sequences of $F(\mathbf{x})$ and $G(\mathbf{x})$, respectively. We will show that $\{I_n\}$ is a numerical Cauchy sequence and therefore possesses a limit. We have

$$\begin{aligned} |I_n - I_m| &= \left| \int_\Omega \{f_n(\mathbf{x}) g_n(\mathbf{x}) - f_m(\mathbf{x}) g_m(\mathbf{x})\}\, d\Omega \right| \\ &= \left| \int_\Omega \{[f_n(\mathbf{x}) - f_m(\mathbf{x})] g_n(\mathbf{x}) + f_m(\mathbf{x})[g_n(\mathbf{x}) - g_m(\mathbf{x})]\}\, d\Omega \right| \end{aligned}$$

and therefore

$$
\begin{aligned}
|I_n - I_m| &\leq \int_\Omega |f_n(\mathbf{x}) - f_m(\mathbf{x})|\, |g_n(\mathbf{x})|\, d\Omega + \int_\Omega |f_m(\mathbf{x})|\, |g_n(\mathbf{x}) - g_m(\mathbf{x})|\, d\Omega \\
&\leq \left(\int_\Omega |f_n(\mathbf{x}) - f_m(\mathbf{x})|^p \, d\Omega \right)^{1/p} \left(\int_\Omega |g_n(\mathbf{x})|^q \, d\Omega \right)^{1/q} \\
&\quad + \left(\int_\Omega |f_m(\mathbf{x})|^p \, d\Omega \right)^{1/p} \left(\int_\Omega |g_n(\mathbf{x}) - g_m(\mathbf{x})|^q \, d\Omega \right)^{1/q} \\
&\to 0 \text{ as } n, m \to \infty
\end{aligned}
$$

since $\{f_n(\mathbf{x})\}$ and $\{g_n(\mathbf{x})\}$ are Cauchy sequences in their respective metrics and

$$
\int_\Omega |f_n(\mathbf{x})|^p \, d\Omega \to \int_\Omega |F(\mathbf{x})|^p \, d\Omega \,, \qquad \int_\Omega |g_n(\mathbf{x})|^q \, d\Omega \to \int_\Omega |G(\mathbf{x})|^q \, d\Omega \,.
$$

So there exists $I = \lim_{n\to\infty} I_n$, which we call the Lebesgue integral

$$
I = \int_\Omega F(\mathbf{x}) G(\mathbf{x}) \, d\Omega \,.
$$

Problem 1.10.4. Show that I is independent of the choice of representatives and hence well-defined. □

Passage to the limit in the inequality

$$
\left| \int_\Omega f_n(\mathbf{x}) g_n(\mathbf{x}) \, d\Omega \right| \leq \left(\int_\Omega |f_n(\mathbf{x})|^p \, d\Omega \right)^{1/p} \left(\int_\Omega |g_n(\mathbf{x})|^q \, d\Omega \right)^{1/q}
$$

shows that Hölder's inequality

$$
\left| \int_\Omega F(\mathbf{x}) G(\mathbf{x}) \, d\Omega \right| \leq \left(\int_\Omega |F(\mathbf{x})|^p \, d\Omega \right)^{1/p} \left(\int_\Omega |G(\mathbf{x})|^q \, d\Omega \right)^{1/q} \tag{1.10.4}
$$

holds for $F(\mathbf{x}) \in L^p(\Omega)$ and $G(\mathbf{x}) \in L^q(\Omega)$ whenever $1/p + 1/q = 1$. Equality holds in Hölder's inequality if and only if $F(\mathbf{x}) = \lambda G(\mathbf{x})$ for some number λ.

Remark 1.10.4. If Ω is unbounded, Hölder's inequality still holds; however, it is not true in general that $L^p(\Omega)$ is a subset of $L^r(\Omega)$ for all $r < p$. □

We conclude by asserting that the properties of the classes in $L^p(\Omega)$ introduced above permit us to deal with the Lebesgue integral as if its integrand were an ordinary function.

Problem 1.10.5. Show that if $f(x)$ and $g(x)$ each belong to $L^2(a, b)$, then $f(x)g(x)$ belongs to $L(a, b)$. Note that we may use lowercase symbols to denote elements of $L^p(\Omega)$. □

1.11 Banach Spaces

Most of the metric spaces we have considered have also been linear spaces. This implies that each pair $x, y \in X$ has a uniquely defined sum $x + y$ such that

1. $x + y = y + x$,
2. $x + (y + z) = (x + y) + z$, and
3. there is a zero element $\theta \in X$ such that $x + \theta = x$;

it also implies that each $x \in X$ has a uniquely defined product by a real (or complex) scalar $\lambda \in \mathbb{R}$ (or $\mathbb{C}$) such that

4. $\lambda(\mu x) = (\lambda\mu)x$,
5. $1x = x$, $0x = \theta$,
6. $x + (-1x) = \theta$,
7. $\lambda(x + y) = \lambda x + \lambda y$,
8. $(\lambda + \mu)x = \lambda x + \mu x$.

We shall denote the zero element of X by 0 instead of θ.

If multiplication by scalars is introduced as multiplication by purely real numbers, the space is called a *real linear space*; if the scalars are in general complex numbers, it is a *complex linear space*.

We could continue to study the general properties of metric spaces, but all spaces of interest to us have a convenient property: their metrics take a special form that can be denoted by

$$d(x, y) = \|x - y\| . \tag{1.11.1}$$

Definition 1.11.1. A function $\|x\|$ is a *norm* on a linear space X if it is real-valued, defined for every $x \in X$, and satisfies the following norm axioms:

N1. $\|x\| \geq 0$, and $\|x\| = 0$ if and only if $x = 0$;

N2. $\|\lambda x\| = |\lambda| \, \|x\|$ for any real (or complex) λ;

N3. $\|x + y\| \leq \|x\| + \|y\|$.

The reader should verify that any metric defined by (1.11.1) satisfies D1–D4, provided that $\|x\|$ satisfies N1–N3. Note that N1 can be formulated as

N1′. $\|x\| = 0$ if and only if $x = 0$.

Indeed, from N2 it follows that $0 = \|0\| = \|x - x\| \leq \|x\| + \|-x\| = 2\|x\|$ and so $\|x\| \geq 0$ for all $x \in X$.

Definition 1.11.2. A linear space X is a *normed space* if, for every $x \in X$, a norm of x satisfying N1–N3 is defined (i.e., $\|x\|$ takes a unique finite value for each $x \in X$). A normed space X is said to be *real* (*complex*) if the scalars λ in the product λx are taken from $\mathbb{R}$ ($\mathbb{C}$).

Each normed space is a metric space, but some linear metric spaces are not normed. For a normed space, we shall use the terminology of the corresponding metric space (and later follow this practice for inner product spaces as well).

It is clear that the sequence spaces c, m, and ℓ^p are normed spaces with $\|\mathbf{x}\| = d(\mathbf{x}, \mathbf{0})$ where $d(\cdot, \cdot)$ is the metric on the respective space. The same is true of the other metric spaces we have considered.

Definition 1.11.3. A complete normed space is a *Banach space*.

Problem 1.11.1. Show that if x and y are elements of a real normed space X, then

$$\|x - y\| \geq \left| \|x\| - \|y\| \right| . \tag{1.11.2}$$

Note that a special case of this result was used in Sect. 1.10 to show that $\{K_n\}$ is a Cauchy sequence. □

Problem 1.11.2. Which of the expressions

(a) $\|f\| = \int_a^b |f(x)|\,dx + \max_{x\in[a,b]} |f'(x)|$,

(b) $\|f\| = |f(a)| + \max_{x\in[a,b]} |f'(x)|$,

(c) $\|f\| = \max_{x\in[a,b]} |f'(x)|$,

can serve as a norm on the set of functions that are continuously differentiable on the interval $[a, b]$? □

Problem 1.11.3. [38] Let X be a normed space over a numerical field $\mathbb{K}$ (= $\mathbb{R}$ or $\mathbb{C}$) and prove the following statements.

(a) If $\{x_n\}$ is a Cauchy sequence and $\alpha \in \mathbb{K}$ is a constant, then $\{\alpha x_n\}$ is a Cauchy sequence.

(b) If $\{x_n\}$ and $\{y_n\}$ are Cauchy sequences in X, then so is $\{x_n + y_n\}$.

(c) If $\alpha_n \to \alpha$ in $\mathbb{K}$ and $x_n \to x$ in X, then $\alpha_n x_n \to \alpha x$ in X as $n \to \infty$.

(d) If $x_n \to x$, then $\|x_n\| \to \|x\|$. Hence the norm is a continuous function.

(e) If $x_n \to x$ and $\|x_n - y_n\| \to 0$, then $y_n \to x$.

(f) If $x_n \to x$, then $\|x_n - y\| \to \|x - y\|$.

(g) If $x_n \to x$ and $y_n \to y$, then $\|x_n - y_n\| \to \|x - y\|$.

(h) If $\{x_n\}$ and $\{y_n\}$ are Cauchy sequences, the sequence $\{\|x_n - y_n\|\}$ converges.

(i) A set $A \subset X$ is bounded if and only if for any sequence $\{x_n\} \subset A$ and any numerical sequence $\{\alpha_n\}$ that tends to zero, the sequence $\{\alpha_n x_n\}$ tends to zero.

(j) The inequality $\|x\| \leq \max\{\|x + y\|, \|x - y\|\}$ holds for all $x, y \in X$. □

Norms in $\mathbb{R}^n$ and $\mathbb{C}^n$. We know the spaces $\mathbb{R}$ and $\mathbb{C}$ are complete, as are $\mathbb{R}^n$ and $\mathbb{C}^n$ because any Cauchy sequence in one of these spaces has a limit in the space if we take the Euclidean norm

$$\|\mathbf{x}\| = \left(\sum_{k=1}^{n} |x_k|^2\right)^{1/2}. \tag{1.11.3}$$

Hence they are Banach spaces. If the x_k in (1.11.3) are the components of $\mathbf{x}$ in a non-canonical basis, the same formula (1.11.3) presents another norm in $\mathbb{R}^n$ or $\mathbb{C}^n$. As ℓ^p and m are normed spaces, the same is true for $\mathbb{R}^n$ or $\mathbb{C}^n$.

Problem 1.11.4. Show that in $\mathbb{R}^n$,

$$\|\mathbf{x}\|_m \equiv \sup_{1\le k\le n} |x_k| = \lim_{p\to\infty} \|\mathbf{x}\|_p \equiv \lim_{p\to\infty} \left(\sum_{k=1}^{n} |x_k|^p\right)^{1/p}.$$

So we can write $\|\cdot\|_m = \|\cdot\|_\infty$. □

We can present even more general norms in $\mathbb{R}^n$ and $\mathbb{C}^n$. Let us introduce a linear transformation

$$z_k = \sum_{m=1}^{n} a_{km} x_m$$

and consider $\|\mathbf{z}\|_p$ for $p \ge 1$. This is a norm on the space of vectors $\mathbf{z}$. When will it be a norm for the initial vectors $\mathbf{x}$? If the matrix (a_{km}) is not degenerate (i.e., if $\det(a_{km}) \ne 0$), we can check the validity of all three norm axioms in terms of x_k. So another type of norm in $\mathbb{R}^n$ or $\mathbb{C}^n$ is

$$\|\mathbf{x}\|_{A,p} = \sum_{k=1}^{n} \left(\left|\sum_{m=1}^{n} a_{km} x_m\right|^p\right)^{1/p} \qquad (p \ge 1).$$

A special case is the norm

$$\|\mathbf{x}\|_{A,\infty} = \max_k \left(\left|\sum_{m=1}^{n} a_{km} x_m\right|\right).$$

Another is the *weighted norm*

$$\|\mathbf{x}\|_w = \left(\sum_{k=1}^{n} h_k |x_k|^p\right)^{1/p}$$

with constants $h_k > 0$.

Problem 1.11.5. (a) Propose a norm in $\mathbb{R}^n$ that does not belong to this family of norms. (b) Verify that if $\alpha_1, \ldots, \alpha_n$ are positive constants, then the following expressions are valid norms on $\mathbb{R}^n$:

$$\|\mathbf{x}\| = \max_{1\le k\le n} (\alpha_k |x_k|), \quad \|\mathbf{x}\| = \sum_{k=1}^{n} \alpha_k |x_k|, \quad \|\mathbf{x}\| = \left(\sum_{k=1}^{n} \alpha_k |x_k|^2\right)^{1/2},$$

where $\mathbf{x} = (x_1, \ldots, x_n)$. □

Thus we have introduced a variety of norms. Is it necessary to prove that the spaces $\mathbb{R}^n$ and $\mathbb{C}^n$ are complete with these norms? The answer is no. To show this, we first introduce

Definition 1.11.4. Two norms $\|\cdot\|_1$ and $\|\cdot\|_2$ defined on a normed space X are *equivalent* if there exist positive constants c_1 and c_2 such that

$$c_1 \|x\|_1 \leq \|x\|_2 \leq c_2 \|x\|_1 \quad \text{for all } x \in X .$$

It is easy to see that norm equivalence is *transitive*: if $\|\cdot\|_1$ is equivalent to $\|\cdot\|_2$, and $\|\cdot\|_2$ is equivalent to $\|\cdot\|_3$, then $\|\cdot\|_1$ is equivalent to $\|\cdot\|_3$.

Theorem 1.11.1. All norms on $\mathbb{R}^n$ ($\mathbb{C}^n$) are equivalent.

The proof will follow from Theorem 1.12.1.

1.12 Norms on *n*-Dimensional Spaces

The natures of the elements of finite dimensional spaces can differ greatly. However, we can put the elements of two spaces having the same finite dimension into a one-to-one correspondence that preserves the algebraic structure of the spaces. In this way we can directly "translate" a result for one space to the corresponding result for another space. Such a linear, bijective correspondence between linear spaces X and Y is called an *isomorphism*. We will use this idea to introduce norms on some n-dimensional spaces. One finite dimensional space is P_n, the space of polynomials having degree less than or equal to n.

Norms on P_n. The space P_n has a basis of monomials $\{1, x, \ldots, x^n\}$, so its dimension is $n + 1$. Setting up a correspondence between the bases of $\mathbb{R}^{n+1}$ and P_n as

$$x^k \leftrightarrow \mathbf{e}_{k+1} \qquad (k = 0, 1, \ldots, n)$$

we obtain an isomorphism T from P_n to $\mathbb{R}^{n+1}$ defined by the formulas

$$\mathbf{a} = (a_0, \ldots, a_n) = T p_n = T\left(\sum_{k=0}^{n} a_k x^k\right).$$

So a polynomial p_n is placed into one-to-one correspondence with a "vector" having components equal to the coefficients of p_n. This transformation preserves algebraic operations in P_n for the "images" in $\mathbb{R}^n$. So based on the norm in $\mathbb{R}^n$, we can introduce the following scale of norms in P_n:

$$\|p_n\|_p = \|T p_n\|_p = \left\|\sum_{k=0}^{n} a_k x^k\right\|_p = \left(\sum_{k=0}^{n} |a_k|^p\right)^{1/p} \qquad (p \geq 1) .$$

In similar fashion we can establish an isomorphism between $\mathbb{R}^n$ and any finite-dimensional space of functions such as the trigonometric polynomials

$$T_n = \frac{a_0}{2} + a_1 \cos x + \cdots + a_n \cos nx + b_1 \sin x + \cdots + b_n \sin nx ,$$

and obtain a normed space.

However, in P_n we can find other norms that do not refer to the isomorphism. The reader can verify (by checking the norm axioms) that the following expressions can serve as norms in P_n:

$$\|p\|_C = \max_{x\in[a,b]} |p(x)| , \quad \|p\|_{C^1} = \max_{x\in[a,b]} (|p(x)| + |p'(x)|), \quad \|p\|_L = \int_a^b |p(x)|\,dx ,$$

for $-\infty < a < b < \infty$. We could continue the list of such norms. Now we should remember that norms are used to measure the distance between the elements of a space. But in functional analysis, they are also used to introduce the limits of sequences in a manner similar to calculus.

The multiplicity of possible norms in P_n prompts us to ask, exactly as for $\mathbb{R}^n$, whether convergence of a sequence of polynomials with respect to one norm implies convergence with respect to another norm. The following theorem answers this question for a general finite dimensional normed space.

Theorem 1.12.1. On an n-dimensional normed space X_n, all norms are equivalent.

Proof. It suffices to show that an arbitrary norm $\|\cdot\|$ is equivalent to a certain norm $\|\cdot\|_m$ that we select as follows. Let $\{\mathbf{e}_1, \ldots, \mathbf{e}_n\}$ be a basis of X_n and take an arbitrary

$$\mathbf{x} = \sum_{k=1}^{n} \alpha_k \mathbf{e}_k .$$

Using the isomorphism $T\mathbf{x} = (\alpha_1, \ldots, \alpha_n)$ with $\mathbb{R}^n$ (or $\mathbb{C}^n$) we take

$$\|\mathbf{x}\|_m = \max_k |\alpha_k| .$$

Let us consider

$$\|\mathbf{x}\| = \left\| \sum_{k=1}^{n} \alpha_k \mathbf{e}_k \right\| \quad \text{on the unit sphere } \|\mathbf{x}\|_m = 1 .$$

By (1.11.2),

$$\left| \left\| \sum_{k=1}^{n} (\alpha_k + \Delta_k)\mathbf{e}_k \right\| - \left\| \sum_{k=1}^{n} \alpha_k \mathbf{e}_k \right\| \right| \le \left\| \sum_{k=1}^{n} \Delta_k \mathbf{e}_k \right\| \le \sum_{k=1}^{n} |\Delta_k| \, \|\mathbf{e}_k\| .$$

This means the function

$$F(\alpha_1, \ldots, \alpha_n) = \left\| \sum_{k=1}^{n} \alpha_k \mathbf{e}_k \right\|$$

is continuous with respect to the arguments $(\alpha_1, \ldots, \alpha_n)$. In $\mathbb{R}^n$ (or $\mathbb{C}^n$) the sphere

$$\left\| \sum_{k=1}^{n} \alpha_k \mathbf{e}_k \right\|_m = 1$$

is compact, hence by Weierstrass' theorem $F(\alpha_1, \ldots, \alpha_n)$ takes its minimum c_1 and maximum c_2 on the sphere. Therefore

$$c_1 \le \left\| \sum_{k=1}^{n} \alpha_k \mathbf{e}_k \right\| \le c_2 \ .$$

Clearly $c_2 < \infty$. Also $c_1 > 0$, because if $c_1 = 0$ then at a point $\mathbf{x}_0$ on the unit sphere we get $\|\mathbf{x}_0\| = 0$ and so $\mathbf{x}_0 = 0$, which is impossible. Now for any $\mathbf{x} \neq \mathbf{0}$, the point $\mathbf{y} = \mathbf{x}/\, \|\mathbf{x}\|_m$ is on the unit sphere, $\|\mathbf{y}\|_m = 1$, and thus

$$c_1 \le \left\| \frac{\mathbf{x}}{\|\mathbf{x}\|_m} \right\| \le c_2 \ .$$

This provides the needed equivalence of the norms. □

We have mentioned other types of norms on P_n, the space of polynomials of order not exceeding n. They are all equivalent. Let us consider one more type of norm.

Problem 1.12.1. Take $n + 1$ distinct points $x_1, \ldots, x_{n+1}$. Show that the correspondence $Tp = (p(x_1), \ldots, p(x_{n+1}))$ is an isomorphism between P_n and $\mathbb{R}^{n+1}$. Demonstrate that

$$\|p\| = \left(\sum_{k=1}^{n+1} |p(x_k)|^r \right)^{1/r} \qquad (r \ge 1)$$

is a norm on P_n. □

Problem 1.12.2. In the norms $\|\cdot\|_p$ on $\mathbb{R}^n$, why is the parameter p restricted to the range $1 \le p < \infty$? □

1.13 Other Examples of Banach Spaces

Let us show that c, m, and ℓ^p are complete in their norms.

1. *The space ℓ^p for* $1 \le p < \infty$. Let $\{\mathbf{x}^{(n)}\}$ be a Cauchy sequence in ℓ^p. That is, for any $\varepsilon > 0$ there exists N such that whenever $n, m > N$,

$$\|\mathbf{x}^{(n)} - \mathbf{x}^{(m)}\|_p = \left(\sum_{k=1}^{\infty} |x_k^{(n)} - x_k^{(m)}|^p \right)^{1/p} < \varepsilon . \tag{1.13.1}$$

For any fixed k it follows that the numerical sequence $\{x_k^{(n)}\}$ is a Cauchy sequence and therefore tends to a limit x_k as $n \to \infty$. From (1.13.1) we find that for any M,

$$\left(\sum_{k=1}^{M} |x_k^{(n)} - x_k^{(m)}|^p \right)^{1/p} < \varepsilon .$$

Now

$$\lim_{m\to\infty} \left(\sum_{k=1}^{M} |x_k^{(n)} - x_k^{(m)}|^p \right)^{1/p} = \left(\sum_{k=1}^{M} |x_k^{(n)} - x_k|^p \right)^{1/p} \le \varepsilon . \tag{1.13.2}$$

By Minkowski's inequality,

$$\begin{aligned} \left(\sum_{k=1}^{M} |x_k|^p \right)^{1/p} &= \left(\sum_{k=1}^{M} |x_k - x_k^{(n)} + x_k^{(n)}|^p \right)^{1/p} \\ &\le \left(\sum_{k=1}^{M} |x_k - x_k^{(n)}|^p \right)^{1/p} + \left(\sum_{k=1}^{M} |x_k^{(n)}|^p \right)^{1/p} \le \varepsilon + R \ \text{ for } n > N . \end{aligned}$$

Here we used the fact that every Cauchy sequence is bounded: $\|\mathbf{x}^{(n)}\|_p \le R$ for all n. As the sequence

$$\left\{ \left(\sum_{k=1}^{M} |x_k|^p \right)^{1/p} \right\}$$

increases monotonically with M and is bounded from above, its limit as $M \to \infty$ exists. Hence $\mathbf{x}^* = (x_1, x_2, \ldots)$ belongs to ℓ^p. A similar limit in (1.13.2),

$$\left(\sum_{k=1}^{\infty} |x_k^{(n)} - x_k|^p \right)^{1/p} \le \varepsilon \ \text{ for all } n > N ,$$

shows that $\mathbf{x}^{(n)} \to \mathbf{x}^*$ in the ℓ^p norm, which means that ℓ^p is complete.

2. *The space m.* Recall that m is the space of bounded sequences $\mathbf{x} = (x_1, x_2, \ldots)$ with the norm

$$\|\mathbf{x}\| = \sup_i |x_i| .$$

Suppose $\{\mathbf{x}^{(n)}\}$ is a Cauchy sequence in m. So for any $\varepsilon > 0$ there is N such that whenever $k, n > N$,

$$\sup_j |x_j^{(k)} - x_j^{(n)}| < \varepsilon .$$

Proceeding as in the case of ℓ^p, we find that for any fixed j, the sequence $\{x_j^{(n)}\}$ is a numerical Cauchy sequence and therefore tends to a limit x_j as $n \to \infty$. Denote $\{x_j\} = \mathbf{x}^*$. Now

$$|x_j| = |x_j - x_j^{(N+1)} + x_j^{(N+1)}| \le |x_j - x_j^{(N+1)}| + |x_j^{(N+1)}| \le \varepsilon + R$$

since $\|\mathbf{x}^{(n)}\| \le R$. So $\sup_j |x_j| \le \varepsilon + R$, and $\mathbf{x}^* \in m$. Next we have

$$\sup_{1\le j\le M} |x_j^{(k)} - x_j^{(n)}| < \varepsilon \quad \text{for all } k, n > N \text{ and any } M\ .$$

Taking $n \to \infty$ we get

$$\sup_{1\le j\le M} |x_j^{(k)} - x_j| \le \varepsilon \quad \text{for all } k > N \text{ and any } M$$

and so

$$\sup_j |x_j^{(k)} - x_j| \le \varepsilon \quad \text{for all } k > N\ .$$

This means that $\mathbf{x}^{(k)} \to \mathbf{x}^*$ in the norm of m, so m is complete.

3. *The space* c. In the space c, which is a subspace of m consisting of those sequences that approach a limit, we also get convergence $\mathbf{x}_k \to \mathbf{x}^*$. We leave it to the reader to show that $\mathbf{x}^* \in c$ so that c is a Banach space. The same is true of its subspace c_0.

Problem 1.13.1. Prove that c is a Banach space. □

Problem 1.13.2. Show that the set

$$S = \{\mathbf{x} = (\xi_1, \xi_2, \ldots) \in \ell^2 : |\xi_k| \le a_k \text{ for } k = 1, 2, \ldots\}$$

is closed. □

Some Banach Spaces of Functions. Now we consider some function spaces.

4. *The space* $C(\Omega)$. With compact $\Omega \subset \mathbb{R}^n$ the space $C(\Omega)$ is a linear space and is clearly a normed space if we set

$$\|f\| = \max_{\mathbf{x}\in\Omega} |f(\mathbf{x})|\ .$$

Because $C(\Omega)$ is complete as a metric space, it is a Banach space. We leave it to the reader to discuss why $L^p(\Omega)$ is a Banach space under the norm

$$\|F\| = \left(\int_\Omega |F(\mathbf{x})|^p \, d\Omega \right)^{1/p} \qquad (p \ge 1).$$

5. *The space* $C^{(k)}(\Omega)$. Let $\Omega \subset \mathbb{R}^n$ be a closed and bounded domain in $\mathbb{R}^n$. The space $C^{(k)}(\Omega)$ consists of functions defined and continuous on Ω and having all their derivatives up to order k continuous on Ω. The norm on $C^{(k)}(\Omega)$ is defined by

$$\|f\| = \max_{\mathbf{x}\in\Omega} |f(\mathbf{x})| + \sum_{|\alpha|\le k} \max_{\mathbf{x}\in\Omega} |D^\alpha f(\mathbf{x})|\ .$$

The reader should check for satisfaction of N1–N3; we proceed to show that the resulting space is complete.

Let $\{f_i\}$ be a Cauchy sequence in $C^{(k)}(\Omega)$. This implies that the sequence $\{f_i\}$ as well as all the sequences $\{D^\alpha f_i\}$ when $|\alpha| \le k$ are Cauchy sequences in $C(\Omega)$. Being uniformly convergent on Ω, each of these sequences has a limit function:

$$\lim_{i\to\infty} f_i(\mathbf{x}) = f(\mathbf{x}), \qquad \lim_{i\to\infty} D^\alpha f_i(\mathbf{x}) = f^\alpha(\mathbf{x}) \quad (|\alpha| \le k)$$

where $f(\mathbf{x})$ and $f^\alpha(\mathbf{x})$ for each $|\alpha| \le k$ are continuous. To complete the verification we must show that

$$D^\alpha f(\mathbf{x}) = f^\alpha(\mathbf{x}).$$

We check this only for $\partial f/\partial x_1$; for the other derivatives it can be done in a similar way. So let

$$\lim_{i\to\infty} \frac{\partial f_i(\mathbf{x})}{\partial x_1} = f^1(\mathbf{x}) = f^1(x_1, x_2, \ldots, x_n).$$

Consider

$$\Delta = f(x_1, x_2, \ldots, x_n) - f(a, x_2, \ldots, x_n) - \int_a^{x_1} f^1(t, x_2, \ldots, x_n)\, dt.$$

Using the identity

$$f_i(x_1, \ldots, x_n) - f_i(a, \ldots, x_n) = \int_a^{x_1} \frac{\partial f_i(t, x_2, \ldots, x_n)}{\partial t}\, dt$$

we have

$$\Delta = [f(x_1, \ldots, x_n) - f_i(x_1, \ldots, x_n)] - [f(a, x_2, \ldots, x_n) - f_i(a, x_2, \ldots, x_n)]$$
$$- \left[\int_a^{x_1} \left(f^1(t, x_2, \ldots, x_n) - \frac{\partial f_i(t, x_2, \ldots, x_n)}{\partial t}\right) dt\right].$$

Each of the terms in square brackets tends to zero uniformly as $i \to \infty$, so $\Delta = 0$ since Δ does not depend on i, i.e.,

$$f(x_1, x_2, \ldots, x_n) - f(a, x_2, \ldots, x_n) = \int_a^{x_1} f^1(t, x_2, \ldots, x_n)\, dt.$$

Thus

$$\frac{\partial f(\mathbf{x})}{\partial x_1} = f^1(\mathbf{x}).$$

6. *The space* $H^{k,\lambda}(\Omega)$. The Hölder space $H^{k,\lambda}(\Omega)$, $0 < \lambda \le 1$, consists of those functions of $C^{(k)}(\Omega)$ whose norms in $H^{k,\lambda}(\Omega)$, defined by

$$\|f\| = \sum_{0 \le |\alpha| \le k} \max_{\mathbf{x}\in\Omega} |D^\alpha f(\mathbf{x})| + \sum_{|\alpha| = k} \sup_{\substack{\mathbf{x},\mathbf{y}\in\Omega \\ \mathbf{x}\ne\mathbf{y}}} \frac{|D^\alpha f(\mathbf{x}) - D^\alpha f(\mathbf{y})|}{|\mathbf{x} - \mathbf{y}|^\lambda},$$

are finite. $H^{k,\lambda}(\Omega)$ is also a Banach space.

1.14 Hilbert Spaces

Just as the notion of distance can be widely extended, so can the notion of vector dot product.

Definition 1.14.1. Let H be a linear space over $\mathbb{C}$. A function (x, y) uniquely defined (and therefore finite) for each pair $x, y \in H$ is an *inner product* on H if it satisfies the following axioms:

P1. $(x, x) \geq 0$, and $(x, x) = 0$ if and only if $x = 0$;

P2. $(x, y) = \overline{(y, x)}$, where the overbar denotes complex conjugation;

P3. $(\lambda x + \mu y, z) = \lambda(x, z) + \mu(y, z)$ whenever $\lambda, \mu \in \mathbb{C}$.

The pair consisting of H and an inner product $(\cdot, \cdot)$ is a (*complex*) *inner product space* (or *pre-Hilbert space*).

We can consider H over $\mathbb{R}$; then the inner product is real-valued, P2 is replaced by

P2′. $(x, y) = (y, x)$,

and H is called a *real* inner product space. If clear from the context, the designation "real" or "complex" shall be omitted.

Let us consider some properties of H. We introduce $\|x\|$ using

$$\|x\| = (x, x)^{1/2}. \tag{1.14.1}$$

To show that we really have a norm, we prove the *Schwarz inequality*.

Theorem 1.14.1 (Schwarz inequality). For any $x, y \in H$, the inequality

$$|(x, y)| \leq \|x\| \, \|y\| \tag{1.14.2}$$

holds. For $x, y \neq 0$, equality holds if and only if $x = \lambda y$.

Proof. If either x or y is zero, there is nothing to show. Let $y \neq 0$ and let λ be a scalar. By P1, we have

$$(x + \lambda y, x + \lambda y) \geq 0\,.$$

But

$$(x + \lambda y, x + \lambda y) = (x, x) + \lambda(y, x) + \overline{\lambda}(x, y) + \lambda\overline{\lambda}(y, y) \equiv A(\lambda)\,.$$

Put $\lambda_0 = -(x, y)/(y, y)$; then

$$A(\lambda_0) = \|x\|^2 - 2\frac{|(x, y)|^2}{\|y\|^2} + \frac{|(x, y)|^2 \, \|y\|^2}{\|y\|^4} \geq 0\,.$$

Inequality (1.14.2) follows directly. □

Now we can verify that the *induced norm* (1.14.1) satisfies N1–N3. Axiom N1 is satisfied by virtue of P1. Axiom N2 is satisfied because

$$\|\lambda x\| = (\lambda x, \lambda x)^{1/2} = (\lambda\bar{\lambda})^{1/2}(x,x)^{1/2} = |\lambda|\,\|x\| \,.$$

Axiom N3 is satisfied because

$$\begin{aligned}\|x+y\|^2 &= (x+y, x+y) = (x,x) + (x,y) + (y,x) + (y,y)\\ &\le \|x\|^2 + \|x\|\,\|y\| + \|x\|\,\|y\| + \|y\|^2 = (\|x\| + \|y\|)^2 \,.\end{aligned}$$

We have shown that an inner product space is a normed space.

Definition 1.14.2. A complete inner product space is a *Hilbert space*.

By analogy with Euclidean space, we say that x is *orthogonal to* y in H if

$$(x, y) = 0 \,.$$

In this case we sometimes write $x \perp y$.

Problem 1.14.1. Show that the *parallelogram equality*

$$\|x+y\|^2 + \|x-y\|^2 = 2(\|x\|^2 + \|y\|^2) \tag{1.14.3}$$

holds for all x, y in an inner product space. □

Let us consider some examples of Hilbert spaces.

1. *The space* ℓ^2. For $\mathbf{x}, \mathbf{y} \in \ell^2$ over $\mathbb{C}$, an inner product is defined by

$$(\mathbf{x}, \mathbf{y}) = \sum_{k=1}^{\infty} x_k \overline{y_k} \,, \tag{1.14.4}$$

where, again, the bar over y_k means complex conjugation. The space ℓ^2 was instrumental in the development of functional analysis; the prototype for all Hilbert spaces, it was introduced by David Hilbert (1862–1943) in a paper devoted to justifying the Dirichlet principle. In ℓ^2 defined over $\mathbb{R}$ instead of $\mathbb{C}$, the inner product is given by

$$(\mathbf{x}, \mathbf{y}) = \sum_{k=1}^{\infty} x_k y_k \,. \tag{1.14.5}$$

2. *The space* $L^2(\Omega)$. Here the inner product is

$$(f, g) = \int_{\Omega} f(\mathbf{x})\overline{g(\mathbf{x})}\, d\Omega \,. \tag{1.14.6}$$

The axioms P1–P3 are readily verified for these spaces. The reader should write down the Schwarz inequality in both cases, along with the inner product for $L^2(\Omega)$ over the reals. It is important to note that the energy spaces introduced earlier are all inner product spaces.

Problem 1.14.2. Write down the Schwarz inequality for use with the spaces ℓ^2 and $L^2(\Omega)$. □

Problem 1.14.3. Prove *Apollonius's identity*:

$$\|z - x\|^2 + \|z - y\|^2 = \tfrac{1}{2}\|x - y\|^2 + 2\left\|z - \tfrac{1}{2}(x + y)\right\|^2$$

for any three elements x, y, z of an inner product space. □

1.15 Factor Spaces

In Sect. 1.5 we encountered a situation where the "distance" between two elements of a metric space for a free elastic object was defined up to a rigid displacement term. For a free membrane, two deflection functions u_1 and u_2 are "equal" under the energy metric if their difference $u_2(x, y) - u_1(x, y) = \text{constant}$. If we really wish to use the energy metric as a metric, we must find a way to remove this ambiguity. Moreover, the energy metric induces both a norm and an inner product on the linear set of all possible membrane deflections, and our approach must preserve a linear structure in the resulting space. An approach that makes mechanical sense is to impose some additional restriction on the functions, obtaining from the class of all functions taking the form $u(x, y)+c$, with indefinite c, just one representative element $u(x, y)+c_0$. To preserve linearity of the new space, c_0 should be defined with a linear restriction such as

$$\iint_\Omega [u(x, y) + c_0]\, dx\, dy = 0 .$$

Although this is a good approach, it does lead to another problem for a membrane whose deflection is restricted by this condition. We will typically use this approach to study equilibrium problems for free bodies. However, in this case we will need a small additional investigation regarding what happens when we remove this artificial geometrical restriction. In an equilibrium problem for a free body, existence of a solution will require a restriction on the set of permissible external loads, viz., that the load should be self-balanced.

Another approach can lead us to the same existence result in the equilibrium problem for a free elastic body. It might seem "more mathematical" but it does have clear mechanical roots. For a membrane, we can "identify" all displacement functions whose differences are constant over Ω. From a mechanical viewpoint, two identified functions $u(x, y) + c_1$ and $u(x, y) + c_2$ describe the same deformed membrane in two coordinate systems shifted along the normal to the membrane. Then we can introduce a new space of classes of such functions that are equal up to a constant, and reformulate an equilibrium problem in this new space. This approach yields existence results equivalent to those obtained by imposing additional restrictions on the deflections. To obtain a well-posed problem, we will still require self-balance conditions on the load. The class of solutions in this approach will have a representative that is a solution of the problem with the restricted deflections.

This idea of identifying the elements that are equal up to an element of some linear set is realized in a structure called a *factor space*. Let X be a normed space and let M be a closed subspace of X, i.e., a linear space $M \subset X$ such that any Cauchy sequence from M has a limit in M. If $x_1, x_2 \in X$, then we say that x_2 is *equivalent* to x_1 *modulo* M if $x_1 - x_2 \in M$. This is sometimes written as $x_2 \equiv x_1 \pmod{M}$ but we will write simply $x_2 \sim x_1$. Note that this relation is

1. *reflexive*, meaning that $x \sim x$ for all $x \in X$;
2. *symmetric* in the sense that $x_2 \sim x_1$ implies $x_1 \sim x_2$;
3. *transitive*: if $x_1 \sim x_2$ and $x_2 \sim x_3$, then $x_1 \sim x_3$.

Hence it is an *equivalence relation* on X.

The *coset* of $x \in X$, denoted $[x]$, is the set of all $y \in X$ such that $x - y \in M$. Note that if $x \neq y$, then $[x]$ and $[y]$ have no common elements. The sum of two cosets $[x_1]$ and $[x_2]$ is defined by the equation

$$[x_1] + [x_2] = [x_1 + x_2] .$$

This definition makes sense because if $y_k \sim x_k$, then $y_1 + y_2 \sim x_1 + x_2$ and therefore $[y_1 + y_2] = [x_1 + x_2]$. Similarly, we introduce the scalar multiple

$$[\alpha x_1] = \alpha[x_1]$$

where $\alpha \in \mathbb{R}$ or $\alpha \in \mathbb{C}$ depending on whether X is a real or complex space. So the collection of cosets $[x]$ becomes a linear space called the *factor space* (or *quotient space*) of X by M. It is denoted by X/M. The subspace M constitutes the zero element of X/M.

The quotient space X/M is a normed space with

$$\|[x]\| = \inf_{m \in M} \|x - m\| .$$

Let us verify satisfaction of the norm axioms. Because $0 \in M$, we find that

$$\|[0]\| = \inf_{m \in M} \|0 - m\| = 0 .$$

Conversely, if $\|[x]\| = 0$ then there is sequence $\{m_k\} \subset M$ such that $\|x - m_k\| \to 0$ as $k \to \infty$. But M is closed, hence the limit x of $\{m_k\}$ is in M. Thus $x \in M$, and so $[x] = [0] = M$. Axiom N1 is verified. Axioms N2 and N3 are also easily verified. Consider N3. For any $x, y \in X$ there are sequences $\{m_k\}$ and $\{n_k\}$ in M such that

$$\|x - m_k\| \to \|[x]\| \quad \text{and} \quad \|y - n_k\| \to \|[y]\| \quad \text{as } k \to \infty .$$

Then

$$\left\|[x+y]\right\| = \inf_{m\in M} \|x+y-m\| \le \|x+y-m_k-n_k\|$$
$$\le \|x-m_k\| + \|y-n_k\| \to \left\|[x]\right\| + \left\|[y]\right\| ,$$

which is the triangle inequality.

It is easy to see that any finite dimensional subspace M is closed in X. This situation will prevail in the linear equilibrium problems for free bodies.

An important result for factor spaces is given by

Theorem 1.15.1. Let X be a Banach space and let M be a closed subspace of X. Then X/M is a Banach space.

Proof. Let $\{[x_k]\}$ be a Cauchy sequence in X/M. To show that $\{[x_k]\}$ has a limit element in X/M, we use the fact (Problem 1.15.1) that if a subsequence of a Cauchy sequence converges to an element, the whole sequence converges to the same element. So we will select a convergent subsequence of $\{[x_k]\}$.

Because $\{[x_k]\}$ is a Cauchy sequence, we can find N_k that

$$\left\|[x_{n+t}] - [x_n]\right\| < 2^{-k} \quad \text{for } n \ge N_k \text{ and any } t > 0 .$$

We can select $N_{k-1} < N_k$ so the subsequence $\{[x_{N_k}]\}$ satisfies the inequality

$$\left\|[x_{N_{k+1}}] - [x_{N_k}]\right\| < 2^{-k} .$$

To establish the theorem, it is enough to prove that $\{[x_{N_k}]\}$ has a limit in X/M. Let us redenote this subsequence as $\{[x_k]\}$ so that

$$\left\|[x_{k+1}] - [x_k]\right\| < 2^{-k} \quad \text{for all } k .$$

Now we choose some special representatives from the elements $[x_{k+1}] - [x_k]$ as follows. For each k, let x_k be an element from the class $[x_k]$. By definition of the norm in X/M, we can find $m_k \in M$ such that

$$\|x_{k+1} - x_k + m_k\| < 2^{-k+1} .$$

Consider

$$z_n = \sum_{k=1}^{n} (x_{k+1} - x_k + m_k) = x_{n+1} - x_1 + \sum_{k=1}^{n} m_k$$

and so

$$x_{n+1} = z_n + x_1 - \sum_{k=1}^{n} m_k . \tag{1.15.1}$$

Let us show that $\{z_n\}$ is a Cauchy sequence in X. Indeed, for any integer $t > 0$

$$\|z_{n+t} - z_n\| \le \|z_{n+t} - z_{n+t-1}\| + \cdots + \|z_{n+1} - z_n\| = \sum_{k=n+1}^{n+t} \|z_k - z_{k-1}\|$$

$$= \sum_{k=n+1}^{n+t} \|x_{k+1} - x_k + m_k\| < \sum_{k=n+1}^{n+t} 2^{-k+1} < 2^{-n+1} \to 0 \text{ as } n \to \infty\,.$$

Because X is a Banach space, $\{z_n\}$ tends to a limit $z \in X$. Now we show that $z + x_1$ is a representative element of the limit of $\{[x_k]\}$ in X/M; using (1.15.1), we get

$$\left\|[x_{n+1}] - [z + x_1]\right\| \le \left\|x_{n+1} - z - x_1 + \sum_{k=1}^{n} m_k\right\| = \|z_n - z\| \to 0 \text{ as } n \to \infty$$

which completes the proof. □

Problem 1.15.1. Show that if a subsequence of a Cauchy sequence has a limit, then the entire sequence must converge to the same limit. □

1.16 Separability

Two sets are said to be *of equal power* if there is a one-to-one correspondence between their elements. Of all sets having infinitely many elements, the set of least power is the set of positive integers.

Definition 1.16.1. A set which is of equal power with the set of positive integers is *countable*.

Roughly speaking, each element of a countable set can be numbered by assigning it a positive-integer index.

Theorem 1.16.1. A countable union of countable sets is countable.

Proof. It suffices to show how we may enumerate the elements of the union. The method is clear from the diagram

$$\begin{array}{ccccccccc}
a_{11} & & a_{12} & & a_{13} & & a_{14} & & \cdots \\
& \swarrow & & \swarrow & & \swarrow & & \swarrow & \\
a_{21} & & a_{22} & & a_{23} & & a_{24} & & \cdots \\
& \swarrow & & \swarrow & & \swarrow & & \swarrow & \\
a_{31} & & a_{32} & & a_{33} & & a_{34} & & \cdots \\
& \swarrow & & \swarrow & & \swarrow & & \swarrow & \\
a_{41} & & a_{42} & & a_{43} & & a_{44} & & \cdots \\
\vdots & \swarrow & & \swarrow & & \swarrow & & &
\end{array}$$

where, for a fixed i, $\{a_{ij}\}$ is a sequence of enumerated elements of the ith set. The first element we choose to enumerate is a_{11}; then we go along the diagonal, enumerating

a_{12} and a_{21}. The next diagonal gives a_{13}, a_{22}, a_{31}. Proceeding in this way, all the elements of all the sets are put into one-to-one correspondence with the sequence of positive integers. □

Corollary 1.16.1. The set of all rational numbers is countable.

Proof. A rational number is represented in the form i/j where i and j are integers. Denoting $a_{ij} = i/j$, we obtain the sequence a_{ij} which satisfies the condition of the theorem. □

Problem 1.16.1. Show that the set of all polynomials with rational coefficients is countable. □

Georg Cantor (1845–1918) proved

Theorem 1.16.2. The set of real numbers of the segment $[0, 1]$ is not countable.

The proof can be found in any textbook on set theory or the theory of functions of a real variable. So the set $[0, 1]$ is not of equal power with the set of positive integers; the points of $[0, 1]$ form a continuum.

It would be beyond our scope to discuss Cantor's theory of sets. Our interests lie in applying the notion of countability to metric spaces. Modern mechanics relies on computational ability. A computer can process only finite sets of numbers, hence can only approximate results to a given decimal accuracy. If x is an arbitrary element of an infinite set X and we want to use a computer to find it, we must be certain that every element of X can be approximated by elements of another set which is finite or, at least, countable. This leads to

Definition 1.16.2. A metric space X is *separable* if it contains a countable subset that is dense in X.

In other words, X is separable if there is a countable set $M \subset X$ such that for every $x \in X$ there is a sequence $\{m_i\}$ in M that approximates x:

$$d(x, m_i) \to 0 \quad \text{as } i \to \infty .$$

An example of a separable metric space is the set of real numbers in $[a, b]$ with the usual metric; here the dense set M is the set of all rational numbers in $[a, b]$. The spaces $\mathbb{R}^n$ and $\mathbb{C}^n$ are obviously separable.

Problem 1.16.2. Prove that ℓ^p for $p \geq 1$, c, and c_0 are separable. □

Lemma 1.16.1. The normed space m is not separable.

Proof. Suppose to the contrary that m is separable. Then there exists a countable set of sequences that is dense in m. Enumerate this set as $(\mathbf{e}_1, \mathbf{e}_2, \ldots)$. So for any $\varepsilon > 0$ and $\mathbf{x} \in m$, we can find $\mathbf{e}_N$ such that $\|\mathbf{x} - \mathbf{e}_N\|_m < \varepsilon$. Let us construct an element of m for which we cannot satisfy the inequality with $\varepsilon = 1/2$. Let $\mathbf{e}_k = (e_1^{(k)}, e_2^{(k)}, \ldots)$ and denote the needed vector by $\mathbf{x}^* = (x_1, x_2, \ldots)$. Choose x_k such that $|x_k| \leq 1$ and $|x_k - e_k^{(k)}| \geq 1$, which is possible for any value of $|e_k^{(k)}|$. By definition of the norm on m, we find that $\|\mathbf{x}^* - \mathbf{e}_k\|_m \geq 1$. Hence m is not separable. □

The set of all polynomials on a closed and bounded domain Ω, equipped with the norm of the space $C(\Omega)$, is separable; the dense subset is the set P_r of all polynomials with rational coefficients. Indeed, approximating the coefficients a_α of an arbitrary polynomial $\sum_\alpha a_\alpha \mathbf{x}^\alpha$ by rational numbers a_{α_r} we can always obtain

$$\max_{\mathbf{x}\in\Omega} \left| \sum_\alpha a_\alpha \mathbf{x}^\alpha - \sum_\alpha a_{\alpha_r} \mathbf{x}^\alpha \right| < \varepsilon \qquad (\mathbf{x}^\alpha \equiv x_1^{\alpha_1} \cdots x_n^{\alpha_n})$$

for any given precision $\varepsilon > 0$.

A nontrivial example of a separable space is provided by the classical Weierstrass theorem, which can be formulated as

Theorem 1.16.3. The set P_r of all polynomials with rational coefficients is dense in $C(\Omega)$, where Ω is a closed and bounded domain in $\mathbb{R}^n$.

Since P_r is countable, the theorem states that $C(\Omega)$ is separable. Now we construct an example of a nonseparable metric space.

Lemma 1.16.2. The set of functions $f(x)$ bounded on $[0, 1]$ and equipped with the norm

$$\|f(x)\| = \sup_{x\in[0,1]} |f(x)|$$

is not separable.

Proof. It suffices to construct a subset M of the space whose elements cannot be approximated by functions from a countable set. Let α be an arbitrary point of $[0, 1]$. The set M is composed of functions defined as follows:

$$f_\alpha(x) = \begin{cases} 1\,, & x \geq \alpha\,, \\ 0\,, & x < \alpha\,. \end{cases}$$

The distance from $f_\alpha(x)$ to $f_\beta(x)$ is

$$\|f_\alpha(x) - f_\beta(x)\| = \sup_{x\in[0,1]} |f_\alpha(x) - f_\beta(x)| = 1 \quad \text{if } \alpha \neq \beta\,.$$

Take a ball B_α of radius 1/3 about $f_\alpha(x)$. If $\alpha \neq \beta$, the intersection $B_\alpha \cap B_\beta$ is empty.

If a countable subset is dense in the space, then each B_α must contain at least one element of this subset; this contradicts Theorem 1.16.2 since the set of balls B_α is of equal power with the continuum. □

We will address the question of separability for the various spaces introduced previously, starting with

Theorem 1.16.4. Let Ω be a closed and bounded domain in $\mathbb{R}^n$. Then $L^p(\Omega)$ is separable for any $1 \leq p < \infty$.

Proof. It suffices to show that P_r, the set of all polynomials with rational coefficients, is dense in $L^p(\Omega)$. We saw (Theorem 1.16.3) that P_r is dense in $C(\Omega)$. Then P_r is dense in the set of all functions continuous on Ω and equipped with the metric of $L^p(\Omega)$. Indeed, if $f(\mathbf{x}) \in C(\Omega)$ then, for a given $\varepsilon > 0$, we can find a polynomial $Q_\varepsilon(\mathbf{x})$ from P_r such that

$$\max_{\mathbf{x}\in\Omega} |f(\mathbf{x}) - Q_\varepsilon(\mathbf{x})| \le \frac{\varepsilon}{(\operatorname{mes}\Omega)^{1/p}} .$$

Then

$$\|f(\mathbf{x}) - Q_\varepsilon(\mathbf{x})\|_{L^p(\Omega)} = \left(\int_\Omega |f(\mathbf{x}) - Q_\varepsilon(\mathbf{x})|^p \, d\Omega\right)^{1/p} \le \left(\frac{\varepsilon^p}{\operatorname{mes}\Omega} \int_\Omega 1 \, d\Omega\right)^{1/p} = \varepsilon .$$

Now let $F(\mathbf{x})$ be an element of $L^p(\Omega)$ and $\{f_n(\mathbf{x})\}$ its representative Cauchy sequence. Each $f_n(\mathbf{x})$ can be approximated by a polynomial $q_n(\mathbf{x})$ from P_r (since $f_n(\mathbf{x})$ is a continuous function) with any accuracy, say $1/n$:

$$\|f_n(\mathbf{x}) - q_n(\mathbf{x})\|_{L^p(\Omega)} < 1/n .$$

This means that the sequence $\{Q_n(\mathbf{x})\} \subset L^p(\Omega)$ such that $Q_n(\mathbf{x})$ contains the stationary sequence $(q_n(\mathbf{x}), q_n(\mathbf{x}), q_n(\mathbf{x}), \ldots)$, tends to $F(\mathbf{x})$ in the norm of $L^p(\Omega)$. As P_r is countable, so is the set of all such $Q_n(\mathbf{x})$. Hence $L^p(\Omega)$ is separable. □

This proof is of a general nature. If P_r is replaced by a countable subset which is dense in a metric space X, and the metric of $L^p(\Omega)$ by the metric of X, the result is an abstract modification of Theorem 1.16.4:

Theorem 1.16.5. The completion of a separable metric space is separable.

Problem 1.16.3. Prove Theorem 1.16.5. □

This theorem allows us to show separability of energy spaces if we prove the following.

Theorem 1.16.6. Let Ω be compact in $\mathbb{R}^n$. Then for any positive integer k, the space $C^{(k)}(\Omega)$ is separable.

We merely sketch the proof. A function $f(\mathbf{x}) \in C^{(k)}(\Omega)$ can be approximated with any accuracy in the norm of $C^{(k)}(\Omega)$ by an infinitely differentiable function $f_1(\mathbf{x})$. This can be done using the averaging technique. We consider the derivative

$$\frac{\partial^{kn} f_1(x_1, \ldots, x_n)}{\partial x_1^k \cdots \partial x_n^k}$$

as an element of $C(\Omega)$ and, to within a prescribed accuracy, approximate it by a polynomial $Q_{kn} \in P_r$. This can be done by virtue of the Weierstrass theorem. Using Q_{kn}, we then construct a polynomial with rational coefficients that approximates $f(\mathbf{x})$ in $C^{(k)}(\Omega)$ within the prescribed accuracy. For this, we choose a point $\mathbf{x}_0 =$

$(x_{10}, \ldots, x_{n0}) \in \Omega$ with rational coordinates. We then choose rational numbers a_α that approximate the values of $D^\alpha f_1(\mathbf{x}_0)$ with some prescribed accuracy. Using these numbers as initial data, we perform successive integrations on the polynomial Q_{kn}:

$$Q_{kn-1}(x_1, x_2, \ldots, x_n) = a_{k-1,k,\ldots,k} + \int_{x_{10}}^{x_1} Q_{kn}(s, x_2, \ldots, x_n)\, ds\,,$$

$$Q_{kn-2}(x_1, x_2, \ldots, x_n) = a_{k-2,k,\ldots,k} + \int_{x_{10}}^{x_1} Q_{kn-1}(s, x_2, \ldots, x_n)\, ds\,,$$

$$\vdots$$

At each stage of integration we get a polynomial with rational coefficients that approximates in $C(\Omega)$ one of the derivatives of $f_1(\mathbf{x})$ to within prescribed accuracy. So the final polynomial approximates $f_1(\mathbf{x})$ and thus $f(\mathbf{x})$ in $C^{(k)}(\Omega)$. We leave it to the reader to complete the proof.

We also need the following almost trivial result.

Theorem 1.16.7. Any subspace E of a separable metric space X is separable.

Proof. Consider a countable set consisting of $(x_1, x_2, \ldots)$ which is dense in X. Let B_{ki} be a ball of radius $1/k$ about x_i. By Theorem 1.16.1, the set of all B_{ki} is countable.

For any fixed k the union $\cup_i B_{ki}$ covers X and thus E. For every B_{ki}, take an element $e_{ki} \in E$ lying in B_{ki} (if one exists). For any $e \in B_{ki} \cap E$, the distance $d(e, e_{ki})$ is less than $2/k$. It follows that the set of all e_{ki} is, on the one hand, countable, and, on the other hand, dense in E. □

Theorem 1.16.7 is significant. We have a limited selection of countable sets of functions with which to demonstrate separability of certain spaces: they are the space P_r of polynomials with rational coefficients, the space of trigonometric polynomials with rational coefficients, and a few others. As a rule, the elements of these spaces do not meet the boundary conditions imposed on functions of, for example, energy spaces. We can circumvent this difficulty by taking a wider space, containing the space of interest, whose separability may be shown. The needed separability is then a consequence of the theorem.

In Sect. 2.1 we will introduce Sobolev spaces $W^{m,p}(\Omega)$ as completions of the spaces $C^{(m)}(\Omega)$ in certain integral norms. As a particular case of Theorems 1.16.5 and 1.16.6, we can get

Lemma 1.16.3. The Sobolev spaces $W^{m,p}(\Omega)$, $p \geq 1$, are separable.

Because all of the energy spaces introduced in Sect. 2.3 and afterwards are subspaces of certain Sobolev spaces, we have

Lemma 1.16.4. All the energy spaces introduced in this book are separable.

In mechanics, series expansions are often employed. As the coefficients in these expansions constitute numerical sequences, it makes sense to examine the separability of the numerical sequence spaces.

In the theory of Fourier expansions we will need the following result.

Lemma 1.16.5. Let S be a set of positive numbers x_α, $\alpha \in A$, such that

$$\sup_A \sum_k x_{\alpha_k} \le C\,, \tag{1.16.1}$$

where the sum is taken over any finite subset of indices $\alpha \in A$ and C is a constant. Then S is countable.

Proof. Write $P_n = [1/n, \infty) \cap S$. The union of all P_n constitutes S. By (1.16.1), each P_n contains no more than $nC + 1$ elements of S, so P_n is a finite set. As a countable set of finite sets $\{P_n\}$ is countable, so is their union S. □

1.17 Compactness, Hausdorff Criterion

The classical Bolzano theorem states that every bounded sequence in $\mathbb{R}^n$ contains a Cauchy subsequence. How does this property depend on the dimension of a space?

Consider, for example, a sequence of elements in ℓ^2:

$$x_1 = (1, 0, 0, 0, \ldots)\,,$$
$$x_2 = (0, 1, 0, 0, \ldots)\,,$$
$$x_3 = (0, 0, 1, 0, \ldots)\,,$$
$$\vdots$$

Since $\|x_i\| = 1$ the sequence is bounded in ℓ^2, but for any pair of distinct elements we get

$$\|x_i - x_j\| = (1^2 + 1^2)^{1/2} = \sqrt{2}\,;$$

hence $\{x_i\}$ does not contain a Cauchy subsequence.

Let us introduce

Definition 1.17.1. A set in a metric space is *sequentially precompact* if every sequence of elements from the set contains a Cauchy subsequence. If the limit elements of these subsequences all belong to the set, then the set is *sequentially compact*. For brevity, we call such sets *precompact* and *compact*, respectively.

A compact set is closed. In these terms, Bolzano's theorem can be reformulated as follows: a bounded subset of $\mathbb{R}^n$ is precompact; a closed and bounded subset of $\mathbb{R}^n$ is compact.[6]

Problem 1.17.1. $\mathbb{R}^n$ itself is not compact under its usual metric. Is there a metric on $\mathbb{R}^n$ such that the whole of $\mathbb{R}^n$ is precompact in this metric? □

---

[6] The sets that we label as *precompact* are often labeled as *compact* in the Russian literature and *totally bounded* in the Western literature.

To establish a criterion for compactness of a set we need

Definition 1.17.2. A finite set E of elements of a metric space X is a *finite ε-net* of a set $M \subset X$ if for every $x \in M$ there exists $e \in E$ such that $d(x, e) < \varepsilon$.

Thus, every element of M lies in one of a finite collection of balls of radius ε if M has a finite ε-net. It is clear that any finite set has a finite ε-net for any $\varepsilon > 0$. In particular, we may have $M = X$.

Now we formulate *Hausdorff's criterion* for compactness.

Theorem 1.17.1 (Hausdorff criterion). A subset of a metric space is precompact if and only if for every $\varepsilon > 0$ there is a finite ε-net for this set.

Proof. (a) *Necessity.* Let M be a precompact subset of a metric space X. The existence of a finite ε-net for any $\varepsilon > 0$ for M is proved by contradiction. Let $\varepsilon_0 > 0$ be such that there is no finite ε_0-net for M: this means that a union of any finite number of balls of radius ε_0 cannot contain all elements of M. Take an element $x_1 \in M$ and a ball B_1 of radius ε_0 about x_1. Since there is no finite ε_0-net for M, there is an element x_2 of M such that $x_2 \notin B_1$. Construct the ball B_2 of radius ε_0 about x_2. Outside $B_1 \cup B_2$ there is a third element x_3 of M — otherwise x_1 and x_2 form an ε_0-net. Continuing to construct a sequence of elements and corresponding balls, we get a sequence $\{x_n\}$ satisfying the condition $d(x_n, x_m) \geq \varepsilon_0$ for $n \neq m$. Therefore $\{x_n\}$ does not contain a Cauchy subsequence, and this contradicts the definition of precompactness.

(b) *Sufficiency.* Suppose that for every $\varepsilon > 0$, there is a finite ε-net of a set $M \subset X$. We must show that M is precompact. Let $\{x_n\}$ be an arbitrary sequence from M: we shall show that we can select a Cauchy subsequence from $\{x_n\}$. For this, take $\varepsilon_1 = 1/1$ and construct a finite ε_1-net for M. One of the balls, say B_1, of radius ε_1 about an element of this finite net must contain an infinite number of elements of $\{x_n\}$. Take one of the latter elements and denote it by x_{i_1}. Next construct a finite ε_2-net with $\varepsilon_2 = 1/2$. One of the balls, say B_2, of radius ε_2 about the elements of this net contains an infinite number of elements of $\{x_n\}$ which belong to B_1. We choose such an element x_{i_2} with subscript $i_2 > i_1$. So $d(x_{i_1}, x_{i_2}) < 2\varepsilon_1 = 2$. Next we construct an ε_3-net with $\varepsilon_3 = 1/3$. There is a ball B_3 of radius $1/3$ about a point of this net that contains an infinite subsequence from $\{x_n\}$ belonging to the intersection of B_1 and B_2. From this subsequence we select an element x_{i_3} with $i_3 > i_2$ and $d(x_{i_2}, x_{i_3}) < 2\varepsilon_2 = 1$. Continuing this procedure, we obtain an infinite sequence $\{x_{i_k}\}$, $i_k < i_{k+1}$, which is a subsequence of $\{x_n\}$. Note that $\{x_{i_k}\}$ belongs to the intersection of the balls $B_1, B_2, \ldots, B_{k-1}$. So for $m > k$, all the elements x_{i_m} lie in the ball B_{k-1}, and thus

$$\left\|x_{i_m} - x_{i_n}\right\| < \frac{2}{k-1} \quad \text{for } n, m > k \,.$$

This means that $\{x_{i_k}\}$ is a Cauchy sequence. □

Note that a closed precompact set is compact. The reader may formulate necessary and sufficient conditions for compactness.

Corollary 1.17.1. A precompact subset S of a metric space M is bounded.

Proof. Take $\varepsilon = 1$ and construct a 1-net. The union of the finite number of unit balls with centers at the nodes of the net covers S. So there is a finite ball of M that contains S. □

In a certain way, the property of compactness is close to the property of separability; namely, we have

Theorem 1.17.2. A precompact metric space M (or a precompact subset of a metric space) is separable.

Proof. Using Theorem 1.17.1 we construct a countable set E which is dense in M, as follows. Take the sequence $\varepsilon_k = 1/k$; let the set $(x_{k1}, x_{k2}, \ldots, x_{kN})$ be a finite ε_k-net of M (N certainly depends on k). The collection of all x_{ki} for all possible k and i, being a countable set, is the needed E; indeed, for every $x \in M$ and any $\varepsilon > 0$, there is a ball B of radius $\varepsilon_k < \varepsilon$ about an x_{ki} such that $x \in B$. Hence there is a sequence in E which converges to x. □

Now we examine an extension of Bolzano's theorem.

Theorem 1.17.3. Every closed and bounded subset of a Banach space X is compact if and only if X has finite dimension.

We recall that X has finite dimension if there is a finite set of elements $x_1, \ldots, x_n$ such that any $x \in X$ can be represented in the form

$$x = \sum_{i=1}^{n} \alpha_i x_i .$$

The least such n is called the *dimension* of X.

Sufficiency of the theorem is proved in a manner similar to that for the classical version of Bolzano's theorem and is omitted. Necessity is a consequence of the following

Lemma 1.17.1 (Riesz). Let M be a closed subspace of a normed space X and suppose $M \neq X$. Then for any ε, $0 < \varepsilon < 1$, there is an element x_ε such that

$$x_\varepsilon \notin M , \qquad \|x_\varepsilon\| = 1 , \quad \text{and} \quad \inf_{y \in M} \|y - x_\varepsilon\| > 1 - \varepsilon .$$

In other words, there exist points on the unit sphere whose distances from M are arbitrarily close to 1.

Proof. Since $M \neq X$ there is an element $x_0 \in X$ such that $x_0 \notin M$. Denote $d = \inf_{y \in M} \|x_0 - y\|$. First we show that $d > 0$. If on the contrary $d = 0$, then there is a sequence $\{y_k\}$, $y_k \in M$, such that $\|x_0 - y_k\| \to 0$ as $k \to \infty$; this means $\lim_{k\to\infty} y_k = x_0$ and so $x_0 \in M$ because M is closed. Thus $d > 0$. By definition of infimum, for any

$\varepsilon > 0$ there exists $y_\varepsilon \in M$ such that $d \leq \|x_0 - y_\varepsilon\| \leq d/(1-\varepsilon/2)$. The needed element is

$$x_\varepsilon = \frac{x_0 - y_\varepsilon}{\|x_0 - y_\varepsilon\|}\ .$$

Indeed $\|x_\varepsilon\| = 1$ and for any $y \in M$ we have

$$\|x_\varepsilon - y\| = \left\| \frac{x_0 - y_\varepsilon}{\|x_0 - y_\varepsilon\|} - y \right\| = \frac{\|x_0 - (y_\varepsilon + \|x_0 - y_\varepsilon\| y)\|}{\|x_0 - y_\varepsilon\|} \geq d/\frac{d}{1-\varepsilon/2} = 1 - \frac{\varepsilon}{2}\ .$$

(Here we used the fact that $y_\varepsilon + \|x_0 - y_\varepsilon\| y \in M$.) □

Proof (of necessity for Theorem 1.17.3.). It suffices to prove that the unit ball about zero is compact only in a finite dimensional Banach space.

Take an element y_1 such that $\|y_1\| = 1$ and denote by E_1 the space spanned by y_1, i.e., the set of all elements of the form αy_1 where $\alpha \in \mathbb{C}$. If $E_1 \neq X$ then, by Lemma 1.17.1, there is an element $y_2 \notin E_1$ such that $\|y_2\| = 1$ and $\|y_1 - y_2\| > 1/2$. Denote by E_2 a linear space spanned by y_1 and y_2. If $E_2 \neq X$ then, by the same lemma, we can find y_3 such that

$$\|y_3\| = 1\ , \qquad \|y_3 - y_1\| > 1/2\ , \qquad \|y_3 - y_2\| > 1/2\ .$$

If X is infinite dimensional the process continues indefinitely and we get a sequence $\{y_k\}$ such that

$$\|y_i - y_j\| > 1/2 \quad \text{if } i \neq j\ .$$

This sequence lying in the unit ball about zero cannot contain a Cauchy subsequence, which contradicts the hypothesis. So the process must terminate and thus $X = E_k$ for some k, i.e., X is finite dimensional. □

Note that we have also proved that any bounded set in a Banach space is precompact if and only if the space is finite dimensional.

In the next section we consider a widely applicable theorem on compactness.

1.18 Arzelà's Theorem and Its Applications

In this section, we denote the Euclidean norm on $\mathbb{R}^n$ by $|\cdot|$.

Theorem 1.18.1 (Arzelà). Let Ω be a closed and bounded (i.e., compact) domain in $\mathbb{R}^n$. A set M of functions continuous on Ω is precompact in $C(\Omega)$ if and only if M satisfies the following pair of conditions:

(i) M is *uniformly bounded*. There is a constant c such that for every $f(\mathbf{x}) \in M$, the inequality $|f(\mathbf{x})| \leq c$ holds for all $\mathbf{x} \in \Omega$.

(ii) M is *equicontinuous*. For any $\varepsilon > 0$ there exists $\delta(\varepsilon) > 0$ such that whenever $\mathbf{x}, \mathbf{y} \in \Omega$ and $|\mathbf{x} - \mathbf{y}| < \delta$, the inequality $|f(\mathbf{x}) - f(\mathbf{y})| < \varepsilon$ holds for every $f(\mathbf{x}) \in M$.

Problem 1.18.1. (a) Does condition (i) mean that M is bounded in $C(\Omega)$? (b) Show that if M contains only a finite number of functions, it is equicontinuous. □

Proof (of Theorem 1.18.1.). (a) *Necessity.* Let M be precompact in $C(\Omega)$. By Theorem 1.17.1, there is a finite 1-net for M; i.e., there is a finite set of continuous functions $g_i(\mathbf{x})$ ($i = 1, \ldots, k$) such that to any $f(\mathbf{x})$ there corresponds $g_i(\mathbf{x})$ for which

$$\|f(\mathbf{x}) - g_i(\mathbf{x})\| = \max_{\mathbf{x}\in\Omega} |f(\mathbf{x}) - g_i(\mathbf{x})| \le 1 .$$

Since the $g_i(\mathbf{x})$ are continuous there is a constant c_1 such that $|g_i(\mathbf{x})| < c_1$ for $i = 1, \ldots, k$, and thus $|f(\mathbf{x})| \le c_1 + 1$. So condition (i) is satisfied.

Let us show (ii). Let $\varepsilon > 0$ be given. By precompactness of M, there is a finite $\varepsilon/3$-net, say $g_i(\mathbf{x})$ for $i = 1, \ldots, m$. Since the $g_i(\mathbf{x})$ are finite in number and equicontinuous on Ω, we can find $\delta > 0$ such that whenever $|\mathbf{x} - \mathbf{y}| < \delta$ then

$$|g_i(\mathbf{x}) - g_i(\mathbf{y})| < \varepsilon/3 \quad \text{for all } i = 1, \ldots, m .$$

For an arbitrary function $f(\mathbf{x})$ from M, there exists $g_r(\mathbf{x})$ such that

$$|f(\mathbf{x}) - g_r(\mathbf{x})| < \varepsilon/3 \quad \text{for all } \mathbf{x} \in \Omega .$$

Let $\mathbf{x}, \mathbf{y} \in \Omega$ be such that $|\mathbf{x} - \mathbf{y}| < \delta$. Then

$$|f(\mathbf{x}) - f(\mathbf{y})| \le |f(\mathbf{x}) - g_r(\mathbf{x})| + |g_r(\mathbf{x}) - g_r(\mathbf{y})| + |g_r(\mathbf{y}) - f(\mathbf{y})| < 3(\varepsilon/3) = \varepsilon$$

and thus condition (ii) is satisfied.

(b) *Sufficiency.* Let M satisfy conditions (i) and (ii) of the theorem. We must show that from any sequence of functions lying in M we can choose a uniformly convergent subsequence. Since Ω is compact in $\mathbb{R}^n$ we can find a finite δ-net for Ω for any $\delta > 0$, say "cubic" close packing. Take $\delta_k = 1/k$ and construct a corresponding finite δ_k-net for Ω. We enumerate all the nodes of these nets successively: first all points of the δ_1-net, then all points of the δ_2-net, and so on. As a result, we get a countable set of points $\{\mathbf{x}_k\}$ which is dense in Ω.

Take a sequence of functions $\{f_k(\mathbf{x})\}$ from M and consider it at $\mathbf{x} = \mathbf{x}_1$. We can choose a convergent subsequence $\{f_{k_1}(\mathbf{x}_1)\}$ from it because the numerical sequence $\{f_k(\mathbf{x}_1)\}$ is bounded. Considering the numerical sequence $\{f_{k_1}(\mathbf{x}_2)\}$, by the same reasoning we can choose a convergent subsequence $\{f_{k_2}(\mathbf{x}_2)\}$ from the latter. The same can be done for $\mathbf{x} = \mathbf{x}_3$, $\mathbf{x} = \mathbf{x}_4$, and so on. On the kth step of this procedure we get a subsequence which is convergent (Cauchy) at $\mathbf{x} = \mathbf{x}_i$ ($i = 1, \ldots, k$).

By construction, the diagonal sequence $\{f_{n_n}(\mathbf{x})\}$ is convergent at every $\mathbf{x} = \mathbf{x}_i$ ($i = 1, 2, 3, \ldots$). Let us show that it is uniformly convergent. By equicontinuity of M, for any $\varepsilon > 0$ we can find $\delta > 0$ such that $|\mathbf{x} - \mathbf{y}| < \delta$ implies

$$|f_{n_n}(\mathbf{x}) - f_{n_n}(\mathbf{y})| < \varepsilon/3 \qquad (n = 1, 2, \ldots) .$$

Let $\tilde{\delta} < \delta$ and take a finite $\tilde{\delta}$-net of Ω with nodes $\mathbf{z}_i$ ($i = 1, \ldots, r$). Since r is finite, for the given ε there exists N such that $n, m > N$ implies

$$|f_{n_n}(\mathbf{z}_i) - f_{m_m}(\mathbf{z}_i)| < \varepsilon/3 \qquad (i = 1, \ldots, r)\,.$$

Let $\mathbf{x}$ be an arbitrary point of Ω and $\mathbf{z}_k$ be the point of the $\tilde{\delta}$-net nearest to $\mathbf{x}$, i.e., $|\mathbf{x} - \mathbf{z}_k| < \tilde{\delta}$. For the above m, n we get

$$\begin{aligned} |f_{n_n}(\mathbf{x}) - f_{m_m}(\mathbf{x})| &\le |f_{n_n}(\mathbf{x}) - f_{n_n}(\mathbf{z}_k)| + |f_{n_n}(\mathbf{z}_k) - f_{m_m}(\mathbf{z}_k)| + |f_{m_m}(\mathbf{z}_k) - f_{m_m}(\mathbf{x})| \\ &< \varepsilon/3 + \varepsilon/3 + \varepsilon/3 = \varepsilon\,. \end{aligned}$$

Therefore $\{f_{n_n}(\mathbf{x})\}$ is uniformly convergent and the proof is complete. □

Problem 1.18.2. Show that a set of functions that all satisfy a Lipschitz condition with the same constant K is equicontinuous. □

Corollary 1.18.1. A set bounded in the space $C^{(1)}(\Omega)$ is precompact in $C(\Omega)$. It is compact if it is closed.

Proof. A set bounded in $C^{(1)}(\Omega)$ is bounded in $C(\Omega)$, i.e., condition (i) of the theorem is satisfied. Fulfillment of (ii) follows from the elementary inequality

$$|f(\mathbf{x}) - f(\mathbf{y})| = \left| \int_0^1 \frac{df(s\mathbf{x} + (1-s)\mathbf{y})}{ds}\, ds \right| \le C|\mathbf{x} - \mathbf{y}| \qquad (1.18.1)$$

which is left for the reader as Problem 1.18.3. This completes the proof. □

Problem 1.18.3. Prove inequality (1.18.1). □

A famous application of Arzelà's theorem is the following local existence theorem, due to Peano, for the Cauchy problem for a system of ordinary differential equations

$$\mathbf{y}'(t) = f(t, \mathbf{y}(t))\,, \quad \mathbf{y}(t_0) = \mathbf{y}_0 \qquad (\mathbf{y} \in \mathbb{R}^n)\,. \qquad (1.18.2)$$

Denote

$$Q(t_0, a, b) = \{(t, \mathbf{y})\colon t_0 \le t \le t_0 + a,\ |\mathbf{y} - \mathbf{y}_0| \le b,\ \mathbf{y} \in \mathbb{R}^n\}\,.$$

Theorem 1.18.2 (Peano). Let $f(t, \mathbf{y})$ be continuous on $Q(t_0, a, b)$ and such that $|f(t, \mathbf{y})| \le m$ on this domain, and let $\alpha = \min\{a, b/m\}$. Then there is a continuous solution $\mathbf{y} = \mathbf{y}(t)$ to the Cauchy problem (1.18.2) on the interval $[t_0, t_0 + \alpha]$.

Proof. On $[t_0 - 1, t_0]$ we define a function $\mathbf{y}_\varepsilon$ by

$$\mathbf{y}_\varepsilon(t) = \mathbf{y}_0 + (t - t_0) f(t_0, \mathbf{y}_0)\,.$$

Its continuation onto $[t_0, t_0 + \alpha]$ is determined by the equation

$$\mathbf{y}_\varepsilon(t) = \mathbf{y}_0 + \int_{t_0}^t f(s, \mathbf{y}_\varepsilon(s - \varepsilon))\, ds\,. \qquad (1.18.3)$$

The process of continuation is done successively onto the segment $[t_0, t_0 + \alpha_1]$, $\alpha_1 = \min\{\alpha, \varepsilon\}$, then onto $[t_0 + \alpha_1, t_0 + 2\alpha_1]$, and so on, until we cover $[t_0, t_0 + \alpha]$. Indeed, if $\varepsilon \le b/m$, then on $[t_0 - \varepsilon, t_0]$ we get

$$|\mathbf{y}_\varepsilon(t) - \mathbf{y}_0| = |t - t_0|\,|f(t_0, \mathbf{y}_0)| \le \varepsilon m \le (b/m)m = b\,. \tag{1.18.4}$$

By (1.18.3), on $[t_0, t_0 + \alpha_1]$ we have

$$|\mathbf{y}_\varepsilon(t) - \mathbf{y}_0| \le \int_{t_0}^{t} |f(s, \mathbf{y}_\varepsilon(s - \varepsilon)|\,ds \le \varepsilon m \le (b/m)m = b$$

because $|t - t_0| \le \varepsilon$, $s \le t$, and $|f(s, \mathbf{y}_\varepsilon(s - \varepsilon))| \le m$. Therefore (1.18.4) also holds on $[t_0, t_0 + \alpha_1]$, and (1.18.3) can be used to define $\mathbf{y}_\varepsilon(t)$ on $[t_0 + \alpha_1, t_0 + 2\alpha_1]$ if $2\alpha_1 < \alpha$. But (1.18.4) continues to hold on this new segment, and we can continue the process as long as $k\alpha_1 < \alpha$. For the maximum value $k = k_0$, the last step will be done on $[t_0 + k_0\alpha_1, t_0 + \alpha]$. This completes the construction of $\mathbf{y}_\varepsilon(t)$ on $[t_0, t_0 + \alpha]$. Moreover, differentiating (1.18.3), we find that on $[t_0, t_0 + \alpha]$

$$|\mathbf{y}'_\varepsilon(t)| = |f(t, \mathbf{y}_\varepsilon(t - \varepsilon))| \le m\,,$$

i.e., the set of all functions $\{\mathbf{y}_\varepsilon(t)\}$ when $\varepsilon \le b/m$, considered on $[t_0, t_0 + \alpha]$, satisfies the conditions of Corollary 1.18.1. Thus we can find a sequence $\{\varepsilon_k\}$ such that $\varepsilon_k \to 0$ as $k \to \infty$ and the sequence $\{\mathbf{y}_{\varepsilon_k}(t)\}$ converges uniformly to a limit function $\mathbf{y}(t)$ on $[t_0, t_0 + \alpha]$.

By equicontinuity of $f(t, \mathbf{y})$, the sequence $\{f(t, \mathbf{y}_{\varepsilon_k}(t)\}$ converges uniformly to $f(t, \mathbf{y}(t))$ on the same segment. Therefore we can take the limit under the integral sign on the right-hand side of (1.18.3); the limit passage yields

$$\mathbf{y}(t) = \mathbf{y}_0 + \int_{t_0}^{t} f(s, \mathbf{y}(s))\,ds\,, \qquad t \in [t_0, t_0 + \alpha]\,.$$

This means $\mathbf{y}(t)$ is a continuous solution to the problem (1.18.2). □

Our subject is now to demonstrate how compactness can be applied to the justification of a finite difference procedure to solve ordinary differential equations. We consider the simplest of these methods, due to Euler, supposing the requirements of Theorem 1.18.2 are met.

Let $\Delta > 0$ be a step of Euler's method for the solution of (1.18.2). Euler's method is defined by the system of equations

$$\frac{\mathbf{z}_{k+1} - \mathbf{z}_k}{\Delta} = f(t_0 + k\Delta, \mathbf{z}_k) \qquad (k = 0, 1, \ldots)\,,$$
$$\mathbf{z}_0 = \mathbf{y}_0\,. \tag{1.18.5}$$

Denote by $\mathbf{y} = \mathbf{y}_\Delta(t)$ the linear interpolation function of the set of pairs $(t_0 + k\Delta, \mathbf{z}_k)$; for $0 \le s \le \Delta$, it is defined by

$$\mathbf{y}_\Delta(t_0 + k\Delta + s) = \frac{\Delta - s}{\Delta}\mathbf{z}_k + \frac{s}{\Delta}\mathbf{z}_{k+1}\,.$$

On each of the segments $[t_0 + k\Delta, t_0 + (k + 1)\Delta]$, $k < \alpha/\Delta$, we get the relation

$$|\mathbf{y}'_\Delta(t)| = |f(t_0 + k\Delta, \mathbf{z}_k)| \le m \tag{1.18.6}$$

from which it follows that

$$|\mathbf{z}_k - \mathbf{z}_0| \equiv |\mathbf{z}_k - \mathbf{y}_0| \le k\Delta m \le \alpha m \le b$$

and thus the system (1.18.5) is solvable until $k < \alpha/\Delta$. Therefore, (1.18.6) holds for all $t \in [t_0, t_0 + \alpha]$ (except t of the form $t_0 + k\Delta$ which, however, does not violate the conditions of Arzelà's theorem) and

$$|\mathbf{y}_\Delta(t) - \mathbf{y}_0| \le b \quad \text{for } t \in [t_0, t_0 + \alpha] \; .$$

Hence the set $\{\mathbf{y}_\Delta(t)\}$ is precompact in $C(t_0, t_0 + \alpha)$ if $\Delta \le \alpha$ since both conditions of Arzelà's theorem are fulfilled. Now we can formulate

Theorem 1.18.3. Assume all conditions of Theorem 1.18.2 are fulfilled and the Cauchy problem (1.18.2) has the unique solution $\mathbf{y} = \mathbf{y}(t)$ in $Q(t_0, a, b)$. Then $\{\mathbf{y}_{\Delta_k}(t)\}$ converges uniformly to $\mathbf{y} = \mathbf{y}(t)$ on $[t_0, t_0 + \alpha]$ as $\Delta_k \to 0$.

Proof. It suffices to show that for any $\varepsilon > 0$ there is only a finite number of functions from the sequence $\{\mathbf{y}_{\Delta_k}(t)\}$ which do not satisfy

$$|\mathbf{y}_{\Delta_k}(t) - \mathbf{y}(t)| \le \varepsilon \quad \text{for all } t \in [t_0, t_0 + \alpha] \; . \tag{1.18.7}$$

Assume to the contrary that there are infinitely many functions from the sequence which do not satisfy (1.18.7) for some $\varepsilon > 0$. Then, using a standard technique from calculus, we can find a point t_1, $t_1 \in [t_0, t_0 + \alpha]$, and a subsequence $\Delta_{k_1} \to 0$ such that

$$|\mathbf{y}_{\Delta_{k_1}}(t_1) - \mathbf{y}(t_1)| > \varepsilon \tag{1.18.8}$$

and the number sequence $\{\mathbf{y}_{\Delta_{k_1}}(t)\}$ is convergent.

As the set $\{\mathbf{y}_{\Delta_{k_1}}(t)\}$ is precompact in $C(t_0, t_0 + \alpha)$ we can choose from it a subsequence $\{\mathbf{y}_{\Delta_{k_2}}(t)\}$ uniformly convergent to $\mathbf{z} = \mathbf{z}(t)$ on $[t_0, t_0 + \alpha]$; by (1.18.8)

$$\mathbf{y}(t) \not\equiv \mathbf{z}(t) \quad \text{on} \quad [t_0, t_0 + \alpha] \; . \tag{1.18.9}$$

Rewriting (1.18.5) in the form

$$\frac{d\mathbf{y}_\Delta(t_0 + k\Delta + s)}{ds} = f(t_0 + k\Delta, \mathbf{y}_\Delta(t_0 + k\Delta)) \qquad (0 \le s \le \Delta)$$

and integrating this with regard to (1.18.5), we get

$$\mathbf{y}_\Delta(t) - \mathbf{y}_0 = \Delta \sum_{i=0}^{k-1} f(t_0 + i\Delta, \mathbf{y}_\Delta(t_0 + i\Delta)) + s f(t_0 + k\Delta, \mathbf{y}_\Delta(t_0 + k\Delta)) \, ,$$

where $t = t_0 + k\Delta + s$, $0 \le s \le \Delta$. The expression on the right is a finite Riemann sum which, under the present conditions, converges as $\Delta = \Delta_{k2} \to 0$ to the integral

$$\int_{t_0}^{t} f(s, \mathbf{z}(s))\, ds\,.$$

So $\mathbf{z}(s)$ satisfies

$$\mathbf{z}(t) - \mathbf{y}_0 = \int_{t_0}^{t} f(s, \mathbf{z}(s))\, ds$$

which is equivalent to the Cauchy problem (1.18.2). By uniqueness of solution to this problem, we have $\mathbf{z}(t) = \mathbf{y}(t)$, which contradicts (1.18.9). □

What can we say about convergence of the first derivatives of $\{\mathbf{y}_\Delta(t)\}$? At the nodes of the Δ-net, $\mathbf{y}_\Delta(t)$ has no derivative, but it has a right-hand derivative at every point (which is discontinuous at the nodes) and it can be shown that the sequence of right-hand derivatives of $\{\mathbf{y}_{\Delta_k}(t)\}$ converges uniformly on $[t_0, t_0 + \alpha)$.

The Euler finite-difference procedure is not used for computer solution of differential equations but there are various finite difference methods, frequently used, for which the problem of convergence is open. A harder question is the justification of a finite difference procedure in a boundary value problem, as it is connected with solvability of the problem and uniqueness of solution. Boundary value problems for partial differential equations and systems are of great interest with respect to the application of finite difference methods, but justification of this applicability is in large part an open problem. Most of the achievements here are for so-called variational difference methods which are close to the finite element method; to justify them one uses the modified technique of energy spaces which is under consideration in this book.

Finally, we state (without proof) the criterion for compactness in $L^p(\Omega)$ where Ω is a closed and bounded domain in $\mathbb{R}^n$.

Theorem 1.18.4. A set M of elements of $L^p(\Omega)$, $1 < p < \infty$, is precompact in $L^p(\Omega)$ if and only if M satisfies the following pair of conditions:

(i) *M is bounded in $L^p(\Omega)$.* There is a constant m such that for every function $f(\mathbf{x})$ from M we have $\|f(\mathbf{x})\|_{L^p(\Omega)} \le m$.

(ii) *M is equicontinuous in $L^p(\Omega)$.* For any $\varepsilon > 0$ we can find $\delta(\varepsilon) > 0$ such that whenever $\mathbf{x}, \mathbf{y} \in \Omega$ and $|\mathbf{x} - \mathbf{y}| < \delta$, the inequality $\|f(\mathbf{x}) - f(\mathbf{y})\|_{L^p(\Omega)} < \varepsilon$ holds for all $f(\mathbf{x}) \in M$. Here $f(\mathbf{x})$ is extended by zero outside Ω.

The proof can be found in Kantorovich [21].

1.19 Theory of Approximation in a Normed Space

In what follows, we consider some minimization problems for functionals — as a rule, energy functionals. We begin with one of the simple problems of this class:

General Problem of Approximation in a Normed Space X. Given $x \in X$ and elements $g_1, \ldots, g_n$ with each $g_i \in X$, find numbers $\lambda_1, \ldots, \lambda_n$ such that the function

$$\phi(\lambda_1, \ldots, \lambda_n) = \left\| x - \sum_{i=1}^{n} \lambda_i g_i \right\|$$

is minimized.

We shall exhibit this and similar problems in the form

$$\phi(\lambda_1, \ldots, \lambda_n) \to \min_{\lambda_1, \ldots, \lambda_n} .$$

This form is shared with the problem of best approximation of a continuous function by an nth order polynomial, a trigonometric polynomial, or a special function. Its solution depends on the norm used to pose the problem.

We suppose that $g_1, \ldots, g_n$ are linearly independent:

$$\text{from the equation } \sum_{i=1}^{n} \lambda_i g_i = 0 \text{ it follows that } \lambda_1 = \cdots = \lambda_n = 0 .$$

Denote by X_n the linear subspace of X spanned by $g_1, \ldots, g_n$.

Theorem 1.19.1. For any $x \in X$ there exists x^*, dependent on x, such that

$$x^* = \sum_{i=1}^{n} \lambda_i^* g_i \quad \text{and} \quad \|x - x^*\| = \min_{\lambda_1, \ldots, \lambda_n} \left\| x - \sum_{i=1}^{n} \lambda_i g_i \right\| .$$

Proof. Consider $\phi(\lambda_1, \ldots, \lambda_n)$ as a function of n variables. The continuity on $\mathbb{R}^n$ (or $\mathbb{C}^n$ if X is a complex space) of this function, and of the function

$$\psi(\lambda_1, \ldots, \lambda_n) = \left\| \sum_{i=1}^{n} \lambda_i g_i \right\|$$

follows from the chain of inequalities

$$\begin{aligned}
\left|\phi(\lambda_1 + \Delta_1, \ldots, \lambda_n + \Delta_n) - \phi(\lambda_1, \ldots, \lambda_n)\right| &= \left| \left\| x - \sum_{i=1}^{n} (\lambda_i + \Delta_i) g_i \right\| - \left\| x - \sum_{i=1}^{n} \lambda_i g_i \right\| \right| \\
&\le \left\| \left[x - \sum_{i=1}^{n} (\lambda_i + \Delta_i) g_i\right] - \left[x - \sum_{i=1}^{n} \lambda_i g_i\right] \right\| \\
&= \left\| \sum_{i=1}^{n} \Delta_i g_i \right\| \le \sum_{i=1}^{n} |\Delta_i| \, \|g_i\|
\end{aligned}$$

where we have used (1.11.2). By continuity, $\psi(\lambda_1, \ldots, \lambda_n)$ assumes a minimum value on the sphere $\sum_{i=1}^{n} |\lambda_i|^2 = 1$ at some point $\tilde{\lambda}_1, \ldots, \tilde{\lambda}_n$ with $\sum_{i=1}^{n} |\tilde{\lambda}_i|^2 = 1$. As the set $\{g_1, \ldots, g_n\}$ is linearly independent we get

$$\left\| \sum_{i=1}^{n} \tilde{\lambda}_i g_i \right\| = \min_{\sum_{i=1}^{n} |\lambda_i|^2 = 1} \left\| \sum_{i=1}^{n} \lambda_i g_i \right\| = d > 0\,.$$

By (1.11.2),

$$\phi(\lambda_1, \ldots, \lambda_n) \geq \left\| \sum_{i=1}^{n} \lambda_i g_i \right\| - \|x\|$$

and, on the domain

$$\left(\sum_{i=1}^{n} |\lambda_i|^2 \right)^{1/2} \geq 3\|x\|/d\,,$$

we have

$$\phi(\lambda_1, \ldots, \lambda_n) \geq (3\|x\|/d)d - \|x\| = 2\,\|x\|\,.$$

Since $\phi(0, \ldots, 0) = \|x\|$, we find that $\phi(\lambda_1, \ldots, \lambda_n)$ has a minimal value at a point $(\lambda_1^*, \ldots, \lambda_n^*)$ of the ball $\left(\sum_{i=1}^{n} |\lambda_i|^2 \right)^{1/2} < 3\|x\|/d$. □

The general problem of approximation is relatively simple in principle — not necessarily in its concrete applications.

Uniqueness of the best approximation is attained in spaces of the next type.

Definition 1.19.1. A normed space is *strictly normed* if from the equality

$$\|x + y\| = \|x\| + \|y\| \qquad (x \neq 0)$$

it follows that $y = \lambda x$ and $\lambda \geq 0$.

The spaces ℓ^p, $L^p(\Omega)$, $W^{k,p}(\Omega)$, when $1 < p < \infty$, are strictly normed. This follows from the properties of Minkowski's inequality (see, for example, Hardy et. al. [14]).

We establish a uniqueness theorem under more general conditions than the existence theorem. Recall Definition 1.6.3.

Theorem 1.19.2. For any element x of a strictly normed space X, there is at most one element of a closed convex set $M \subset X$ that minimizes the functional $F(y) = \|x - y\|$ on the set M.

Proof. Suppose to the contrary that $F(y)$ has two minimizers y_1 and y_2:

$$\|x - y_1\| = \|x - y_2\| = \inf_{y \in M} \|x - y\| \equiv d\,. \tag{1.19.1}$$

If $x \in M$ then $x = y_1 = y_2$. Let $x \notin M$ so that $d > 0$. By convexity of M, the element $(y_1 + y_2)/2$ belongs to M. So

$$\left\| x - \frac{y_1 + y_2}{2} \right\| \geq d\,.$$

On the other hand,

$$d \le \left\| x - \frac{y_1 + y_2}{2} \right\| = \left\| \frac{x - y_1}{2} + \frac{x - y_2}{2} \right\| \le \frac{1}{2}\|x - y_1\| + \frac{1}{2}\|x - y_2\| = d\,.$$

Therefore

$$\left\| \frac{x - y_1}{2} + \frac{x - y_2}{2} \right\| = \left\| \frac{x - y_1}{2} \right\| + \left\| \frac{x - y_2}{2} \right\|.$$

Because X is strictly normed, it follows that

$$x - y_1 = \lambda(x - y_2) \qquad (\lambda \ge 0)$$

so $\|x - y_1\| = \lambda\|x - y_2\|$. From (1.19.1) we get $\lambda = 1$, hence $y_1 = y_2$. □

Lemma 1.19.1. An inner product space is strictly normed.

Proof. Let $\|x + y\| = \|x\| + \|y\|$, $x \neq 0$. Then

$$\|x + y\|^2 = (\|x\| + \|y\|)^2\,.$$

This can be rewritten (for a complex space) in the form

$$\|x\|^2 + 2\,\mathrm{Re}(x, y) + \|y\|^2 = \|x\|^2 + 2\,\|x\|\,\|y\| + \|y\|^2 \tag{1.19.2}$$

so $\mathrm{Re}(x, y) = \|x\|\,\|y\|$. By the Schwarz inequality, we obtain $\mathrm{Im}(x, y) = 0$ and thus

$$(x, y) = \|x\|\,\|y\| \tag{1.19.3}$$

(in the real case, this equality comes directly from (1.19.2)). Theorem 1.14.1 gives $y = \lambda x$ and, placing this into (1.19.3), we get $\lambda \ge 0$. □

In a Hilbert space, we can combine Theorems 1.19.2 and 1.19.1 as follows.

Theorem 1.19.3. For any element x of a Hilbert space H, there is a unique element of a closed convex set M that minimizes the functional $F(y) = \|x - y\|$ on M.

Proof. Uniqueness was proved in Theorem 1.19.2. We show existence of a minimizer. Let $\{y_k\}$ be a minimizing sequence of $F(y)$, i.e.,

$$\lim_{k\to\infty} F(y_k) = \lim_{k\to\infty} \|x - y_k\| = \inf_{y\in M} \|x - y\|\,.$$

(By definition of infimum, such a sequence exists.) As M is closed it suffices to show that $\{y_k\}$ is a Cauchy sequence. For this, we write down the parallelogram equality for elements $x - y_i$ and $x - y_j$:

$$\|2x - y_i - y_j\|^2 + \|y_i - y_j\|^2 = 2\left(\|x - y_i\|^2 + \|x - y_j\|^2\right),$$

so

$$\|y_i - y_j\|^2 = 2\left(\|x - y_i\|^2 + \|x - y_j\|^2\right) - 4\left\| x - \frac{y_i + y_j}{2} \right\|^2. \tag{1.19.4}$$

Since $\|x - y_k\|^2 = d^2 + \varepsilon_k$, where $\varepsilon_k \to 0$ as $k \to \infty$, from (1.19.4) it follows that

$$\|y_i - y_j\|^2 \le 2(d^2 + \varepsilon_i + d^2 + \varepsilon_j) - 4d^2 = 2(\varepsilon_i + \varepsilon_j) \to 0$$

as $i, j \to \infty$. □

All requirements of Theorem 1.19.3 are met if M is a closed linear subspace of H. But this case is so important that we treat it separately in Sect. 1.20.

Problem 1.19.1. Show that ℓ^2 is a strictly normed space. □

1.20 Decomposition Theorem, Riesz Representation

Let x be an arbitrary element of a Hilbert space H. Let M be a closed linear subspace of H, and let m be the unique (by Theorem 1.19.3) minimizer of $F(y)$ on M:

$$\|x - m\| = \inf_{v \in M} \|x - v\| .$$

Taking a fixed element $v \in M$, we consider a real-valued function

$$f(\alpha) = \|x - m - \alpha v\|^2$$

of a real variable α. This function takes its minimum value at $\alpha = 0$, so

$$\left.\frac{df}{d\alpha}\right|_{\alpha=0} = 0 .$$

Direct calculation gives

$$\left.\frac{df}{d\alpha}\right|_{\alpha=0} = \frac{d}{d\alpha}(x - m - \alpha v, x - m - \alpha v)\bigg|_{\alpha=0} = -2\,\mathrm{Re}(x - m, v) = 0 .$$

Replacing v by iv, we get $\mathrm{Im}(x - m, v) = 0$ so that

$$(x - m, v) = 0 . \tag{1.20.1}$$

It follows that $x - m$ is orthogonal to every $v \in M$.

Definition 1.20.1. An element n of a Hilbert space H is *orthogonal to a subspace* M of H, if n is orthogonal to every element of M. Two subspaces M and N of H are *mutually orthogonal* (denoted $M \perp N$) if any $n \in N$ is orthogonal to M and any $m \in M$ is orthogonal to N.

We say that there is an *orthogonal decomposition* of a Hilbert space H into M and N if M and N are mutually orthogonal subspaces of H and any element $x \in H$ can be uniquely represented in the form

$$x = m + n \qquad (m \in M,\ n \in N) . \tag{1.20.2}$$

Relation (1.20.1) leads to the *decomposition theorem* for a Hilbert space:

Theorem 1.20.1 (decomposition). Assume that M is a closed subspace of a Hilbert space H. Then there is a closed subspace N of H that is orthogonal to M and such that H can be uniquely decomposed into the orthogonal sum of M and N, i.e., any $x \in H$ has the unique representation (1.20.2).

Proof. Denote by N the set of all elements of H such that any $n \in N$ is orthogonal to every $m \in M$ (we suppose that $M \neq H$). The set N is not empty as contains the element $x - m$, cf. (1.20.1). It is seen that N is a subspace of H; indeed, if $n_1, n_2 \in N$, i.e.

$$(n_1, m) = (n_2, m) = 0 \quad \text{for all } m \in M\ ,$$

then $(\lambda_1 n_1 + \lambda_2 n_2, m) = 0$ for any numbers λ_1, λ_2 and any $m \in M$. Moreover, N is closed: every Cauchy sequence $\{n_k\} \subset N$ has a limit element $n \in N$ because

$$(n, m) = \lim_{k\to\infty} (n_k, m) = 0 \quad \text{for all } m \in M\ .$$

At the beginning of this section, we constructed for an arbitrary element $x \in H$ its projection m on M in such a way that $n = x - m$ is, by (1.20.1), orthogonal to M. So the representation (1.20.2) holds.

It remains to prove uniqueness. Assume that for some x there are two such representations:

$$x = m_1 + n_1 \quad \text{and} \quad x = m_2 + n_2 \qquad (m_i \in M, n_i \in N)\ .$$

Then $m_1 + n_1 = m_2 + n_2$ or

$$m_1 - m_2 = n_1 - n_2\ .$$

Multiplying both sides of this equality by $m_1 - m_2$ and then by $n_1 - n_2$ in H, we get $\|m_1 - m_2\|^2 = 0$ and $\|n_1 - n_2\|^2 = 0$, respectively. This completes the proof. □

Theorem 1.20.1 has widespread applications. One is the *Riesz representation theorem* in a Hilbert space, which is of great importance in what follows. The roots of the theorem are in the unique representation of a linear[7] homogeneous function $F(\mathbf{x}) = a_1x_1 + \cdots + a_nx_n$ on $\mathbb{R}^n$, in the form of a dot product: $F(\mathbf{x}) = \mathbf{a} \cdot \mathbf{x}$ where $\mathbf{a} = (a_1, \ldots, a_n)$ and $\mathbf{x} = (x_1, \ldots, x_n)$.

Theorem 1.20.2 (Riesz representation). Let $F(x)$ be a continuous linear functional given on a Hilbert space H. There is a unique element $f \in H$ such that

$$F(x) = (x, f) \quad \text{for all } x \in H\ . \tag{1.20.3}$$

Moreover, $\|F\| = \|f\|$.

[7] The term *linear operator*, of which the term *linear functional* is a special case, is formally defined on p. 73. Roughly, a linear functional F is a real- (complex-) valued mapping that satisfies the relation $F(\alpha x + \beta y) = \alpha F(x) + \beta F(y)$ for all real (complex) scalars α, β and all elements x, y of the domain.

Proof. Consider the *null space* or *kernel* of F, defined as the set M of all elements satisfying

$$F(x) = 0\,. \tag{1.20.4}$$

By linearity of $F(x)$, any finite linear combination $\sum_{i=1}^{n} \lambda_i m_i$ of elements $m_i \in M$ belongs to M; if $\{m_k\} \subset M$ is a Cauchy sequence in H, then by continuity of $F(x)$ we see that $x = \lim_{k\to\infty} m_k$ satisfies (1.20.4), i.e., $x \in M$. Therefore M is a closed subspace of H.

By Theorem 1.20.1, there is a closed subspace N of H which is orthogonal to M, and any $x \in H$ can be uniquely represented as $x = m + n$ where $m \in M$, $n \in N$, and $(m, n) = 0$.

Let us show that N is one-dimensional, i.e., that any $n \in N$ has the form $n = \alpha n^*$ for some fixed $n^* \in N$. Let n_1 and n_2 be two arbitrary elements of N. Then the element $n_3 = F(n_1)n_2 - F(n_2)n_1$ belongs to N. On the other hand

$$F(n_3) = F(n_1)F(n_2) - F(n_2)F(n_1) = 0\,,$$

which means that $n_3 \in M$. So $n_3 = 0$, hence N is one-dimensional.

Take $n \in N$ and define n_0 by $n_0 = n/\|n\|$. Any $x \in H$ can be represented as $x = m + \alpha n_0$ ($m \in M$) where $\alpha = (x, n_0)$. It follows that

$$F(x) = F(m + \alpha n_0) = F(m) + \alpha F(n_0) = \alpha F(n_0) = F(n_0)(x, n_0) = (x, \overline{F(n_0)}n_0)\,.$$

Denoting $\overline{F(n_0)}n_0$ by f, we obtain (1.20.3).

Suppose there are two representers f_1 and f_2, i.e., $F(x) = (x, f_1) = (x, f_2)$ so that

$$(x, f_1 - f_2) = 0\,.$$

Putting $x = f_1 - f_2$, we get $\|f_1 - f_2\|^2 = 0$, hence $f_1 = f_2$. Finally, the equality $\|F\| = \|f\|$ follows from the definition of $\|F\|$ and the Schwarz inequality. □

Although carried out in a complex Hilbert space, this proof remains valid for a real Hilbert space. The sense of the theorem is that any continuous linear functional given on a Hilbert space can be uniquely identified with an element of the same space. Since for a fixed $f \in H$ the inner product (x, f) is a continuous linear functional on H, we have a one-to-one correspondence between the set H' of all continuous linear functionals given on a Hilbert space H (called the *conjugate space* to H) and H itself.

1.21 Linear Operators and Functionals

We have already used the notions of operator and functional, with the understanding that the reader has surely encountered them in other subject areas. Let us present a formal definition valid for metric and normed spaces.

Definition 1.21.1. Let X and Y be metric spaces. A correspondence $x \mapsto y = A(x)$, where $x \in X$ and $y \in Y$, is an *operator* if to each $x \in X$ there corresponds no more than one $y \in Y$. We say that A *acts from* X *to* Y. If $Y = X$, we say that A *acts in* X. The set of all x for which there exists a corresponding $y = A(x)$ is the *domain* of A, denoted $D(A)$. The image of $D(A)$ is the *range* of A, denoted $R(A)$. If $R(A) \subseteq \mathbb{R}$ or $\mathbb{C}$, we refer to A as a *real* or *complex functional*, respectively.

Remark 1.21.1. In the applied literature, transformations of a function $u(x, y)$ like

$$A(u)(x, y) = \nabla^2 u(x, y) \quad \text{or} \quad B(u)(x, y) = \iint_S k(x, y, \xi, \chi)\, u(\xi, \chi)\, d\xi d\chi$$

are usually called operators. In functional analysis, A and B become operators only when we clearly specify their domains and ranges. Any change in the domain yields a new operator, even if the expression for the mapping rule remains the same. If, for example, we enlarge the domain of A, we get a new operator called an *extension* or *continuation* of A. □

In accordance with the classical definition of function continuity, we say that A is *continuous* at $x_0 \in X$ if for any $\varepsilon > 0$ there exists $\delta > 0$ (dependent on ε) such that $d(A(x), A(x_0)) < \varepsilon$ whenever $d(x, x_0) < \delta$. If A is continuous at every point of an open domain M, it is said to be continuous in M.

If X and Y are linear spaces, we can consider a class of linear operators A from X to Y.

Definition 1.21.2. A *linear operator* A satisfies

$$A(\lambda_1 x_1 + \lambda_2 x_2) = \lambda_1 A(x_1) + \lambda_2 A(x_2)$$

for all $x_1, x_2 \in X$ and any (real or complex) numbers λ_1, λ_2. In this case the image $A(x)$ of a point x is usually denoted by Ax.

In this section, from now on, we let X and Y be normed spaces. The definition of continuity of an operator is changed in an evident way (recall relation (1.11.1)). For a linear operator A we have $Ax - Ax_0 = A(x - x_0)$, so A is continuous in the whole space X if and only if it is continuous at a single point $x_0 \in X$, say $x_0 = 0$. This brings us to the next result.

Theorem 1.21.1. Let X and Y be normed spaces. A linear operator A from X to Y is continuous if and only if there is a constant c such that for every $x \in X$

$$\|Ax\| \le c\,\|x\| . \tag{1.21.1}$$

The infimum of all such constants c is called the *norm* of A, denoted $\|A\|$.

Proof. We need only show continuity of A at $x = 0$. If (1.21.1) holds, then this continuity is clear by definition. Conversely, suppose A is continuous at $x = 0$. Take $\varepsilon = 1$; by definition there exists $\delta > 0$ such that $\|Ax\| < 1$ whenever $\|x\| < \delta$. For every nonzero $x \in X$, the norm of $x^* = \delta x/(2\,\|x\|)$ is

$$\|x^*\| = \|\delta x/(2\,\|x\|)\| = \delta/2 < \delta\,,$$

so $\|Ax^*\| < 1$. Since A is linear, we have

$$\frac{\delta}{2\,\|x\|}\,\|Ax\| < 1 \quad \text{or} \quad \|Ax\| < \frac{2}{\delta}\,\|x\|\,.$$

This is (1.21.1) with $c = 2/\delta$. □

Problem 1.21.1. Note that, by definition,

1. $\|Ax\| \le \|A\|\,\|x\|$ for all $x \in X$, and
2. for every $\varepsilon > 0$ there exists x_ε such that $\|Ax_\varepsilon\| > (\|A\| - \varepsilon)\,\|x_\varepsilon\|$.

Use these two properties to show that $\|A\| = \sup_{\|x\|\le 1} \|Ax\|$. □

Definition 1.21.3. A linear operator A satisfying relation (1.21.1) is a *bounded linear operator.*

Theorem 1.21.2. Let X and Y be normed spaces. A linear operator A from X to Y is continuous if and only if for every sequence $\{x_n\}$ such that $x_n \to 0$, the sequence $Ax_n \to 0$ as $n \to \infty$.

Proof. It is clear that for a continuous operator, $Ax_n \to 0$ if $x_n \to 0$ as $n \to \infty$. We prove the converse. Let $Ax_n \to 0$ for every sequence $\{x_n\}$ such that $x_n \to 0$. Suppose to the contrary that A is not continuous (i.e., is not bounded). Then there exists a sequence $\{x_n\}$ such that $\|x_n\| \le 1$ but $\|Ax_n\| \to \infty$. We can assume (why?) that $\|Ax_n\| \ge n$. Consider the sequence $y_n = x_n/\sqrt{n}$. It is clear that $\|y_n\| \to 0$ but $\|Ay_n\| \ge \sqrt{n}$ so Ay_n does not tend to 0. This contradiction completes the proof. □

Thus, for a linear operator acting from X to Y we can introduce two other equivalent definitions of continuity: the first uses the property of boundedness (inequality (1.21.1)), and the second, defined by Theorem 1.21.2, uses the notion of *sequential continuity*.

Let us consider some examples of operators.

1. The differentiation operator is d/dx. It is clear that it is bounded from $C^{(1)}(-\infty,\infty)$ to $C(-\infty,\infty)$.

2. A differential operator

$$Au = \sum_{|\alpha|\le m} a_\alpha D^\alpha u(\mathbf{x})$$

with constant coefficients a_α is bounded (hence continuous) from $C^{(m)}(\overline{\Omega})$ to $C(\overline{\Omega})$ and from $W^{m,2}(\Omega)$ to $L^2(\Omega)$. The reader should construct an example where it is not continuous.

3. The integral operator defined by

$$Bu(x) = \int_0^x u(s)\,ds$$

is continuous from $C(0,1)$ to $C(0,1)$. It is left as an exercise for the reader to determine whether it is continuous from $C(0,1)$ to $C^{(1)}(0,1)$.

Problem 1.21.2. Verify the statements made in these examples. □

We mention that various authors employ other terms instead of "operator." We sometimes find instead the terms transformation, map, mapping, or simply function.

Problem 1.21.3. Prove that for an operator (not necessarily linear), *sequential continuity* is equivalent to continuity as defined on p. 73. That is, show that an operator f, acting from X to Y, is continuous if and only if for every sequence $\{x_n\} \subset X$ such that $x_n \to x_0$ in X, it follows that $f(x_n) \to f(x_0)$ in Y. □

Problem 1.21.4. Show that the inner product is a continuous function with respect to each of its arguments. □

Problem 1.21.5. Suppose f is a continuous operator from the space X onto the space Y (i.e., $f(X) = Y$). Show that if A is dense in X, then $f(A)$ is dense in Y. □

Problem 1.21.6. Suppose f and g are continuous operators acting from X to Y. Show that the set $S = \{x \in X : f(x) = g(x)\}$ is closed in X. □

Problem 1.21.7. Let A be a linear operator from a linear space X to a linear space Y. (a) Show that A maps convex subsets of $D(A)$ into convex subsets of Y. (b) Show that the inverse image of a convex set lying in $R(A)$ is a convex set in $D(A)$. □

Problem 1.21.8. Let X, Y be linear spaces and A a linear operator from X to Y. Show that if $\{x_1, \ldots, x_n\}$ is a linearly dependent subset of $D(A)$, then $\{Ax_1, \ldots, Ax_n\}$ is a linearly dependent subset of $R(A)$. □

Problem 1.21.9. [38] Show that each of the following expressions defines a bounded linear operator acting between the spaces indicated. Find or estimate the norm of each.

(a) $Af(x) = \int_0^x f(\xi)\, d\xi$, acting in $C(0,1)$;

(b) $Af(x) = f(x)$, acting from $C(-1,1)$ to $C(0,1)$;

(c) $Af(x) = x \int_0^1 f(x)\, dx$, acting in $L^2(0,1)$;

(d) $Af(x) = f(x)$, acting from $C^{(1)}(a,b)$ to $C(a,b)$;

(e) $Af(x) = f'(x)$, acting from $C^{(1)}(a,b)$ to $C(a,b)$;

(f) $Af(x) = \int_0^x f(\xi)\, d\xi$, acting in $L^2(0,1)$;

(g) $Af(x) = \frac{1}{2}[f(x) + f(-x)]$, acting in $C(-1,1)$.

Show that each of the following expressions defines a bounded linear functional on the space indicated:

(h) $\phi(f) = \int_{-1}^{1} x f(x)\,dx$ on $L^2(-1,1)$;

(i) $\phi(f) = \int_{-1}^{1} f(x)\,dx - f(0)$ on $C(-1,1)$. □

1.22 Space of Linear Continuous Operators

Let X and Y be normed spaces. In Sect. 1.21 we saw that a linear operator A defined on X is continuous if and only if there is constant c such that

$$\|Ax\| \le c\,\|x\| \quad \text{for all } x \in X\,.$$

The infimum of all such constants is $\|A\|$, the norm of A.

Consider the set $L(X, Y)$ of continuous linear operators from X to Y; it is clearly a linear space.

Lemma 1.22.1. $L(X, Y)$ is a normed space.

Proof. We need to check only that the norm axioms N1–N3 are satisfied by the operator norm introduced above. $\|A\|$ is clearly non-negative. If $\|A\| = 0$, then $\|Ax\| = 0$ for all $x \in X$, i.e., $A = 0$. Conversely, if $A = 0$ then $\|A\| = 0$. Hence N1 is satisfied. It is obvious that N2 is satisfied. The chain of inequalities

$$\|(A + B)x\| = \|Ax + Bx\| \le \|Ax\| + \|Bx\| \le \|A\|\,\|x\| + \|B\|\,\|x\|$$

shows that $\|A + B\| \le \|A\| + \|B\|$. Hence $\|A\|$ also satisfies N3. □

As in any normed space, the notion of convergence applies in $L(X, Y)$.

Definition 1.22.1. A sequence $\{A_n\}$ of continuous linear operators *converges uniformly* to an operator A if $\|A_n - A\| \to 0$ as $n \to \infty$.

This terminology makes sense because by the inequality

$$\|A_n x - Ax\| \le \|A_n - A\|\,\|x\|\,, \tag{1.22.1}$$

we have $A_n x \to Ax$ as $n \to \infty$ for all $x \in X$ simultaneously as $\|A_n - A\| \to 0$.

Theorem 1.22.1. If X is a normed space and Y is a Banach space, then $L(X, Y)$ is a Banach space.

Proof. Let $\{A_n\}$ be a Cauchy sequence in $L(X, Y)$, i.e.,

$$\|A_{n+m} - A_n\| \to 0 \;\text{ as } n \to \infty\,, \quad m > 0\,.$$

We must show that there is a continuous operator A such that $A_n \to A$ in the operator norm as $n \to \infty$. For any $x \in X$, $\{A_n x\}$ is also a Cauchy sequence because

$$\|A_{n+m}x - A_n x\| \le \|A_{n+m} - A_n\| \, \|x\| \; ;$$

since Y is a Banach space, there exists $y \in Y$ such that $A_n x \to y$. This unique correspondence between $x \in X$ and $y \in Y$ constitutes an operator A with $y = Ax$. By properties of the limit and the linearity of A_n, the operator A is linear. Since $\{A_n\}$ is a Cauchy sequence, the sequence of norms $\{\|A_n\|\}$ is bounded, say $\|A_n\| \le c$, and so

$$\|Ax\| = \lim_{n\to\infty} \|A_n x\| \le \limsup_{n\to\infty} \|A_n\| \, \|x\| \le c \, \|x\| \, .$$

That is, A is continuous and the proof is finished. □

Problem 1.22.1. Verify that A is linear and that $A_n \to A$ in the operator norm. □

In a Banach space $L(X, Y)$ we can introduce series $\sum_{n=1}^{\infty} A_n$ and define their sums by

$$S = \lim_{k\to\infty} \sum_{n=1}^{k} A_n \, .$$

Such a series is said to be *absolutely convergent* if the numerical series $\sum_{n=1}^{\infty} \|A_n\|$ is convergent.

Problem 1.22.2. Suppose $A_n \in L(X, Y)$ where Y is a Banach space. Show that if $\sum_{n=1}^{\infty} A_n$ is absolutely convergent, then it is convergent. □

If X is a Banach space, we denote the space $L(X, X)$ by $L(X)$. In $L(X)$ we can introduce the product of operators A, B by

$$ABx = A(Bx) \, .$$

The product possesses the usual properties of a numerical product except commutativity; we have

$$A(B + C) = AB + AC \, , \qquad (AB)C = A(BC) \, ,$$
$$(A + B)C = AC + BC \, , \qquad IA = AI = A \, ,$$

where I is the identity operator. The product is also a continuous operator, because

$$\|ABx\| \le \|A\| \, \|Bx\| \le \|A\| \, \|B\| \, \|x\|$$

gives

$$\|AB\| \le \|A\| \, \|B\| \, .$$

Problem 1.22.3. Show that if $A_n \to A$ and $B_n \to B$ in the operator norm, where $A_n, B_n \in L(X)$, then $A_n B_n \to AB$. □

Note that $L(X)$ is a non-commutative ring. Denoting by A^k the product $A \cdot A \cdots A$ (k times), we can introduce functions of operators in $L(X)$, for example,

$$e^A = I + \sum_{k=1}^{\infty} \frac{1}{k!} A^k .$$

Now we introduce another type of convergence of linear operators, an analogue of pointwise convergence for a sequence of ordinary functions.

Definition 1.22.2. A sequence $\{A_n\} \subset L(X, Y)$ is *strongly convergent* to $A \in L(X, Y)$ if, whenever $x \in X$,

$$\|A_n x - Ax\| \to 0 \text{ as } n \to \infty .$$

In this case A is the *strong limit* of $\{A_n\}$.

By (1.22.1), an operator sequence $\{A_n\}$ uniformly convergent to A is also strongly convergent to A; for any x,

$$\|A_n x - Ax\| \le \|A_n - A\| \, \|x\| \to 0 \text{ as } n \to \infty .$$

However, the strong convergence of a sequence of linear operators, which is sometimes called *pointwise convergence*, does not imply its uniform convergence. An example is presented in Sect. 2.8 in the Fourier series theory. Let $g_1, g_2, \ldots$ be an orthonormal basis in a Hilbert space H, which means that $(g_k, g_m) = \delta_{km}$ and for $x \in H$ we have $x = \sum_{k=1}^{\infty} (x, g_k) g_k$. We introduce the sequence of *projection operators* P_n by the equality

$$P_n x = \sum_{k=1}^{n} (x, g_k) g_k .$$

Problem 1.22.4. Show that P_n is a bounded linear operator with $\|P_n\| = 1$. □

By the results of Sect. 2.8, $P_n x \to x$ as $n \to \infty$ and so $\{P_n\}$ is strongly convergent to the unit operator I. However,

$$P_{n+m} x - P_n x = \sum_{k=n+1}^{n+m} (x, g_k) g_k$$

and so for $x = g_{n+1}$ we get

$$\|P_{n+m} g_{n+1} - P_n g_{n+1}\| = \left\| \sum_{k=n+1}^{n+m} (g_{n+1}, g_k) g_k \right\| = \|(g_{n+1}, g_{n+1}) g_{n+1}\| = 1 .$$

This means $\{P_n\}$ does not converge uniformly on the unit ball $\|x\| \le 1$.

For the strong convergence of an operator sequence we will establish the important Banach–Steinhaus principle (Theorem 1.24.2).

The space of linear continuous functionals, i.e., the space $L(X, \mathbb{R})$ or $L(X, \mathbb{C})$ (depending on whether X is a real or complex normed space) is often denoted by X' (or X'_s) and is called the *dual space* to X. Since $\mathbb{R}$ and $\mathbb{C}$ are Banach spaces, by Theorem 1.22.1, so is X'. The Hahn–Banach theorem (Theorem 1.30.1) shows that X' is not empty and, moreover, contains linear continuous functionals with useful properties.

1.23 Uniform Boundedness Theorem

Now we study some properties of the strong convergence of continuous linear operators in a Banach space X. First we establish an interesting property of a Banach space given by the Baire category theorem (Theorem 1.23.1).

Definition 1.23.1. A set $S \subset X$ is *nowhere dense in* X if its closure $\overline{S}$ does not contain any open ball of nonzero radius of X entirely.

In other words, a ball $B(x_0, r) = \{x \in X: \|x - x_0\| < r\}$ with any $r > 0$ and any $x_0 \in X$ contains at least one point that is not in $\overline{S}$.

Definition 1.23.2. A set $V \subset X$ is *of the first category* if it can be represented as a countable union of sets that are nowhere dense in X. Otherwise it is *of the second category.*

The notion of category can be applied to the whole space as well. We shall prove

Theorem 1.23.1 (Baire). A (nonempty) Banach space X is of the second category.

Proof. Suppose to the contrary that X is of the first category, which means X is the union of some sets $S_1, S_2, \ldots$ each of which is nowhere dense in X:

$$X = \cup_{k=1}^{\infty} S_k . \tag{1.23.1}$$

We will show the existence of a point $x_0 \in X$ that is not an element of any $\overline{S_k}$, which will contradict (1.23.1). We start to construct the point.

By Definition 1.23.1, there exists $x_1 \in X$ that does not belong to $\overline{S_1}$. Because x_1 is not an accumulation point of S_1, there is a closed ball

$$\overline{B}(x_1, r_1) = \{x \in X: \|x - x_1\| \le r_1\}$$

centered at x_1 with radius $0 < r_1 < 1$ that does not intersect with S_1 and, moreover, with $\overline{S_1}$. Again by Definition 1.23.1, the open ball

$$B(x_1, r_1) = \{x \in X: \|x - x_1\| < r_1\}$$

contains a point x_2 that does not belong to $\overline{S_2}$. As x_2 is not an accumulation point of S_2 there is a closed ball

$$\overline{B}(x_2, r_2) = \{x \in X: \|x - x_2\| \le r_2\}$$

of radius $0 < r_2 < 1/2$ that is inside $B(x_1, r_1)$ and such that $\overline{B}(x_2, r_2)$ does not have any point of $\overline{S_2}$. Clearly, $\overline{B}(x_2, r_2)$ also does not intersect $\overline{S_1}$.

On the kth step we obtain an element x_k and a ball $B(x_k, r_k)$ of radius $0 < r_k < 1/k$ such that

(1) $\overline{B}(x_k, r_k) \subset B(x_{k-1}, r_{k-1})$, and

(2) $\overline{B}(x_k, r_k) \cap \overline{S_m}$ is empty for all $m \leq k$.

Continuing this process, we get an infinite sequence of balls $B(x_k, r_k)$ such that

$$B(x_1, r_1) \supset B(x_2, r_2) \supset \cdots \supset B(x_k, r_k) \supset \cdots ,$$

$r_k < 1/k$, $\overline{B}(x_k, r_k)$ does not intersect $\overline{S_1} \cup \cdots \cup \overline{S_k}$, and $\overline{B}(x_k, r_k) \subset B(x_{k-1}, r_{k-1})$.

The centers of the closed balls satisfy

$$||x_k - x_{k+m}|| < 1/k \;\text{ for all } m \geq 1 , \tag{1.23.2}$$

so $\{x_k\}$ is a Cauchy sequence. Since X is complete, $x_k \to x_0$ for some $x_0 \in X$. Letting $m \to \infty$ in (1.23.2) we get

$$||x_k - x_0|| \leq 1/k .$$

So $x_0 \in \overline{B}(x_k, r_k)$ and thus x_0 does not belong to $\overline{S_k}$ for any k. This contradicts (1.23.1). □

Corollary 1.23.1. Let X be a Banach space and

$$X = \cup_{k=1}^{\infty} S_k .$$

At least one of sets $\overline{S_k}$ contains an open ball of X.

This follows from the fact that at least one of the sets $\overline{S_k}$ is not nowhere dense in X and hence contains an open ball.

The next theorem is called the *principle of uniform boundedness* or the *resonance theorem.*

Theorem 1.23.2 (uniform boundedness). Let X be a Banach space and Y a normed space. Suppose $A_i \in L(X, Y)$ for each $i \in I$, where I is an uncountable index set. Further, suppose that for each $x \in X$, $\sup_{i \in I} ||A_i x||$ is bounded. Then $\sup_{i \in I} ||A_i||$ is bounded; in other words, the family A_i is uniformly bounded in the norm.

Proof. For integer k, we introduce a set $S_k \subset X$ as

$$S_k = \left\{ x \in X : \sup_{i \in I} ||A_i x|| \leq k \right\} .$$

A_i is a continuous operator, so $S_{ik} = \{x \in X : ||A_i x|| \leq k\}$ is a closed set. S_k is the intersection of all the S_{ik}, hence S_k is also closed. As $\sup_{i \in I} ||A_i x||$ is bounded for any $x \in X$, we have

$$X = \cup_{k=1}^{\infty} S_k .$$

By Corollary 1.23.1, one of the sets (S_k), say S_m, contains an open ball

$$B(x^*, r) = \{z \in X : ||z - x^*|| < r\} ,$$

centered at x^* and having radius $r > 0$.

For any $x \in X$, the element

$$z_x = x^* + rx/(2\|x\|) \in B(x^*, r)$$

and so $\|A_i z_x\| \le m$, i.e.

$$\left\| A_i x^* + A_i\left(\frac{r}{2\|x\|}x\right)\right\| \le m\,.$$

Using the inequality

$$\|a\| = \|(a+b) - b\| \le \|a+b\| + \|b\|$$

we get

$$\left\| A_i\left(\frac{r}{2\|x\|}x\right)\right\| \le m + \|A_i x^*\| \le 2m$$

(here we have used the fact that $\|A_i x^*\| \le m$ as $x^* \in B(x^*, r)$). Multiplying both sides of the inequality by $2\|x\|/r$ we find that

$$\|A_i x\| \le \frac{4m}{r}\|x\|$$

so $\|A_i\| \le 4m/r$ for all $i \in I$. This completes the proof. □

A consequence of Theorem 1.23.2 is

Corollary 1.23.2. Let X be a Banach space and Y a normed space. Let the sequence $\{A_n\} \subset L(X, Y)$ be such that for each $x \in X$ there exists $\lim_{n\to\infty} A_n x = Ax$. Then A is a linear continuous operator and

$$\|A\| \le \liminf_{n\to\infty} \|A_n\|\,. \tag{1.23.3}$$

Proof. Because $A_n x \to Ax$, the sequence $\{\|A_n x\|\}$ is bounded for any $x \in X$. So we can apply Theorem 1.23.2 and conclude that $\sup_n \|A_n\|$ is finite. It is easy to verify that A is a linear operator. Next, for any $x \in X$,

$$\|Ax\| = \lim_{n\to\infty} \|A_n x\| \le \liminf_{n\to\infty} \|A_n\|\,\|x\|$$

so we see that (1.23.3) is valid (Problem 1.23.1). □

Problem 1.23.1. Complete the details of the proof of (1.23.3). □

Remark 1.23.1. Corollary 1.23.2 does *not* imply that $\|A_n - A\| \to 0$. However, the pointwise convergence of $\{A_n x\}$ at every point $x \in X$ implies the existence of a linear continuous operator A for which $A_n x \to Ax$ at any point. If Y is a Banach space, then for the existence of $Ax = \lim_{n\to\infty} A_n x$ it suffices to assume that $\{A_n x\}$ is a Cauchy sequence for every x.

Theorem 1.23.2 and its corollary hold for the families of continuous linear functionals on a Banach space X. □

1.24 Banach–Steinhaus Principle

In applications, we frequently encounter situations where a differential operator, initially defined on a set of smooth functions that is dense in some space, must be extended to the larger space. For linear operators, a similar situation is considered in the following two theorems.

Theorem 1.24.1. Let A be a linear operator with domain $D(A)$ dense in a normed space X, acting to a Banach space Y, and bounded on $D(A)$:

$$\|Ax\| \le M\,\|x\| \quad \text{for all } x \in D(A)\,.$$

Let us call the infimum of the constants M on $D(A)$ the norm of A and denote it $\|A\|$. There is a continuation of A, denoted A_c, defined on X and such that

(i) $A_c \in L(X, Y)$;

(ii) $A_c x = Ax$ for every $x \in D(A)$;

(iii) $\|A_c\| = \|A\|$.

Proof. If $x \in D(A)$ then $A_c x = Ax$. Take $x_0 \notin D(A)$. The value Ax_0 is defined as follows. As $D(A)$ is dense in X, there exists $\{x_n\} \subset D(A)$ such that $\|x_n - x_0\| \to 0$. The sequence $\{Ax_n\}$ is a Cauchy sequence because

$$\|Ax_n - Ax_m\| \le \|A\|\,\|x_n - x_m\| \to 0 \quad \text{as } n, m \to \infty\,;$$

hence, by the completeness of Y, there exists $\lim_{n\to\infty} Ax_n$ which does not depend on the choice of sequence $\{x_n\}$. Let $A_c x_0 = \lim_{n\to\infty} Ax_n$. As

$$\|Ax_n\| \le \|A\|\,\|x_n\|\,,$$

the limit as $n \to \infty$ gives

$$\|A_c x_0\| \le \|A\|\,\|x_0\|\,.$$

This inequality holds for any $x_0 \in X$. Since $A_c x = Ax$, we have $\|A_c\| = \|A\|$, so (iii) is valid. □

Problem 1.24.1. Show that $\lim_{n\to\infty} Ax_n$ does not depend on the choice of $\{x_n\}$. □

Now we prove the *Banach–Steinhaus principle.*

Theorem 1.24.2 (Banach–Steinhaus). Let $\{A_n\}$ be a sequence of continuous linear operators in $L(X, Y)$, X being a normed space and Y a Banach space, such that

(i) $\|A_n\| \le M$ for all n;

(ii) there exists $\lim_{n\to\infty} A_n x$ for all x in a dense subspace X^* of X.

Then $\{A_n\}$ converges strongly to a continuous linear operator A, i.e., for every $x \in X$ we have $\|A_n x - Ax\| \to 0$ as $n \to \infty$.

Proof. The linear operator A, defined on X^* by the relation

$$Ax = \lim_{n\to\infty} A_n x \qquad (x \in X^*)$$

is bounded on X^* by (i); indeed,

$$\|Ax\| = \lim_{n\to\infty} \|A_n x\| \le M\,\|x\| \qquad (x \in X^*)\,.$$

Using the construction of Theorem 1.24.1 we can extend A to the entire space X with preservation of norm. Let us denote this continuation again by A.

Now we show that $\lim_{n\to\infty} A_n x_0 = Ax_0$ for any $x_0 \in X$. Let $\{x_m\} \subset X^*$ be such that

$$\|x_m - x_0\| \to 0 \ \text{ as } m \to \infty\,.$$

Then $Ax_0 = \lim_{m\to\infty} Ax_m$. On the other hand,

$$\begin{aligned}\|Ax_0 - A_n x_0\| &= \|Ax_0 - Ax_m + Ax_m - A_n x_m + A_n x_m - A_n x_0\| \\ &\le \|Ax_0 - Ax_m\| + \|Ax_m - A_n x_m\| + \|A_n x_m - A_n x_0\| \\ &\le \|A\|\,\|x_0 - x_m\| + \|Ax_m - A_n x_m\| + \|A_n\|\,\|x_m - x_0\| \\ &\le 2M\,\|x_0 - x_m\| + \|Ax_m - A_n x_m\|\,.\end{aligned}$$

Given $\varepsilon > 0$, we can choose m_1 such that $\|x_0 - x_{m_1}\| < \varepsilon/(3M)$; fixing this m_1, by (ii) there exists n_1 such that

$$\|Ax_{m_1} - A_n x_{m_1}\| < \varepsilon/3 \ \text{ for all } n > n_1\,.$$

Hence we get

$$\|Ax_0 - A_n x_0\| < \varepsilon \ \text{ for all } n > n_1\,,$$

and this completes the proof. □

1.25 Closed Operators and the Closed Graph Theorem

Broader than the class of continuous linear operators is the class of closed linear operators.

Definition 1.25.1. A linear operator A acting from a Banach space X to a Banach space Y is *closed* if for any sequence $\{x_n\} \subset D(A)$ such that $x_n \to x$ and $Ax_n \to y$, it follows that $x \in D(A)$ and $y = Ax$.

A nearly obvious property of a closed operator is given by

Lemma 1.25.1. The set $N(A) = \{x \in X\colon Ax = 0\}$ is a closed subspace of the Banach space X.

Proof. By linearity of A, the set $N(A)$ is a subspace. Suppose $\{x_n\} \subset N(A)$ converges to x. As $Ax_n = 0$, the sequence $\{Ax_n\}$ converges to zero in Y. By Definition 1.25.1 we have $Ax = 0$, so $x \in N(A)$ and $N(A)$ is a closed subspace of X. □

By definition, a continuous linear operator having domain X is closed. There are closed linear operators that are not continuous, however. An example is the derivative operator d/dt acting from $C[0, 1]$ to $C[0, 1]$. Indeed, let $x_n(t) \to x(t)$ and $x_n'(t) \to y(t)$, both uniformly (i.e., in $C[0, 1]$), so $x(t)$ and $y(t)$ are continuous functions on $[0, 1]$. In order to show that $x'(t) = y(t)$, i.e., that Definition 1.25.1 is fulfilled for d/dt, we start with the Newton–Leibnitz formula

$$x_n(t) = x_n(0) + \int_0^t \frac{dx_n(s)}{ds}\, ds\,.$$

Passage to the limit as $n \to \infty$ yields

$$x(t) = x(0) + \int_0^t y(s)\, ds$$

from which it follows that $x'(t) = y(t)$. Unboundedness of d/dt is seen from its action on the sequence $\{t^n\}$.

Similarly, a general differential operator A given by

$$Af(\mathbf{x}) = \sum_{|\alpha|\le n} c_\alpha(\mathbf{x}) D^\alpha f(\mathbf{x})$$

with smooth coefficients $c_\alpha(\mathbf{x})$, acting in $C(\overline{\Omega})$, is closed.

Let us consider closed operators from another standpoint. We begin with the definition of the product $X \times Y$ of Banach spaces X, Y over the same scalar field. The elements of $X \times Y$ are ordered pairs (x, y) where $x \in X$ and $y \in Y$. This is a linear space with operations defined as follows:

$$(x_1, y_1) + (x_2, y_2) = (x_1 + x_2, y_1 + y_2)\,, \qquad \alpha(x, y) = (\alpha x, \alpha y)\,.$$

Moreover, $X \times Y$ is a Banach space under the norm

$$\|(x, y)\| = (\|x\|^2 + \|y\|^2)^{1/2}.$$

Definition 1.25.2. The subset $\{(x, Ax) \colon x \in D(A)\}$ of $X \times Y$ is the *graph* of a linear operator A acting from $D(A) \subset X$ to Y.

The following is equivalent to Definition 1.25.1:

Definition 1.25.3. A linear operator A acting from $D(A) \subset X$ to Y is *closed* if its graph is a closed linear subspace of $X \times Y$.

Problem 1.25.1. Prove that Definitions 1.25.1 and 1.25.3 are equivalent. □

We have denoted the domain of a linear operator A by $D(A)$. Clearly $D(A)$ is a linear subspace of X. In general, it is not closed. However over $D(A)$ we can introduce another norm related to A in such a way that $D(A)$ with this norm becomes a complete space.

Theorem 1.25.1. Let X, Y be Banach spaces and A a closed linear operator from X to Y. The domain $D(A)$ of A with the norm

$$\|x\|_A = \|x\| + \|Ax\|$$

is a Banach space.

Proof. Let $\{x_n\} \subset D(A)$ be a Cauchy sequence with respect to the norm $\|\cdot\|_A$. This means that $\{x_n\}$ is a Cauchy sequence in X and simultaneously $\{Ax_n\}$ is a Cauchy sequence in Y. As X, Y are Banach spaces, there exist $x \in X$ and $y \in Y$ such that $x_n \to x$ in X and $Ax_n \to y$ in Y as $n \to \infty$. Because A is closed, we have $Ax = y$ and so $\|x_n - x\|_A \to 0$ as $n \to \infty$. □

It is instructive to consider this theorem for the operator $A = d/dt$ given in $C[0, 1]$. In this case

$$\|x\|_A = \max_{t\in[0,1]} |x(t)| + \max_{t\in[0,1]} \left|\frac{dx(t)}{dt}\right|.$$

Hence on $D(A)$ the norm $\|\cdot\|_A$ is the norm of $C^{(1)}[0, 1]$, and it is clear that $D(A)$ becomes the set of continuously differentiable functions on $[0, 1]$ with the norm of $C^{(1)}[0, 1]$. So $D(A)$ is $C^{(1)}[0, 1]$, which is a Banach space.

We say that a linear operator A acting from $D(A) \subset X$ to Y has a *closed extension* if the closure of the graph $G(A)$ in $X \times Y$ is the graph of a linear operator, say B, acting from $D(B) \subset X$ to Y.

Lemma 1.25.2. A linear operator A acting from a Banach space X to a Banach space Y has a closed extension if and only if from the condition

(c) $\{x_n\} \subset D(A)$ is an arbitrary sequence such that $x_n \to 0$ and $Ax_n \to y$

it follows that $y = 0$.

Proof. The necessity of condition (c) follows from the definition of a closed operator. To prove its sufficiency, we construct an operator B, called the *least closed extension* of A, as follows. An element x belongs to $D(B)$ if and only if there is a sequence $\{x_n\} \subset D(A)$ such that $x_n \to x$ and $Ax_n \to y$ as $n \to \infty$; for this x, we define by $y = Bx$. By condition (c) of the theorem, y is uniquely defined by x so B is an operator. Linearity of B is evident.

We will show that B is closed. Let $\{u_n\} \subset D(B)$ be a sequence such that $u_n \to u$ and $Bu_n \to v$. We must prove that $Bu = v$. By definition of B, there is a sequence $\{x_n\} \subset D(A)$ such that $\|x_n - u_n\| < 1/n$ and $\|Ax_n - Bu_n\| < 1/n$. Hence $x_n \to u$ and $Ax_n \to v$ as $n \to \infty$. Thus, by definition of B, we have $u \in D(B)$ and $Bu = v$. □

Consider, for example, an extension of the operator

$$A = \sum_{|\alpha| \le k} c_\alpha(\mathbf{x}) D^\alpha$$

acting in $L^2(\Omega)$. Suppose $c_\alpha(\mathbf{x}) \in C^{(k)}(\Omega)$ and Ω is compact in $\mathbb{R}^n$. We define the domain of A as the set $L^2(\Omega) \cap C^{(k)}(\Omega)$. The range of A lies in $L^2(\Omega)$. We now try the condition of the lemma to show that A has a closed extension. Let $\{u_n(\mathbf{x})\} \subset D(A)$ be such that $\|u_n\|_{L^2(\Omega)} \to 0$ and $\|Au_n(\mathbf{x}) - v(\mathbf{x})\|_{L^2(\Omega)} \to 0$ as $n \to \infty$. Let $\varphi(\mathbf{x}) \in C_0^{(k)}(\Omega)$, where $C_0^{(k)}(\Omega)$ is the subspace of $C^{(k)}(\Omega)$ consisting of functions that together with all their derivatives up to the order k vanish on the boundary $\partial\Omega$. Integration by parts gives

$$\int_\Omega \varphi(\mathbf{x}) A u_n(\mathbf{x})\, d\Omega = \int_\Omega u_n(\mathbf{x}) \sum_{|\alpha| \le k} (-1)^{|\alpha|} D^\alpha [c_\alpha(\mathbf{x})\varphi(\mathbf{x})]\, d\Omega$$

(the boundary terms vanish because $\varphi(\mathbf{x})$ and $D^\alpha\varphi(\mathbf{x})$ are zero on $\partial\Omega$). By definition of $L^2(\Omega)$, passage to the limit as $n \to \infty$ gives

$$\int_\Omega \varphi(\mathbf{x}) v(\mathbf{x})\, d\Omega = \int_\Omega 0 \cdot \sum_{|\alpha| \le k} (-1)^{|\alpha|} D^\alpha [c_\alpha(\mathbf{x})\varphi(\mathbf{x})]\, d\Omega = 0 .$$

Since $C_0^k(\Omega)$ is dense in $L^2(\Omega)$, it follows that $v(\mathbf{x}) = 0$ as an element of $L^2(\Omega)$. Hence there is a closed extension of the differential operator A.

The procedure we have used is an approach that can lead us to generalized derivatives; it is equivalent to the approach due to S.L. Sobolev.

Theorem 1.25.2. If A is a closed linear operator and its inverse A^{-1} exists, then A^{-1} is also closed.

Proof. The graph of A^{-1} can be obtained from the graph of A by rearrangement: $(x, Ax) \mapsto (Ax, x)$. Hence $G(A^{-1})$ is a closed set in $Y \times X$ and A^{-1} is a closed operator from Y to X. □

A useful result is

Theorem 1.25.3. Let A be a closed linear operator from X to Y, where X and Y are Banach spaces. Assume there is a set $M \subset D(A)$ which is dense in X, and a positive constant c such that

$$\|Ax\| \le c\,\|x\| \quad \text{for all } x \in M .$$

Then A is continuous in X.

Proof. Take $x_0 \in X$. Since M is dense in X, there is a sequence $\{x_n\} \subset M$ such that

$$\|x_n - x_0\| < 1/n .$$

By the condition of the theorem, we get

$$\|Ax_{k+m} - Ax_k\| \le c\,\|x_{k+m} - x_k\| \le c\,(\|x_{k+m} - x_0\| + \|x_k - x_0\|)$$
$$\le 2c/k \to 0 \text{ as } k \to \infty\,.$$

So $\{Ax_k\}$ is a Cauchy sequence which, by completeness of Y, has a limit y. By closedness of A, we get $Ax_0 = y$. On the other hand,

$$\|Ax_0\| = \lim_{k\to\infty} \|Ax_k\| \le \lim_{k\to\infty} c\,\|x_k\| = c\,\|x_0\|\,,$$

which completes the proof. □

We now formulate Banach's *closed graph theorem.*

Theorem 1.25.4 (closed graph). Let X, Y be Banach spaces and let A be a closed linear operator from X to Y with domain $D(A) = X$. Then A is continuous on X.

Proof. To prove that A is a continuous operator we will use the following statement:

If there exist positive constants r, C such that $\|x\| < r$ implies $\|Ax\| < C$, then A is continuous. (*)

Indeed, for an arbitrary $x \in X$ we introduce $z = rx/(2\,\|x\|)$ for which $\|z\| = r/2$ and so $\|Az\| < C$. But this is

$$\left\| A\left(\frac{r}{2\,\|x\|}x\right)\right\| < C$$

and so

$$\|Ax\| < \frac{2C}{r}\,\|x\|\,.$$

Thus A is bounded.

So we will prove the validity of (*) for A in two steps.

Step 1. For integer k we introduce the set

$$S_k = \{x \in X\colon\ \|Ax\| < k\}\,.$$

Since $D(A) = X$, we have

$$X = \cup_{k=1}^{\infty} S_k\,.$$

As a Banach space, X is of the second category and so one of the sets $\overline{S_k}$, say $\overline{S_m}$, contains an open ball $B(x^*, r) = \{x \in X\colon\ \|x - x^*\| < r\}$ of a radius $r > 0$. Since S_m is dense in $B(x^*, r)$, we can choose $x_0 \in S_m$ so close to x^* that there exists $B(x_0, r/2) \subset B(x^*, r) \subset \overline{S_m}$.

Now we show that

$$B(0, r/2) \subset \overline{S_{2m}}\,. \tag{1.25.1}$$

Indeed, let $x \in B(0, r/2)$. Then $x + x_0 \in B(x_0, r/2)$ and so $x + x_0 \in \overline{S_m}$. As $x_0 \in \underline{S_M}$ by definition of S_m we get that $-x_0 \in S_m$. So we get that $x + x_0 \in \overline{S_m}$ and $x_0 \in \overline{S_m}$. By the definition of S_m, the sum of two elements of $\overline{S_m}$ is an element of $\overline{S_{2m}}$, so $x = (x + x_0) + (-x_0)$ is an element of $\overline{S_{2m}}$, thus (1.25.1) is valid.

By linearity of A, from (1.25.1) it follows that for any $a > 0$

$$B(0, a) \subset \overline{S_b} \text{ with } b = \frac{2m}{R}a\,, \quad R = \frac{r}{2}\,. \tag{1.25.2}$$

Step 2. Now we show that for any fixed ε such that $0 < \varepsilon < 1$

$$B(0, R) \subset S_{4m/(1-\varepsilon)}\,. \tag{1.25.3}$$

This means that for any $x \in B(0, R)$ we have $\|Ax\| < 4m/(1 - \varepsilon)$ and by (*) it shall follow that A is continuous.

So we prove (1.25.3). Let $x \in B(0, R)$. First we construct a sequence $\{x_k\}$ that belongs to $B(0, R) \cap S_{2m}$ and converges to x, which will also lie in the set $S_{4m/(1-\varepsilon)}$.

By (1.25.1), S_{2m} is dense in $B(0, R)$, so there is $x_1 \in B(0, R) \cap S_{2m}$ such that $\|x - x_1\| < \varepsilon R$ with $0 < \varepsilon < 1$ and $\|Ax_1\| \le 2m$. By (1.25.2), $x - x_1 \in \overline{S_{b_1}}$, where $b_1 = 2m\varepsilon$.

Now we take successively $x_2, x_3, \ldots$ such that $x_k \in B(0, R) \cap S_{2m}$ and $\|x - x_k\| < \varepsilon^k R$. By (1.25.2), $x - x_k \in \overline{S_{b_k}}$ with $b_k = 2m\varepsilon^k$. Moreover,

$$\|x_k - x_{k-1}\| = \|(x - x_{k-1}) - (x - x_k)\| \le \|x - x_{k-1}\| + \|x - x_k\| < (\varepsilon^{k-1} + \varepsilon^k)R\,.$$

By (1.25.2),

$$\|Ax_k - Ax_{k-1}\| = \|A(x_k - x_{k-1})\| \le 2m(\varepsilon^{k-1} + \varepsilon^k)\,.$$

Let us show that all $x_k \in S_{4m/(1-\varepsilon)}$. Indeed,

$$\begin{aligned}
\|Ax_k\| &= \|(Ax_k - Ax_{k-1}) + \cdots + (Ax_2 - Ax_1) + Ax_1\| \\
&\le \|(Ax_k - Ax_{k-1}\| + \cdots + \|(Ax_2 - Ax_1)\| + \|Ax_1\| \\
&\le 2m((\varepsilon^k + \varepsilon^{k-1}) + \cdots + (\varepsilon^2 + \varepsilon) + 1) \\
&< 4m(\varepsilon^k + \varepsilon^{k-1} + \cdots + 1) < \frac{4m}{1 - \varepsilon}\,.
\end{aligned} \tag{1.25.4}$$

Now we show that $\{Ax_k\}$ is a Cauchy sequence:

$$\begin{aligned}
\|Ax_{k+n} - Ax_k\| &= \|(Ax_{k+n} - Ax_{k+n-1}) + \cdots + (Ax_{k+1} - Ax_k)\| \\
&\le \|Ax_{k+n} - Ax_{k+n-1}\| + \cdots + \|Ax_{k+1} - Ax_k\| \\
&\le 2m((\varepsilon^{k+n} + \varepsilon^{k+n-1}) + \cdots + (\varepsilon^{k+1} + \varepsilon^k)) \\
&< 4m(\varepsilon^{k+n} + \cdots + \varepsilon^k) < \frac{4m\varepsilon^k}{1 - \varepsilon} \to 0 \text{ as } k \to \infty\,.
\end{aligned}$$

As Y is a Banach space, there exists $y = \lim_{k\to\infty} Ax_k$. We know that A is closed and $x_k \to x$ so $y = Ax$. By (1.25.4), $\|Ax\| < 4m/(1 - \varepsilon)$ and so we have obtained the needed constants from condition (*). □

1.26 Inverse Operator

We are now interested in solving an *operator equation*

$$Ax = y \tag{1.26.1}$$

where A is a linear operator, y is a given element of a normed space Y, and x is an unknown element of a normed space X.

If for any $y \in Y$ equation (1.26.1) has no more than one solution $x \in X$, then the correspondence from Y to X defined by (1.26.1) is an operator; this operator is called the *inverse* to A and is denoted A^{-1}. It is clear that $D(A^{-1}) = R(A)$ and $R(A^{-1}) = D(A)$.

Theorem 1.26.1. Let A be a linear operator from a normed space X to a normed space Y. The operator A^{-1} exists if and only if the equation $Ax = 0$ has the unique solution $x = 0$. The operator A^{-1}, if it exists, is also linear.

Problem 1.26.1. Supply the proof. □

In terms of the *null space* (or *kernel*) of A, defined as the set

$$N(A) = \{x \in X : Ax = 0\} ,$$

we see that A^{-1} exists if and only if $N(A) = \{0\}$.

We are interested not only in the solvability of (1.26.1), but whether its solution x depends continuously on y. Because A^{-1} is linear, continuity implies the existence of a constant m, independent of y, such that

$$\|x\| \le m \|y\| \quad \text{for all } y \in R(A) .$$

A simple result is

Theorem 1.26.2. The operator A^{-1} is bounded on $R(A)$ if and only if there is a positive constant c such that

$$\|Ax\| \ge c \|x\| \quad \text{for all } x \in D(A) . \tag{1.26.2}$$

Proof. Necessity. Suppose A^{-1} exists and is bounded on $R(A)$. There is a constant $m > 0$ such that $\|A^{-1}y\| \le m\|y\|$. Denoting $y = Ax$ and $c = 1/m$, we get (1.26.2).

Sufficiency. From (1.26.2) it follows that the equation $Ax = 0$ has the unique solution $x = 0$, i.e., A^{-1} exists. Putting $x = A^{-1}y$ in (1.26.2), we get $\|A^{-1}y\| \le (1/c)\|y\|$ for all $y \in R(A)$. This completes the proof. □

Definition 1.26.1. We call A *continuously invertible* if $A^{-1} \in L(Y, X)$.

Theorem 1.26.2 implies

Theorem 1.26.3. An operator A is continuously invertible if and only if (i) $R(A) = Y$ and (ii) there is a positive constant c such that (1.26.2) holds.

Let us consider some examples.

1. We begin with the Fredholm equation with degenerate kernel:

$$u(t) - \lambda \int_a^b \sum_{k=1}^n \varphi_k(t)\, \psi_k(s)\, u(s)\, ds = f(t)\,, \tag{1.26.3}$$

where λ is a parameter. Assume that all $\varphi_k(t)$, $\psi_k(t)$, and $f(t)$ are of class $C(a, b)$. What can we say about the inverse of the operator A_F given by

$$(A_F u)(t) = u(t) - \lambda \int_a^b \sum_{k=1}^n \varphi_k(t)\, \psi_k(s)\, u(s)\, ds\,,$$

acting in $C(a, b)$? If (1.26.3) is solvable, its solution has the form

$$u(t) = f(t) + \sum_{k=1}^n c_k \varphi_k(t)\,.$$

Putting this into (1.26.3), we get

$$\sum_{k=1}^n c_k \varphi_k(t) + f(t) - \lambda \sum_{k=1}^n \varphi_k(t) \int_a^b \psi_k(s) \Big(\sum_{i=1}^n c_i \varphi_i(s) + f(s) \Big) ds = f(t)\,.$$

If the system $\varphi_1(t), \ldots, \varphi_n(t)$ is linearly independent, we obtain the linear algebraic system

$$c_k - \lambda \sum_{i=1}^n c_i \int_a^b \varphi_i(s)\, \psi_k(s)\, ds = \lambda \int_a^b f(s)\, \psi_k(s)\, ds \qquad (k = 1, \ldots, n)\,,$$

whose solution by Cramer's rule is

$$c_k = \frac{D_k(\lambda, f)}{D(\lambda)} \qquad (k = 1, \ldots, n)\,.$$

Hence

$$u(t) = f(t) + \sum_{k=1}^n \frac{D_k(\lambda, f)}{D(\lambda)} \varphi_k(t)\,. \tag{1.26.4}$$

This solution is valid if $D(\lambda) \neq 0$; then from (1.26.4) we see that

$$\max_{t \in [a,b]} |u(t)| \leq m(\lambda) \max_{t \in [a,b]} |f(t)|\,,$$

with a constant $m(\lambda)$ independent of $f(t)$. In terms of norms this is

$$\|u\| \leq m(\lambda)\, \|f\|\,,$$

which means that $A_F^{-1} \in L(C(a, b))$ if $D(\lambda) \neq 0$ and $\|A_F^{-1}\| \leq m(\lambda)$.

Suppose $D(\lambda) = 0$. Since $D(\lambda)$ is a polynomial in λ of order n, it has no more than n distinct zeros λ_i. For $\lambda = \lambda_i$, there is a nontrivial solution $c_1^{(i)}, \dots, c_n^{(i)}$ to the system

$$c_k - \lambda_i \sum_{j=1}^{n} c_j \int_a^b \varphi_j(s)\psi_k(s)\,ds = 0 \qquad (k = 1, \dots, n)\,.$$

Because the equation $A_F u = 0$ has a nonzero solution, A_F^{-1} does not exist. These $\{\lambda_i\}$ comprise the spectrum of the integral operator A_F.

2. Now we consider a simple boundary value problem

$$u''(t) = f(t)\,, \qquad u(0) = u_0\,, \qquad u(1) = u_1\,, \tag{1.26.5}$$

with given $f(t) \in C(0,1)$ and constants u_0, u_1. Its solution is

$$u(t) = \int_0^t \int_0^s f(s_1)\,ds_1\,ds + u_0 + \left(u_1 - u_0 - \int_0^1 \int_0^s f(s_1)\,ds_1\,ds\right)t\,. \tag{1.26.6}$$

In terms of operators, equations (1.26.5) define an operator B whose domain is $C^{(2)}(0,1)$ and range is the space whose elements are pairs consisting of a function $f(t) \in C(0,1)$ and a vector (u_0, u_1). From (1.26.6) it follows that B^{-1} exists and is bounded.

A useful consequence of the closed graph theorem is

Theorem 1.26.4. If A is a closed linear operator mapping a Banach space X onto a Banach space Y (i.e., $R(A) = Y$) and $N(A) = \{0\}$, then A^{-1} exists and is continuous on Y.

Proof. A^{-1} exists because $N(A) = \{0\}$, and it is closed by Theorem 1.25.2. Because $D(A^{-1}) = Y$, continuity of A^{-1} follows from Theorem 1.25.4. □

We recall that a continuous operator is also closed, so Theorem 1.26.4 holds for $A \in L(X, Y)$ as well.

It is interesting to note that despite the visual difference between the closed graph theorem, Theorem 1.26.4, and the corresponding result for a continuous operator, the three results are equivalent (i.e., starting with any one of them we can derive the other two).

Problem 1.26.2. Let X be a Banach space with respect to each of two norms $\|x\|_1$ and $\|x\|_2$. Show that if for every $x \in X$ there is a positive constant c_1 not depending on x such that $\|x\|_1 \le c_1\|x\|_2$, then there is a positive constant c_2 such that $\|x\|_2 \le c_2\|x\|_1$. That is, show that the norms $\|x\|_1$ and $\|x\|_2$ are equivalent. □

The preceding problem might seem to provide a convenient tool for proving norm equivalence. However, proving that X is a Banach space in two norms is not likely to be easier than proving equivalence of the norms directly.

1.27 Lax–Milgram Theorem

An important application of the Riesz representation theorem is

Theorem 1.27.1 (Lax–Milgram). Let $a(u, v)$ be a bilinear form in $u, v \in H$, a real Hilbert space (i.e., $a(u, v)$ is linear in each variable u, v), such that for all $u, v \in H$

(i) $|a(u, v)| \le M \|u\| \|v\|$,

(ii) $|a(u, u)| \ge \alpha \|u\|^2$,

with positive constants M, α that do not depend on u, v. Then there exists a continuous linear operator A having the properties

(a) the range of A is the whole of H;

(b) A has a continuous inverse and $\|A^{-1}\| \le 1/\alpha$;

(c) for all $u, v \in H$,

$$a(u, v) = (Au, v) . \tag{1.27.1}$$

Proof. Fix $u \in H$. By (i), the form $a(u, v)$ is a linear continuous functional in v. By Theorem 1.20.2 there is a uniquely defined element g such that $a(u, v) = (v, g) = (g, v)$. The correspondence $u \mapsto g$, by which g is uniquely determined by u, constitutes an operator A with $g = Au$. Thus we have established (1.27.1).

Let us prove the stated properties of this operator. The linearity of $a(u, v)$ in u implies the linearity of A (Problem 1.27.1, below). By (i) of the theorem, we have

$$|a(u, v)| = |(Au, v)| \le M \|u\| \|v\| \quad \text{for all } u, v \in H .$$

Putting $v = Au$ we get

$$|(Au, Au)| = \|Au\|^2 \le M \|u\| \|Au\| \quad \text{for all } v \in H$$

and so $\|Au\| \le M \|u\|$. This means that A is continuous.

By the Schwarz inequality, (1.27.1), and (ii), it follows that

$$\|Au\| \|u\| \ge |(Au, u)| = |a(u, u)| \ge \alpha \|u\|^2$$

so

$$\|Au\| \ge \alpha \|u\| . \tag{1.27.2}$$

This means that on its range $R(A)$ the operator A has a continuous inverse A^{-1} such that $\|A^{-1}\| \le 1/\alpha$. Here we have used Theorem 1.26.2.

It remains to demonstrate that $R(A) = H$. First, $R(A)$ is a closed subspace of H. Indeed, if $\{Ax_n\}$ is a Cauchy sequence then, by (1.27.2), $\{x_n\}$ is also a Cauchy sequence converging to x_0. By continuity of A we have $Ax_n \to Ax_0$, hence $R(A)$ is closed. If $R(A) \ne H$, there exists $v_0 \ne 0$ such that $v_0 \perp R(A)$, hence $(Au, v_0) = 0$ for all $u \in H$. In particular $(Av_0, v_0) = 0$ and (ii) yields $v_0 = 0$. This contradiction completes the proof. □

Problem 1.27.1. Verify that A is linear. □

Corollary 1.27.1. Suppose $\Phi(v)$ is a continuous linear functional in H, and $a(u, v)$ is the bilinear form of the theorem. Then there is a unique $u_0 \in H$ such that

$$a(u_0, v) = \Phi(v) \tag{1.27.3}$$

for all $v \in H$.

Proof. By Theorem 1.20.2 there is a unique representation $\Phi(v) = (v, f) = (f, v)$. So by Theorem 1.27.1 we can rewrite (1.27.3) as $(Au, v) = (f, v)$. Since v is arbitrary it follows that $Au = f$. Denoting $u_0 = A^{-1}f$, we define the element with the needed properties. □

The Lax–Milgram theorem is traditionally used to demonstrate existence and uniqueness of weak solutions for boundary value problems. We shall take an alternative approach based on the direct use of the Riesz representation theorem and energy norms. In physical applications $a(u, v)$ is defined by the expressions for the energy functionals and is normally symmetric so that $a(u, v) = a(v, u)$. With the conditions of the Lax–Milgram theorem, $a(u, v)$ becomes another inner product in the Hilbert space, under which the space remains a Hilbert space. In this book we prefer to use this new inner product as the principal inner product, so the operator A of the Lax–Milgram theorem reduces to the unit operator. The approaches are equivalent, but we prefer the energy approach; it bears a deeper relation to the nature of the problems, as all the transformations and characteristics are obtained in terms of the strain energy of a physical object under consideration.

1.28 Open Mapping Theorem

It is not so obvious that a linear continuous operator $A \in L(X, Y)$ maps any open set in X onto an open set in Y. It is even less obvious that a closed operator has a similar property, given by the following

Theorem 1.28.1 (open mapping). Let X and Y be Banach spaces, and let A be a linear closed operator from X to Y such that $R(A) = Y$. Let S be an open set in $D(A)$ with norm $\|\cdot\|_A$ (see Theorem 1.25.1). Then the image $A(S)$ of S is an open set in Y.

Proof. We will use Theorem 1.26.4. One of its conditions is that $N(A) = \{0\}$, so we will introduce another operator $\tilde{A}$ related to A that has the needed properties. By Theorem 1.25.1, $D(A)$, the domain of A, with norm $\|\cdot\|_A$ is complete. Lemma 1.25.1 states that the null space of A is a closed subspace of $D(A)$, hence by Theorem 1.15.1 the factor space $D(A)/N(A)$ with the norm induced by $\|\cdot\|_A$ is complete. On $D(\tilde{A}) = D(A)/N(A)$ we introduce the needed operator $\tilde{A}$ whose range is Y as follows:

$$\tilde{A}[x] = Ax\,,$$

where x is a representative of the coset $[x]$. From the properties of A, it follows that the operator $\tilde{A}$ is also linear and closed from $D(\tilde{A})$ onto Y. Furthermore, $N(\tilde{A}) =$

$\{[0]\}$. By Theorem 1.26.4, $\tilde{A}$ has a continuous inverse and there is a constant c such that

$$\|\tilde{A}^{-1}y\|_A \le c\,\|y\| \quad \text{for all } y \in Y\,. \tag{1.28.1}$$

Let S be an open set in $D(A)$ and y_0 an arbitrary point from $A(S)$. There exists $x_0 \in S$ such that $y_0 = Ax_0$. As S is open, there is an open ball $B(x_0, \varepsilon) \subset S$ for some sufficiently small $\varepsilon > 0$. With regard for (1.28.1), taking $y \in Y$ such that $\|y - y_0\| < \delta$ with $\delta = \varepsilon/c$, we get

$$\|\tilde{A}^{-1}(y - y_0)\|_A \le c\,\|y - y_0\| < \varepsilon\,.$$

Denote $[z] = \tilde{A}^{-1}(y - y_0) \in D(\tilde{A})$. For $[z]$ we have $\|[z]\|_A < \varepsilon$ and $\tilde{A}([z]) = y - y_0$. From $[z]$ we can choose a representative $z \in D(A)$ such that $\|z\|_A < \varepsilon$. Moreover, $Az = y - y_0$. Let $x = z + x_0$. We see that $x \in D(A)$. Moreover, $\|x - x_0\|_A = \|z\|_A < \varepsilon$ and so $x \in B(x_0, \varepsilon) \subset S$ and, besides, $Ax = y$. So any y such that $\|y - y_0\| < \delta$ is in the image $A(S)$ of S. This means $A(S)$ is an open set in Y. □

Corollary 1.28.1. Assume X and Y are Banach spaces, $A \in L(X, Y)$, and $R(A) = Y$. The image $A(S)$ of an open set $S \in X$ is open in Y.

Proof. Now $D(A) = X$. As A is a bounded operator, there is a constant c_1 such that for any $x \in X$ we get $\|Ax\| \le c_1\,\|x\|$. It follows that

$$\|x\| \le \|x\|_A = \|x\| + \|Ax\| \le (1 + c_1)\,\|x\|$$

for all $x \in X$. The norm $\|\cdot\|_A$ is equivalent to the initial norm of X, and so an open set S in X is also open in $D(A) = X$. By the theorem, $A(S)$ is open in Y. □

1.29 Dual Spaces

We have considered the space $L(X, Y)$ where X and Y are normed spaces. Since $\mathbb{R}$ and $\mathbb{C}$ are Banach spaces, so are spaces of the form $L(X, \mathbb{R})$ and $L(X, \mathbb{C})$. The space of linear continuous functionals on X is called the *dual space* to X and is denoted by X' or X_s. We wish to emphasize that

Theorem 1.29.1. X', the dual space to a real or complex normed space X, is a Banach space.

By the Riesz representation theorem, we can identify the dual to a Hilbert space H with the same space H. In some books on functional analysis, to underline the duality of X and X', the linear continuous functional on X is written using inner-product-type notation: $\langle x, F\rangle$ or $\langle x\,|\,F\rangle$.

Let $F \in X'$, where X' is the dual to a normed space X. We know that the set of all $x \in X$ such that $Fx = 0$ is called the null space or kernel of F.

Theorem 1.29.2. The kernel K of $F \in X'$ is a closed linear subspace of X having codimension 1, which means that there is a one-dimensional space $\{x_0\}$ spanned by an element x_0 such that $X = K + \{x_0\}$.

Proof. By linearity of F, the set $Fx = 0$, denoted by K, is a linear subspace of X. If $\{x_k\} \subset K$ and $x_k \to x$, then $Fx_k \to Fx = 0$ by continuity of F. Suppose $F \neq 0$. As $K \neq X$ there is $x_0 \notin K$ from X such that $Fx_0 \neq 0$. We show that $x \notin K$ takes the form $x = \alpha x_0$, hence the set of all elements of X that do not belong to K, supplemented with zero, is a one-dimensional subspace $\{x_0\}$ of X. Indeed, the combination $(Fx)x_0 - (Fx_0)x$ belongs to K because

$$F((Fx)x_0 - (Fx_0)x) = (Fx)(Fx_0) - (Fx_0)(Fx) = 0\,.$$

Since the kernel of F is K, a closed subspace, which has zero as its only common point with the set of $x \notin K$, we have

$$x = \frac{Fx}{Fx_0}x_0\,.$$

This completes the proof. □

Proving the Riesz theorem, we have a version of this theorem in a Hilbert space. Recall that the general form of a linear continuous functional in $\mathbb{R}^n$ is

$$F\mathbf{x} = \sum_{k=1}^{n} x_k F\mathbf{e}_k\,,$$

where $(\mathbf{e}_1, \ldots, \mathbf{e}_n)$ is a basis of $\mathbb{R}^n$. Supposing this basis is canonical and denoting $a_k = F\mathbf{e}_k$, we see that F is a linear function $F\mathbf{x} = a_1x_1 + \cdots + a_nx_n$ which can be represented using the dot product:

$$F\mathbf{x} = \mathbf{x}\cdot\mathbf{a}\,, \qquad \mathbf{a} = (a_1, \ldots, a_n)\,. \tag{1.29.1}$$

Note that the question of continuity of a linear functional in $\mathbb{R}^n$ does not arise, as F can be discontinuous only when the a_k can be infinite.

For $\mathbb{C}^n$ the general form of F is also (1.29.1), but as we wish to use the inner product in $\mathbb{C}^n$ we transform it to

$$F\mathbf{x} = \mathbf{x}\cdot\bar{\mathbf{a}}\,, \qquad \bar{\mathbf{a}} = (\bar{a}_1, \ldots, \bar{a}_n)$$

supposing $\mathbf{x}, \mathbf{a}$ to be represented in a canonical basis of $\mathbb{C}^n$.

Examples of dual spaces. Now we wish to consider some linear continuous functionals on other spaces. Let us examine the dual $(\ell^p)'$ to the sequence space ℓ^p.

Theorem 1.29.3. The dual space to ℓ^p, $1 < p < \infty$, is isometrically isomorphic to ℓ^q with $1/q = 1 - 1/p$. In short, we have $(\ell^p)' = \ell^q$.

Proof. First note that, as a dual space to a normed space, ℓ^q is complete. Hence ℓ^p is also complete, as we have established directly.

Clearly

$$Fx = \sum_{k=1}^{\infty} x_k a_k \tag{1.29.2}$$

is a linear functional in ℓ^p. Let $\mathbf{a} = (a_1, a_2, \ldots) \in \ell^q$. By Hölder's inequality,

$$\left| \sum_{k=1}^{\infty} x_k a_k \right| \leq \|\mathbf{x}\|_p \|\mathbf{a}\|_q$$

and so F is bounded in ℓ^p and moreover,

$$\|F\| \leq \|\mathbf{a}\|_q \ . \tag{1.29.3}$$

Our goal is to show that any linear continuous functional in ℓ^p can be represented in the form (1.29.2) with $\mathbf{a} \in \ell^q$.

Denote $\mathbf{e}_1 = (1, 0, 0, \ldots)$, $\mathbf{e}_2 = (0, 1, 0, 0 \ldots)$, and so on; in other words, let $\mathbf{e}_k$ be the sequence consisting of zero elements in all positions except the kth position where its value is 1: $e_k = (0, \ldots, 0, 1, 0, 0, \ldots)$. Then $\mathbf{x} = (x_1, x_2, \ldots)$ can be represented in the form

$$\mathbf{x} = \sum_{k=1}^{\infty} x_k \mathbf{e}_k \ .$$

Note that the sequence of partial sums for this representation,

$$\mathbf{s}_n = \sum_{k=1}^{n} x_k \mathbf{e}_k \ ,$$

converges to $\mathbf{x}$ in ℓ^p because

$$\|\mathbf{x} - \mathbf{s}_n\|_p = \left\| \sum_{k=n+1}^{\infty} x_k \mathbf{e}_k \right\|_p = \left(\sum_{k=n+1}^{\infty} |x_k|^p \right)^{1/p} \to 0 \ \text{ as } n \to \infty \ .$$

Consider

$$F\mathbf{s}_n = F\left(\sum_{k=1}^{n} x_k \mathbf{e}_k \right) = \sum_{k=1}^{n} x_k F\mathbf{e}_k \ .$$

By continuity of F we have

$$F\mathbf{x} = \lim_{n \to \infty} F\mathbf{s}_n = \sum_{n=1}^{\infty} x_n F\mathbf{e}_n \ .$$

We have obtained the needed form of $F\mathbf{x}$, which is $\sum_{k=1}^{\infty} x_k a_k$ where $a_k = F\mathbf{e}_k$. Now we should demonstrate that $\mathbf{a} = (a_1, a_2, \ldots)$ belongs to ℓ^q. For this, we take a special element $\hat{\mathbf{x}} = (\hat{x}_1, \hat{x}_2, \ldots)$ with

$$\hat{x}_k = \begin{cases} |a_k|^{q-2}\overline{a_k}\,, & a_k \neq 0\,, \\ 0\,, & a_k = 0\,. \end{cases}$$

As $p = q/(q-1)$, for this $\hat{\mathbf{x}}$ we get

$$\|\hat{\mathbf{s}}_n\|_p^p = \sum_{k=1}^n |\hat{x}_k|^p = \sum_{k=1}^n |a_k|^{p(q-1)} = \sum_{k=1}^n |a_k|^q$$

(to avoid division by zero, we delete terms with $a_k = 0$ from the sums). In ℓ^p it follows that

$$|F\hat{\mathbf{s}}_n| \leq \|F\| \Big(\sum_{k=1}^n |a_k|^q \Big)^{1/p}.$$

Next, by the definition of $\hat{x}_k$ we get

$$F\hat{\mathbf{s}}_n = \sum_{k=1}^n \hat{x}_k a_k = \sum_{k=1}^n |a_k|^q$$

and so

$$\sum_{k=1}^n |a_k|^q \leq \|F\| \Big(\sum_{k=1}^n |a_k|^q \Big)^{1/p}.$$

Since $1/q = 1 - 1/p$, we get

$$\Big(\sum_{k=1}^n |a_k|^q \Big)^{1/q} \leq \|F\|$$

and, by arbitrariness of n,

$$\Big(\sum_{k=1}^\infty |a_k|^q \Big)^{1/q} \leq \|F\|\,. \tag{1.29.4}$$

So $\mathbf{a} \in \ell^q$, which completes the proof. However, from (1.29.3) and (1.29.4) we also conclude that $\|F\| = \|\mathbf{a}\|_q$. □

We will not prove the similar looking result

Theorem 1.29.4. The dual space to $L^p(V)$, $1 < p < \infty$, is $L^q(V)$ with $1/q = 1 - 1/p$.

We will, however, establish

Theorem 1.29.5. The dual to ℓ^1 is the space m.

Proof. For a functional $F\mathbf{x} = \sum_{k=1}^\infty x_k a_k$ where $\|\mathbf{a}\|_m = \sup_k |a_k|$ is finite, we have

$$|F\mathbf{x}| = \Big| \sum_{k=1}^\infty x_k a_k \Big| \leq \sup_k |a_k| \sum_{k=1}^\infty |x_k| = \|\mathbf{a}\|_m \, \|\mathbf{x}\|_1$$

so $F\mathbf{x} = \sum_{k=1}^\infty x_k a_k$ is bounded and, moreover,

$$\|F\| \leq \|\mathbf{a}\|_m . \tag{1.29.5}$$

Now we show that any $F \in (\ell^1)'$ can be represented in the needed form. Indeed, for any $\mathbf{x} \in \ell^1$ the partial sums

$$\mathbf{s}_n = \sum_{k=1}^{n} x_k \mathbf{e}_k \to \mathbf{x} \text{ as } n \to \infty .$$

Continuity of F gives

$$F\mathbf{x} = \sum_{k=1}^{\infty} x_k \, F\mathbf{e}_k = \sum_{k=1}^{\infty} x_k a_k$$

where $a_k = F\mathbf{e}_k$. Take $\mathbf{x} = \mathbf{e}_k$:

$$|F\mathbf{e}_k| = |a_k| \leq \|F\| \, \|\mathbf{e}_k\|_1 .$$

Because $\|\mathbf{e}_k\|_1 = 1$, for any k we have $|a_k| \leq \|F\|$. So $\|\mathbf{a}\|_m = \sup_k |a_k| \leq \|F\|$. We conclude that $\mathbf{a} \in m$. The last inequality and (1.29.5) imply that $\|F\| = \|\mathbf{a}\|_m$. □

1.30 Hahn–Banach Theorem

Although we have considered the representation of linear continuous functionals in some normed spaces, we still have not asked whether linear continuous functionals exist in a general normed space. This and some special problems for linear functionals are addressed by the Hahn–Banach theorem. The theorem shows that a linear functional continuous on a subspace can be extended to the entire normed space with preservation of the norm. We will present it in a restricted but still quite general form.

Theorem 1.30.1 (Hahn–Banach). Let X be a separable normed real space, let S be a subspace of X, and let f be a bounded linear functional on S:

$$|fx| \leq \|f\| \, \|x\| \qquad (x \in S) .$$

There is a linear continuous functional F defined on X, called an *extension* of f, such that $Fx = fx$ for all $x \in S$ and $\|F\| = \|f\|$.

Proof of Hahn–Banach theorem. We prove the present version of the theorem via three lemmas.

We suppose $S \neq X$ and so there is $x_0 \in X$ such that $x_0 \notin S$, $x_0 \neq 0$. The set of all elements of the form $y = x + tx_0$, where $x \in S$ and $t \in \mathbb{R}$ are arbitrary, constitutes a linear space S_1. Note that the representation of an element of S_1, which is $x + tx_0$, is unique. Indeed, if for some y we have two representations $y = x + tx_0$ and $y = x' + t'x_0$, then $x - x' = (t' - t)x_0$. If $t' \neq t$ then $x_0 = (x - x')/(t' - t) \in S$, which contradicts the fact that $x_0 \notin S$. But if $t' = t$, then $x - x' = 0$ and this implies uniqueness of the representation.

The first two lemmas show how to extend f to F on S_1 with preservation of norm.

Lemma 1.30.1. There is a functional F extending f to S_1 such that

$$Fy \le \|f\|\,\|y\| \qquad (y \in S_1)\,. \tag{1.30.1}$$

Proof. We have $fx \le \|f\|\,\|x\|$ on S. The lemma states that we can extend f to F while preserving this inequality, i.e., (1.30.1) will hold for all $y \in S_1$ while $Fx = fx$ for all $x \in S$. As the extension F should be linear, we have

$$Fy = F(x + tx_0) = Fx + tFx_0 = fx + tFx_0 \qquad (x \in S)\,.$$

Finding an appropriate numerical value Fx_0, we define F on S_1 completely. For $y = x + tx_0$, inequality (1.30.1) takes the form

$$F(x + tx_0) = fx + tFx_0 \le \|f\|\,\|x + tx_0\| \qquad (x \in S,\ t \in \mathbb{R})\,. \tag{1.30.2}$$

From this, we will derive the constraints for the value of Fx_0.

Let $t > 0$. From (1.30.2) we get

$$Fx_0 \le \frac{1}{t}(\|f\|\,\|x + tx_0\| - fx) = \|f\|\left\|\frac{x}{t} + x_0\right\| - f\left(\frac{x}{t}\right).$$

Because $x/t = z_1$ is an arbitrary element of S, we have

$$Fx_0 \le \|f\|\,\|z_1 + x_0\| - fz_1 \ \text{ for all } z_1 \in S\,.$$

For $t < 0$, from (1.30.2) we get

$$Fx_0 \ge \frac{1}{t}(\|f\|\,\|x + tx_0\| - fx) = -\|f\|\left\|-\frac{x}{t} - x_0\right\| + f\left(-\frac{x}{t}\right).$$

So denoting $z_2 = -x/t$, which is still arbitrary, we get the inequality

$$Fx_0 \ge -\|f\|\,\|z_2 - x_0\| + fz_2\,.$$

Together with the previous constraint, this yields

$$-\|f\|\,\|z_2 - x_0\| + fz_2 \le Fx_0 \le \|f\|\,\|z_1 + x_0\| - fz_1\,. \tag{1.30.3}$$

If for all $z_1, z_2 \in S$ we can find Fx_0 satisfying (1.30.3), i.e., if

$$-\|f\|\,\|z_2 - x_0\| + fz_2 \le \|f\|\,\|z_1 + x_0\| - fz_1 \qquad (z_1, z_2 \in S)\,, \tag{1.30.4}$$

then F has the properties announced in the lemma. Indeed, if (1.30.3) holds, then repeating the transformations in the reverse order, we find that F satisfies (1.30.1) as needed.

Consider (1.30.4). It is equivalent to

$$f(z_1 + z_2) \le \|f\| \left(\|z_1 + x_0\| + \|z_2 - x_0\| \right) \qquad (z_1, z_2 \in S),$$

but this is valid because

$$\begin{aligned} f(z_1 + z_2) &\le \|f\| \, \|z_1 + z_2\| \\ &= \|f\| \, \|(z_1 + x_0) + (z_2 - x_0)\| \\ &\le \|f\| \left(\|z_1 + x_0\| + \|z_2 - x_0\| \right). \end{aligned}$$

So (1.30.4) holds for all $z_1, z_2 \in S$. This means that

$$m_1 = \sup_{z_2} \left(-\|f\| \, \|z_2 - x_0\| + f z_2 \right) \le \inf_{z_1} \left(\|f\| \, \|z_1 + x_0\| - f z_1 \right) = m_2 .$$

Thus we complete the proof by taking $F x_0 = m_1$. □

Lemma 1.30.2. There is a functional F extending f to S_1 such that

$$|Fx| \le \|f\| \, \|x\| \text{ on } S_1 \tag{1.30.5}$$

and the norm of F on S_1 is $\|f\|$.

Proof. Let F be constructed in the previous lemma. Changing y to $-x$ in (1.30.1), we get $F(-x) \le \|f\| \, \|x\|$, i.e., $Fx \ge -\|f\| \, \|x\|$. Together with (1.30.1) this implies (1.30.5), so $\|F\| \le \|f\|$. As on S the norm of the functional $F = f$ is $\|f\|$, its norm on S_1 cannot be less than $\|f\|$, so $\|F\| = \|f\|$ on S_1. □

Lemma 1.30.3. Theorem 1.30.1 is valid.

Proof. We have shown how to extend f from S to S_1, which is the linear space of all linear combinations of elements of S and x_0, while preserving the needed properties. Now, extending f to the entire space X, we will use the fact that X is separable.

By separability, X contains a countable dense subset C. Let us select a sequence of points from C and construct a sequence of subspaces S_k as follows:

Choose $x_1 \notin S_1$.
Let S_2 be the space of all linear combinations of x_1 with the elements of S_1.
Choose $x_2 \notin S_2$.
Let S_3 be the space of all linear combinations of x_2 with the elements of S_2.
Choose $x_3 \notin S_3$.

$\vdots$

Note that for each k, the space S_k is a linear space that extends the previous space S_{k-1} by one dimension. Now, step by step, we extend f to S_2, S_3, and so on, as described in the above lemmas, preserving $\|f\|$ each time. As a result, we extend f to a linear functional on a dense subspace of X; on this subspace, the norm of the extended functional is $\|f\|$.

Recall that by Theorem 1.24.1, a bounded linear operator A from X to Y with domain $D(A)$ dense in X can be extended to X with preservation of the boundedness constant. Changing A to f and Y to $\mathbb{R}$ in the formulation of the theorem, we complete the proof. □

Now we consider the Hahn–Banach theorem in a complex normed space.

Theorem 1.30.2. Let X be a separable complex normed space, S a subspace of X, and f a linear functional with norm $\|f\|$ on S:

$$\|f\| = \sup_{\substack{x \in S \\ \|x\| \le 1}} |fx| .$$

The functional f can be extended to a linear functional F such that $Fx = fx$ on S and $\|F\| = \|f\|$ on the whole space X.

Proof. We reduce the proof to the use of the Hahn–Banach theorem on a real normed space. For this, we split f into the "real" part functional f_r and the "imaginary" part functional f_i on S:

$$fx = f_R x + i f_I x ; \qquad f_R x = \operatorname{Re} fx , \qquad f_I x = \operatorname{Im} fx .$$

We then establish their properties on S by considering a "real" normed space S_R, constructed by taking all the elements of S and making it a linear space while permitting only the product of the elements of S by real numbers. On S_R, the functionals f_R and f_I are linear. For example, for any $\alpha_k \in \mathbb{R}$ and $x_k \in S$ we get

$$\begin{aligned} f_R(\alpha_1 x_1 + \alpha_2 x_2) &= \operatorname{Re} f(\alpha_1 x_1 + \alpha_2 x_2) \\ &= \alpha_1 \operatorname{Re} f x_1 + \alpha_2 \operatorname{Re} f x_2 \\ &= \alpha_1 f_R x_1 + \alpha_2 f_R x_2 . \end{aligned}$$

Next, using the definition of f_R and f_I, on S_R we get

$$f(ix) = f_R(ix) + i f_I(ix) , \qquad i fx = i f_R x - f_I x .$$

Comparing the real and imaginary parts of these, from $f(ix) = ifx$ it follows that

$$f_R(ix) = -f_I x , \qquad f_I(ix) = f_R x . \tag{1.30.6}$$

These two equalities are equivalent. Next, S_R contains along with any x the element ix and $\|x\| = \|ix\|$. On S_R we have

$$\|f_R\| = \sup_{\substack{ix \in S_R \\ \|ix\| \le 1}} |f_R x| = \sup_{\substack{x \in S_R \\ \|x\| \le 1}} |f_I x| = \|f_I\| \ .$$

As $\|f\| = (\|f_R\|^2 + \|f_I\|^2)^{1/2}$, we have

$$\|f_R\| = \|f_I\| = \frac{1}{\sqrt{2}} \|f\| \ .$$

Thus we have two bounded linear real functionals f_R, f_I on S_R. We apply the real version of the Hahn–Banach Theorem 1.30.1 to f_R and extend the functional with preservation of norm to a linear real-valued functional F_R on the whole space X, which is considered as a "real" normed space. Using (1.30.6) we introduce the extension of f_I by $F_I x = -F_R(ix)$. Finally, the extension of f is

$$F = F_R + iF_I \ .$$

Clearly, on X considered as a real linear space, the functional F is linear. Taking into account the relations (1.30.6), which hold when we change f_R, f_I to F_R, F_I respectively, we get

$$F(ix) = F_R(ix) + iF_I(ix) = -F_I x + iF_R x = iFx \ .$$

We see that F is also linear in the complex space X.

On S we find that $Fx = fx$, so F is a continuation of f. Finally,

$$\|F\| = \sup_{\|x\| \le 1} |Fx| = \sup_{\|x\| \le 1} (|F_R x|^2 + |F_I x|^2)^{1/2} \le \|f\| \ .$$

This completes the proof. □

The Hahn–Banach theorem in a nonseparable case can be proved using Zorn's lemma (see, for example, [44]).

1.31 Consequences of the Hahn–Banach Theorem

We consider some simple but important consequences of the Hahn–Banach theorem.

Theorem 1.31.1. Let $x_0 \neq 0$ be an element of a Banach space X. There is a linear continuous functional F on X such that $\|F\| = 1$ and $Fx_0 = \|x_0\|$.

Proof. Let S be the subspace of X spanned by the element x_0, i.e., the set of elements $x = \alpha x_0$ for $\alpha \in \mathbb{K}$. We define a functional on S by

$$fx = f(\alpha x_0) = \alpha \, \|x_0\| \ .$$

Linearity of f on S is obvious, as is the fact that $fx_0 = \|x_0\|$. Moreover,

$$\|f\| = \sup_{|\alpha|\,\|x_0\|\le 1} |\alpha|\,\|x_0\| = 1\,.$$

By the Hahn–Banach theorem, we extend f to F with the domain X, obtaining the desired functional. □

Corollary 1.31.1. If $x_1 \in X$ is such that $fx_1 = 0$ for any linear continuous functional f on X, then $x_1 = 0$.

This follows from the existence of a linear continuous functional F such that $Fx_1 = \|x_1\|$.

Theorem 1.31.2. Let S be a subspace of a Banach space X, and let $x_0 \notin S$ be an element of X. Let

$$d = \inf_{x\in S} \|x_0 - x\| > 0$$

be the distance from x_0 to S. Then there is a linear continuous functional F on X such that $Fx_0 = \|x_0\|$, $Fx = 0$ for any $x \in S$, and $\|F\| = 1$.

Proof. In the proof of the Hahn–Banach theorem, we saw that for any $y \in X$, the representation $y = x+tx_0$ determines the element $x \in S$ and the numerical parameter t uniquely. The equality

$$fy = f(x + tx_0) = td$$

defines a unique functional f on $S_1 = S + \{x_0\}$. For $x \in S$, which is when $t = 0$, we get $fx = 0$. Hence f is linear on S_1. As S is a subspace, for arbitrary $t \neq 0$ we get

$$d = \inf_{x\in S} \|x_0 - x\| = \inf_{x\in S} \|x_0 - x/t\| = \inf_{x\in S} \|x_0 + x/t\|\ ,$$

so

$$|f(x + tx_0)| = |t|d = |t| \inf_{x\in S} \left\| x_0 + \frac{x}{|t|} \right\| \le \|tx_0 + x\|$$

which states that $\|f\| \le 1$. Finally, for any $\varepsilon > 0$ we can find $x_1 \in S$ such that $\|x_0 - x_1\| < d + \varepsilon$. But $f(x_0 - x_1) = fx_0 = d$ and so

$$\frac{|f(x_0 - x_1)|}{\|x_0 - x_1\|} > \frac{d}{d+\varepsilon} \to 1 \text{ as } \varepsilon \to 0$$

from which it follows that $\|f\| = 1$. So by the Hahn–Banach theorem we can extend f to F given on X with the needed properties. □

1.32 Contraction Mapping Principle

In the particular case $X = Y$, an operator A is said to be *acting in* X. Many problems of mechanics can be formulated as equations of the form

$$x = A(x)\,, \tag{1.32.1}$$

where A acts in a metric space X. A solution to (1.32.1) is called a *fixed point* of A. We denote the image of x under A as $A(x)$, since A can be a nonlinear operator. In the Introduction we saw two different problems whose solutions were in a certain sense similar. There are many other such problems; their general similarity is captured in the following definition.

Definition 1.32.1. An operator A acting in a metric space X is a *contraction* in X if for every pair $x, y \in X$ there is a number $0 < q < 1$ such that

$$d(A(x), A(y)) \leq q\, d(x, y) . \tag{1.32.2}$$

The following central theorem is known as the *contraction mapping principle* or *Banach's principle of successive approximations.*

Theorem 1.32.1 (contraction mapping). Let A be a contraction operator (with a constant $q < 1$) in a complete metric space X. Then:

(i) A has a unique fixed point $x_* \in X$;

(ii) For any initial approximation $x_0 \in X$, the sequence of successive approximations

$$x_{k+1} = A(x_k) \qquad (k = 0, 1, 2, \ldots) \tag{1.32.3}$$

converges to x_*; the rate of convergence is estimated by

$$d(x_k, x_*) \leq \frac{q^k}{1-q}\, d(x_0, x_1) . \tag{1.32.4}$$

Remark 1.32.1. Note that X need not be a linear space. Banach's principle works if X is a closed subset of a metric space such that $A(X) \subset X$ and A is a contraction operator in X. □

Proof (of Theorem 1.32.1.). We first show uniqueness of a fixed point of A. Assuming the existence of two such points x_1 and x_2 with

$$x_1 = A(x_1) \quad \text{and} \quad x_2 = A(x_2) ,$$

we get

$$d(x_1, x_2) = d(A(x_1), A(x_2)) \leq q\, d(x_1, x_2) ;$$

as $q < 1$, this implies $d(x_1, x_2) = 0$. Hence $x_1 = x_2$.

Now take an element $x_0 \in X$ and consider the iterative procedure (1.32.3). For $d(x_n, x_{n+m})$, we successively obtain

$$\begin{aligned} d(x_n, x_{n+m}) &= d(A(x_{n-1}), A(x_{n+m-1})) \\ &\le q\, d(x_{n-1}, x_{n+m-1}) \\ &= q\, d(A(x_{n-2}), A(x_{n+m-2})) \\ &\le q^2 d(x_{n-2}, x_{n+m-2}) \\ &\vdots \\ &\le q^n d(x_0, x_m)\,. \end{aligned}$$

But

$$\begin{aligned} d(x_0, x_m) &\le d(x_0, x_1) + d(x_1, x_2) + d(x_2, x_3) + \cdots + d(x_{m-1}, x_m) \\ &\le d(x_0, x_1) + q\, d(x_0, x_1) + q^2 d(x_0, x_1) + \cdots + q^{m-1} d(x_0, x_1) \\ &= (1 + q + q^2 + \cdots + q^{m-1})\, d(x_0, x_1) \\ &= \frac{1 - q^m}{1 - q}\, d(x_0, x_1) \le \frac{1}{1 - q}\, d(x_0, x_1) \end{aligned}$$

so

$$d(x_n, x_{n+m}) \le \frac{q^n}{1 - q}\, d(x_0, x_1)\,. \tag{1.32.5}$$

It follows that $\{x_n\}$ is a Cauchy sequence. Since X is complete there is an element $x_* \in X$ such that

$$x_* = \lim_{n\to\infty} x_n = \lim_{n\to\infty} A(x_{n-1})\,.$$

Let us estimate $d(x_*, A(x_*))$:

$$\begin{aligned} d(x_*, A(x_*)) &\le d(x_*, x_n) + d(x_n, A(x_*)) \\ &= d(x_*, x_n) + d(A(x_{n-1}), A(x_*)) \\ &\le d(x_*, x_n) + q\, d(x_{n-1}, x_*) \to 0 \ \text{ as } n \to \infty\,. \end{aligned}$$

Thus

$$x_* = A(x_*)$$

and x_* is a fixed point of A.

Passage to the limit as $m \to \infty$ in (1.32.5) gives the estimate (1.32.4). □

Problem 1.32.1. Use one of the intermediate estimates of the proof to establish that

$$d(x_k, x_*) \le q^k d(x_0, x_*)\,.$$

Why is this estimate of less practical value than (1.32.4)? □

Let us denote

$$A^N(x) = \underbrace{A(A(\cdots(A(x))\cdots))}_{N \text{ times}}\,.$$

Problem 1.32.2. Let

$$Ay(t) = \int_0^t g(t-\tau)y(\tau)\,d\tau\,,$$

with $g(t) \in C(0,T)$, be an operator acting in $C(0,T)$. (1) Prove that A^N is a contraction operator for some integer N. Similar operators appear in the theory of viscoelasticity. (2) Is the statement valid if $g(t) \in L^p(0,T)$, $p > 1$? □

Corollary 1.32.1. Let A^N, for some N, be a contraction operator in a complete metric space X. Then the operator A has a unique fixed point x_* to which the sequence of successive approximations (1.32.3) converges independently of choice of initial approximation $x_0 \in X$ and with rate

$$d(x_i, x_*) \le \frac{q^{i/N-1}}{1-q} \max\{d(x_0, x_N), d(x_1, x_{N+1}), \ldots, d(x_{N-1}, x_{2N-1})\}\,.$$

Proof. The operator A^N meets all requirements of Theorem 1.32.1, so the equation

$$x = A^N(x) \tag{1.32.6}$$

has a unique solution x_* and we have $x_* = A^N(x_*)$. An application of A to both sides of (1.32.6) gives

$$A(x_*) = A(A^N(x_*)) = A^N(A(x_*))\,.$$

Hence $A(x_*)$ is also a solution to (1.32.6). From uniqueness of solution of (1.32.6), it follows that

$$x_* = A(x_*)\,,$$

i.e., equation (1.32.1) is solvable. Noting that any fixed point of A is a fixed point of A^N, we get uniqueness of solution of (1.32.1). Finally, the whole sequence of successive approximations can be constructed by taking elements from each of the N subsequences

$$\begin{aligned}
&\{x_0, A^N(x_0), A^{2N}(x_0), \ldots\}\,,\\
&\{A(x_0), A^{N+1}(x_0), A^{2N+1}(x_0), \ldots\}\,,\\
&\vdots\\
&\{A^{N-1}(x_0), A^{2N-1}(x_0), A^{3N-1}(x_0)\,, \ldots\}\,.
\end{aligned}$$

For each of these subsequences, the estimate (1.32.4) is valid: we have

$$d(A^{kN}(x_0), x_*) \le \frac{q^k}{1-q}\, d(x_0, A^N(x_0)) ,$$

$$d(A^{kN+1}(x_0), x_*) \le \frac{q^k}{1-q}\, d(A(x_0), A^{N+1}(x_0)) ,$$

$$\vdots$$

$$d(A^{kN+N-1}(x_0), x_*) \le \frac{q^k}{1-q}\, d(A^{N-1}(x_0), A^{2N-1}(x_0)) .$$

Replacing kN by i in the first inequality, $kN + 1$ by i in the second inequality, and so on, and remembering that $x_k = A^k(x_0)$, we get

$$d(x_i, x_*) \le \frac{q^{i/N}}{1-q}\, d(x_0, x_N) \qquad (i = kN),$$

$$d(x_i, x_*) \le \frac{q^{(i-1)/N}}{1-q}\, d(x_1, x_{N+1}) \qquad (i = kN + 1),$$

$$\vdots$$

$$d(x_i, x_*) \le \frac{q^{(i-N+1)/N}}{1-q}\, d(x_{N-1}, x_{2N-1}) \qquad (i = kN + (N - 1))$$

and the desired estimate follows. □

We consider some applications of Banach's principle to linear algebraic systems more general that those treated in the Introduction. We wish to solve the system

$$x_i = \sum_{j=1}^{\infty} a_{ij} x_j + c_i . \tag{1.32.7}$$

The corresponding operator A is defined by

$$\mathbf{y} = A\mathbf{x} , \qquad \mathbf{y} = (y_1, y_2, \ldots) , \qquad y_i = \sum_{j=1}^{\infty} a_{ij} x_j + c_i .$$

Our treatment of this system depends on the space in which we seek a solution. If we take $X = m$, the space of bounded sequences with metric

$$d(\mathbf{x}, \mathbf{y}) = \sup_i |x_i - y_i| ,$$

then we find that A is a contraction operator if

$$q = \sup_i \sum_{j=1}^{\infty} |a_{ij}| < 1 \quad \text{and} \quad \mathbf{c} = (c_1, c_2, \ldots) \in m . \tag{1.32.8}$$

So we can solve (1.32.7) by iterating from any initial approximation in m.

Another restriction on the infinite matrix (a_{ij}) appears if we consider successive approximations in ℓ^p, $p > 1$. Here we get

$$d(A\mathbf{x}, A\mathbf{y}) = \left(\sum_{i=1}^{\infty} \left| \sum_{j=1}^{\infty} a_{ij}(x_j - y_j) \right|^p \right)^{1/p} .$$

Applying Hölder's inequality we obtain

$$d(A\mathbf{x}, A\mathbf{y}) \leq \left[\sum_{i=1}^{\infty} \left(\sum_{j=1}^{\infty} |a_{ij}|^r \right)^{p/r} \sum_{j=1}^{\infty} |x_j - y_j|^p \right]^{1/p} = \left[\sum_{i=1}^{\infty} \left(\sum_{j=1}^{\infty} |a_{ij}|^r \right)^{p/r} \right]^{1/p} d(\mathbf{x}, \mathbf{y})$$

where $1/r + 1/p = 1$. So A is a contraction operator in ℓ^p if $\mathbf{c} \in \ell^p$ and

$$q = \left[\sum_{i=1}^{\infty} \left(\sum_{j=1}^{\infty} |a_{ij}|^r \right)^{p/r} \right]^{1/p} < 1 , \qquad r = \frac{p}{p-1} . \tag{1.32.9}$$

Now we can solve the system (1.32.7) by an iterative procedure in ℓ^p.

The values of q from (1.32.8) and (1.32.9) are the corresponding operator norms for the matrix portion of the operator A defined by the equations

$$y_i^* = \sum_{j=1}^{\infty} a_{ij} x_j$$

in m and ℓ^p, respectively (prove this).

In a similar way, we can extend the result of the Introduction for the system (2) to more general systems of integral equations in different spaces. We leave this to the reader as an exercise.

Finally, we formulate

Theorem 1.32.2. Assume that X and Y are Banach spaces, $A \in L(X, Y)$ is continuously invertible, and $B \in L(X, Y)$ is such that $\|B\| < \|A^{-1}\|^{-1}$. Then $A + B$ has an inverse $(A + B)^{-1} \in L(Y, X)$ and

$$\|(A + B)^{-1}\| \leq (\|A^{-1}\|^{-1} - \|B\|)^{-1} . \tag{1.32.10}$$

Proof. The equation

$$(A + B)x = y \tag{1.32.11}$$

can be reduced to

$$x - Cx = x_0 , \quad C = -A^{-1}B , \quad x_0 = A^{-1}y .$$

By hypothesis $\|C\| \leq \|A^{-1}\| \|B\| < 1$. Hence, by the contraction mapping principle, this equation has a unique solution x^* for any $y \in Y$ which can be found by iteration:

$$x_{k+1} = x_0 + Cx_k$$

or

$$x_k = (I + C + \cdots + C^k)x_0 .$$

Existence of the unique solution to (1.32.11) means that the inverse to $A + B$ exists and its domain is Y.

We now obtain (1.32.10). From $x = A^{-1}Ax$ it follows that

$$\|x\| \leq \|A^{-1}\| \, \|Ax\| ,$$

and so

$$\|Ax\| \geq \|A^{-1}\|^{-1}\|x\| .$$

For any $y \in Y$, (1.32.11) shows that

$$\|y\| = \|(A + B)x\| \geq \|Ax\| - \|Bx\| \geq \|A^{-1}\|^{-1}\|x\| - \|B\| \, \|x\| = (\|A^{-1}\|^{-1} - \|B\|) \, \|x\| .$$

From this (1.32.10) follows, and the proof is complete. □

We shall see Banach's principle applied to problems in plasticity. When applicable to such problems, it is a convenient and useful tool.

What can we say about the method relative to its use in numerical computation? The advantages of iterative procedures are well known; in particular, numerical error at each iterative step does not degrade the solution as a whole. However, the convergence rate does depend on q: with $q = 0.5$ or greater, say, convergence may be too slow to carry out many iterations on a complicated system of equations. In such cases it may be possible to transform the iterative procedure in order to speed up convergence (e.g., Seidel's method).

1.33 Topology, Weak and Weak* Topologies

We conclude the chapter with a small excursion into topology that will permit us to consider weak convergence from another point of view and to touch on some theoretical questions that arise in applications.

We introduced an open set in a normed space as a set containing, together with each point, some open ball centered at the point. The collection of all open sets is said to constitute a *topology* of the normed space. Now we extend this notion to a general topological space.

We say that a collection τ of subsets of X, called *open sets* in X, constitutes a *topology* in X if it includes the following sets:

1. the empty set $\emptyset$ and the whole set X;
2. any union of the subsets of τ;
3. the intersection of any finite collection of open sets.

The pair (X, τ), consisting of X along with this system τ, is called a *topological space*. When the topology on X is understood, the topological space can be denoted by X.

We consider only *Hausdorff spaces* having the property that for any two points $x_1 \neq x_2 \in X$ there are nonintersecting open sets S_1, S_2 such that $x_1 \in S_1$ and $x_2 \in S_2$.

As in a metric space, we call a set N a *neighborhood* of $x \in X$ if there is an open set $S \in \tau$ such that $x \in S \subset N$. We also call x a *limit* (or *accumulation*) *point* of a set $M \subset X$ if any neighborhood of x contains a point of M distinct from x. Finally, we say that $M \subset X$ is *closed* if it contains all its limit points.

Note that any subset M of X becomes a topological space with the *relative topology* (induced by τ) if we define its topology as the system of all of its subsets taking the form $M \cap S$ with $S \subset \tau$.

We also can introduce a *basis of the topology* as a collection of open sets such that any set from τ can be represented as a union of sets from the basis.

Finally, we note that a metric space is a topological space. The system of all its open balls constitutes a basis of its topology. The same holds for a normed space.

Below, we will introduce a "weak topology" in a normed space, which differs from the topology defined by the basis consisting of open balls. We wish to compare two topologies on the same space X. We say that topology τ_1 is *finer* than τ_2 (equivalently, that τ_2 is *coarser* than τ_1) if the collection τ_1 includes all sets from τ_2.

In a normed space, the topology is introduced by the unique functional called the norm. In a linear (real or complex) space X, we can introduce a topology using a family of seminorms.

A real-valued functional $p(x)$ defined for each $x \in X$ is called a *seminorm* if it satisfies the following two axioms for any $\alpha \in \mathbb{R}$ (or $\mathbb{C}$) and $x, y \in X$:

(1) $p(\alpha x) = |\alpha|\, p(x)$,

(2) $p(x + y) \leq p(x) + p(y)$.

From (1) it follows that $p(0) = 0$, and from (1) and (2) that $p(x) \geq 0$. So one difference between a seminorm and a norm is that $p(x) = 0$ does not imply $x = 0$. If $p(x) = 0$ does imply $x = 0$, then the seminorm is a norm in X. We also have

$$|p(x) - p(y)| \geq |p(x - y)| \,.$$

Let us show how to construct the topology X given by a family of seminorms $(p_i(x),\ i \in I)$. First we define a neighborhood of zero defined by a seminorm $p_i(x)$, that is

$$B_{i,c} = \{x \in X\colon\ p_i(x) < c\} \,.$$

Each $B_{i,c}$ is an open set by definition. Any $x_0 \in B_c$ has some neighborhood defined by $p_i(x)$ that belongs to $B_{i,c}$. Indeed, let $p_i(x_0) = b < c$. The needed neighborhood of x_0 is the set of points

$$z = x_0 + B_{i,(c-b)/2} = \{z \in X\colon\ z = x_0 + x,\ p_i(x) < (c - b)/2\} \,.$$

Because

$$p_i(z) = p_i(x_0 + x) \le p(x_0) + p_i(x) \le b + (c - b)/2 < c\,,$$

we have $z \in B_{i,c}$. So $B_{i,c}$ contains some neighborhood of each interior point as was introduced in the general definition of an open set in a topological space. This is valid for any $i \in I$.

Next we can prove that any union of constructed open sets, as well as the intersection of a finite number of such sets, is also open. So all sets of the form $x_0 + B_{i,c}$ for all $x_0 \in X$, $i \in I$, $c > 0$ constitute a basis of the topology. We also can prove it is a Hausdorff topology if for any $x^* \in X$ there is a seminorm $p_\lambda(x)$ from the family $(p_i(x),\ i \in I)$ such that $p_\lambda(x^*) \ne 0$.

Now we are ready to discuss

1.34 Weak Topology in a Normed Space X

We redenote a linear continuous functional on a normed space X by

$$\langle x, x^* \rangle\,,$$

where $x \in X$ and $x^* \in X'$, a linear continuous functional. In our earlier notation this functional should be written as $x^*(x)$; the present notation is more symmetric and reflects the Riesz representation theorem for a linear continuous functional in a Hilbert space.

For $x^* \in X'$ we introduce a seminorm by

$$p_{x^*}(x) = |\langle x, x^* \rangle|\,.$$

The family of all these seminorms defines a basis of the weak topology in X, denoted by $\tau(X, X')$. A set $S \subset X$ is open in the weak topology if for any $x_0 \in S$ there are $\varepsilon > 0$ and a finite set of seminorms $p_{x_1^*}(x), \dots, p_{x_n^*}(x)$ such that $V(x_0) \subset S$,

$$V(x_0) = \{x \in X\colon\ p_{x_1^*}(x - x_0) < \varepsilon, \dots, p_{x_n^*}(x - x_0) < \varepsilon\}\,.$$

We recall that the kernel of a linear continuous functional x^* in X, i.e.,

$$N(x^*) = \{x\colon\ \langle x, x^* \rangle = 0,\ x \in X\}\,,$$

has codimension one, and so the set $\{x \in X\colon\ p(x^*) < \varepsilon\}$ contains an infinite dimensional subspace of X if X is infinite dimensional. The intersection of a finite number of such sets, one of which is presented by $V(0)$, also contains an infinite dimensional subspace of X, which means that an open ball in a normed space X cannot be open in the weak topology. However an open set of X in the weak topology is open in the strong topology defined by the norm. So the weak topology of X is coarser than the strong topology.

It makes sense to note that a linear functional is still continuous at any point of X if and only if it is continuous at zero. Any linear continuous functional x^* from X' is continuous in the weak topology, as follows from the trivial inequality

$$|\langle x, x^*\rangle| \le p(x^*) .$$

Using the notation $\langle x, x^*\rangle$ for a linear continuous functional, we can introduce a seminorm on X', the dual space to X, using

$$p_x(x^*) = |\langle x, x^*\rangle| .$$

The family of all seminorms of this form defines a basis of a so-called *weak* topology* in X', denoted by $\tau(X', X)$.

We recall that in a Hilbert space the second dual space $(X')'$ is identified with X. A similar situation can happen with a Banach space; in this case, X is said to be *reflexive*. In the latter case, the weak topology in X' is exactly the topology referred to here as the weak* topology.

We note without proof that in a normed space, a closed linear subspace is also closed in the weak topology.

We shall consider weak convergence in a normed space in some detail.

Definition 1.34.1. A sequence $\{x_k\}$ *converges weakly* to x_0 if

$$\lim_{k\to\infty} \langle x_k, x^*\rangle = \langle x_0, x^*\rangle$$

for all $x^* \in X'$. We denote this by

$$\operatorname*{w-lim}_{k\to\infty} x_k = x_0 \quad \text{or} \quad x_k \rightharpoonup x_0 \ \text{ as } k \to \infty .$$

Theorem 1.34.1. The weak limit x_0 of a sequence $\{x_k\}$, if it exists, is unique.

Proof. Suppose there exist two weak limits x_{01} and x_{02} of $\{x_k\}$ such that

$$\langle x_{01} - x_{02}, x^*\rangle = 0$$

for any linear continuous functional x^*. By a corollary to the Hahn–Banach theorem, there is a linear continuous functional $x^* \in X'$ such that

$$\langle x_{01} - x_{02}, x^*\rangle = \|x_{01} - x_{02}\| .$$

This yields a contradiction if $x_{01} \neq x_{02}$. □

Every strongly convergent sequence is weakly convergent. As we saw in a Hilbert space, a sequence of orthonormal elements converges weakly to zero but does not converge strongly to zero, so weak convergence does not imply strong convergence.

Definition 1.34.2. We call $\{x_k\}$ a *weak Cauchy sequence* if for every $x^* \in X'$ the sequence $\{\langle x_k, x^*\rangle\}$ is a numerical Cauchy sequence.

Theorem 1.34.2. In a normed space X, a weak Cauchy sequence $\{x_k\}$ is bounded. If $\{x_k\}$ converges weakly to x_0, then

$$\|x_0\| \leq \liminf_{k\to\infty} \|x_k\| . \tag{1.34.1}$$

Proof. We reduce this to Corollary 1.23.2. Recall that the dual space X' is complete. Consider $\langle x_k, x'\rangle$ as the value of an operator (functional) defined by x_k over X'. By the definition of weak continuity, for each $x' \in X'$ the numerical sequence $\{\langle x_k, x'\rangle\}$ converges to some number that defines some linear functional $\langle x_0, x'\rangle$. By Corollary 1.23.2, the set of norms of the functionals is bounded, i.e., $\sup_k \|x_k\| \leq c$, and (1.34.1) follows from the corollary. □

A reflexive Banach space X has two important properties:

1. X is *weakly complete*: any weakly Cauchy sequence has a weak limit in X;
2. a bounded set in X is weakly sequentially compact; that is, any bounded sequence has a weakly convergent subsequence (*Alaoglu's theorem*).

1.35 Conclusion and Further Reading

In the next chapter, we will apply the results of this chapter to mechanical problems.

Functional analysis borrowed many ideas from analysis, topology, the calculus of variations, the theory of differential equations, measure theory, etc. However, we can say that functional analysis itself — as a separate part of mathematics — arose in the books by S. Banach [4] along with papers by a great many individuals. A couple of early and interesting treatises on functional analysis are those by Riesz [30] and von Neumann [29]. Although modern textbooks on functional analysis are abundant, the contents of a given book may strongly reflect the interests of its authors. Works that complement the present book include [31], [28], [15], and [18].

Chapter 2
Mechanics Problems from the Functional Analysis Viewpoint

In the past, an engineer could calculate mechanical stresses and strains using a pencil and a logarithmic slide rule. Modern mechanical models, on the other hand, are nonlinear, and even the linear models are complicated. Numerical methods in structural dynamics cannot be applied without computers running specialized programs. However, a researcher should have a solid grasp of the equations that underlie a numerical model and the types of results that can be expected. New models appear in mechanics on a regular basis. Some of these, when written out in detail, can span multiple pages and are clearly beyond pencil-and-paper approaches. Although functional analysis does not provide a detailed picture of the results to be expected from a complicated model, it can answer questions regarding whether the problem is mathematically well-posed (e.g., whether a solution exists and is unique). It may also indicate whether the results can be obtained by a general computer program or whether a special program, based on a knowledge of the general properties of the model, is required. In other words, functional analysis can provide valuable insight even to those who rely heavily on numerical approaches.

2.1 Introduction to Sobolev Spaces

In his famous book [33], S.L. Sobolev (1908–1989) introduced some normed spaces that now bear his name; they are denoted by $W^{m,p}(\Omega)$. The norm in $W^{m,p}(\Omega)$ is

$$\|u\| = \left(\int_\Omega \sum_{|\alpha|\le m} |D^\alpha u|^p \, d\Omega \right)^{1/p}, \tag{2.1.1}$$

where m is an integer, $p \ge 1$, and Ω is compact in $\mathbb{R}^n$. (The D^α notation was introduced on p. 19.) Indeed this is a norm on the set of functions that possess continuous derivatives up to order m on Ω, with satisfaction of axiom N3 ensured by Minkowski's inequality (1.3.6). The completion of the resulting normed space is the Banach space $W^{m,p}(\Omega)$.

For Ω a segment $[a,b]$, the spaces $W^{m,p}(a,b)$ were introduced by Stefan Banach (1892–1945) in his dissertation. Our interest in Sobolev spaces is clear, as the elements of each of our energy spaces will belong to $W^{m,2}(\Omega)$ for some m.

Problem 2.1.1. (a) Demonstrate that if $f(x) \in W^{1,2}(a,b)$ and $g(x) \in C^{(1)}(a,b)$, then $f(x)g(x) \in W^{1,2}(a,b)$. (b) Show that $Af(x) = f(x)$ defines a bounded linear operator acting from $W^{1,2}(0,1)$ to $L^2(0,1)$. □

Generalized Notions of Derivative. For $u \in L^p(\Omega)$, K.O. Friedrichs (1901–1982) [12] introduced the notion of *strong derivative*, calling $v \in L^p(\Omega)$ a strong derivative $D^\alpha(u)$ if there is a sequence $\{\varphi_n\} \subset C^{(\infty)}(\Omega)$ such that

$$\int_\Omega |u(\mathbf{x}) - \varphi_n(\mathbf{x})|^p \, d\Omega \to 0 \quad \text{and} \quad \int_\Omega |v(\mathbf{x}) - D^\alpha \varphi_n(\mathbf{x})|^p \, d\Omega \to 0 \ \text{ as } n \to \infty .$$

Since $C^{(\infty)}(\Omega)$ is dense in any $C^{(k)}(\Omega)$, an element of $W^{m,p}(\Omega)$ has all strong derivatives up to the order m lying in $L^p(\Omega)$.

Another version of generalized derivative was proposed by Sobolev. He used, along with the classical integration by parts formula, an idea from the classical calculus of variations: if

$$\int_\Omega u(\mathbf{x})\, \varphi(\mathbf{x})\, d\Omega = 0$$

for all infinitely differentiable functions $\varphi(\mathbf{x})$ having compact support in an open domain Ω, then $u(\mathbf{x}) = 0$ almost everywhere (everywhere if $u(\mathbf{x})$ is to be continuous). We say that a function $\varphi(\mathbf{x})$, infinitely differentiable in Ω, has *compact support* in $\Omega \subset \mathbb{R}^n$ if the closure of the set

$$M = \{\mathbf{x} \in \Omega : \varphi(\mathbf{x}) \neq 0\}$$

is a compact set in Ω. In the sense introduced by Sobolev, $v \in L^p(\Omega)$ is called a *weak derivative* $D^\alpha u$ of $u \in L^p(\Omega)$ if for every infinitely differentiable function $\varphi(\mathbf{x})$ with compact support in Ω we have

$$\int_\Omega u(\mathbf{x})\, D^\alpha \varphi(\mathbf{x})\, d\Omega = (-1)^{|\alpha|} \int_\Omega v(\mathbf{x})\, \varphi(\mathbf{x})\, d\Omega . \tag{2.1.2}$$

The two definitions of generalized derivative are equivalent [33]. The proof would exceed the scope of our presentation, and the same is true for some other facts of this section.

Imbedding Theorems. The results presented next are particular cases of *Sobolev's imbedding theorem*. Later we will give proofs of some particular results when studying certain energy spaces for elastic models which turn out to be subspaces of Sobolev spaces. The theorems are formulated in terms of imbedding operators, a notion elucidated further in Sect. 2.2.

We assume that the compact set $\Omega \subset \mathbb{R}^n$ satisfies the *cone condition*. This means there is a finite circular cone in $\mathbb{R}^n$ such that any point of the boundary of Ω can be touched by the vertex of the cone while the cone lies fully inside Ω. This is the condition under which Sobolev's imbedding theorem is proved. We denote by Ω_r an r-dimensional piecewise smooth hypersurface in Ω. (This means that, at any point of smoothness, in a local coordinate system, it is described by functions having all derivatives continuous up to order m locally, if we consider $W^{m,p}(\Omega)$.)

The theory of Sobolev spaces is a substantial branch of mathematics (see Adams [1], Lions and Magenes [25], etc.). We formulate only what is needed for our purposes. This is Sobolev's imbedding theorem with some extensions:

Theorem 2.1.1 (Sobolev–Kondrashov). The imbedding operator of the Sobolev space $W^{m,p}(\Omega)$ to $L^q(\Omega_r)$ is continuous if one of the following conditions holds:

(i) $n > mp$, $r > n - mp$, $q \le pr/(n - mp)$;
(ii) $n = mp$, q is finite with $q \ge 1$.

If $n < mp$, then the space $W^{m,p}(\Omega)$ is imbedded into the Hölder space $H^\alpha(\overline{\Omega})$ when $\alpha \le (mp - n)/p$, and the imbedding operator is continuous.

The imbedding operator of $W^{m,p}(\Omega)$ to $L^q(\Omega_r)$ is compact (i.e., takes every bounded set of $W^{m,p}(\Omega)$ into a precompact set of the corresponding space[1]) if

(i) $n > mp$, $r > n - mp$, $q < pr/(n - mp)$ or
(ii) $n = mp$ and q is finite with $q \ge 1$.

If $n < mp$ then the imbedding operator is compact to $H^\alpha(\overline{\Omega})$ when $\alpha < (mp - n)/p$.

Note that this theorem provides imbedding properties not only for functions but also for their derivatives:

$$u \in W^{m,p}(\Omega) \implies D^\alpha u \in W^{m-k,p}(\Omega) \;\text{ for } |\alpha| = k\,.$$

Also available are stricter results on the imbedding of Sobolev spaces on Ω into function spaces given on manifolds Ω_r of dimension less than n; however, such *trace theorems* require an extended notion of Sobolev spaces.

Let us formulate some useful consequences of Theorem 2.1.1.

Theorem 2.1.2. Let γ be a piecewise differentiable curve in a compact set $\Omega \subset \mathbb{R}^2$. For any finite $q \ge 1$, the imbedding operator of $W^{1,2}(\Omega)$ to the spaces $L^q(\Omega)$ and $L^q(\gamma)$ is continuous (and compact), i.e.,

$$\max\{\|u\|_{L^q(\Omega)}, \|u\|_{L^q(\gamma)}\} \le m\, \|u\|_{W^{1,2}(\Omega)} \tag{2.1.3}$$

with a constant m which does not depend on $u(\mathbf{x})$.

Theorem 2.1.3. Let $\Omega \subset \mathbb{R}^2$ be compact. If $\alpha \le 1$, the imbedding operator of $W^{2,2}(\Omega)$ to $H^\alpha(\overline{\Omega})$ is continuous; if $\alpha < 1$, it is compact. For the first derivatives, the imbedding operator to $L^q(\Omega)$ and $L^q(\gamma)$ is continuous (and compact) for any finite $q \ge 1$.

---
[1] The notion of compact operator will be explored in Chap. 3.

Theorem 2.1.4. Let γ be a piecewise smooth surface in a compact set $\Omega \subset \mathbb{R}^3$. The imbedding operator of $W^{1,2}(\Omega)$ to $L^q(\Omega)$ when $1 \le q \le 6$, and to $L^p(\gamma)$ when $1 \le p \le 4$, is continuous; if $1 \le q < 6$ or $1 \le p < 4$, respectively, then it is compact.

We merely indicate how such theorems are proved. We will establish similar results for the beam problem (see (2.3.5)) and for the clamped plate problem (see (2.3.21)) via integral representations of functions from certain classes. In like manner, Sobolev's original proof is given for Ω a union of bounded star-shaped domains. A domain is called *star-shaped with respect to a ball B* if any ray with origin in B intersects the boundary of the domain only once. For a domain Ω which is bounded and star-shaped with respect to a ball B, a function $u(\mathbf{x}) \in C^{(m)}(\Omega)$ can be represented in the form

$$u(\mathbf{x}) = \sum_{|\alpha|\le m-1} x_1^{\alpha_1} \cdots x_n^{\alpha_n} \int_B K_\alpha(\mathbf{y})\, u(\mathbf{y})\, d\Omega + \int_\Omega \frac{1}{|\mathbf{x}-\mathbf{y}|^{n-m}} \sum_{|\alpha|=m} K_\alpha(\mathbf{x},\mathbf{y})\, D^\alpha u(\mathbf{y})\, d\Omega_{\mathbf{y}} \tag{2.1.4}$$

where $K_\alpha(\mathbf{y})$ and $K_\alpha(\mathbf{x},\mathbf{y})$ are continuous functions. Investigating properties of the integral terms on the right-hand side of the representation (2.1.4), Sobolev formulated his results; later they were extended to more general domains.

Another method is connected with the Fourier transformation of functions. In the case of $W^{m,2}(\Omega)$, it is necessary to extend functions of $C^{(m)}(\overline{\Omega})$ outside Ω in such a way that they belong to $C^m(\mathbb{R}^n)$ and $W^{m,2}(\mathbb{R}^n)$. Then using the Fourier transformation

$$\hat{u}(\mathbf{y}) = (2\pi)^{-n/2} \int_{\mathbb{R}^n} e^{-i\mathbf{x}\cdot\mathbf{y}} u(\mathbf{x})\, dx_1 \cdots dx_n$$

along with the facts that

$$\|u(\mathbf{x})\|_{L^2(\mathbb{R}^n)} = \|\hat{u}(\mathbf{y})\|_{L^2(\mathbb{R}^n)}$$

and

$$\widehat{D^\alpha u}(\mathbf{x}) = (iy_1)^{\alpha_1} \cdots (iy_n)^{\alpha_n}\, \hat{u}(\mathbf{y})$$

for $u \in L^2(\mathbb{R}^n)$, we can present the norm in $W^{m,2}(\Omega)$ in the form

$$\|u(\mathbf{x})\|^2_{W^{m,2}(\mathbb{R}^n)} = \sum_{|\alpha|\le m} \|y_1^{\alpha_1} \cdots y_n^{\alpha_n}\, \hat{u}(\mathbf{y})\|^2_{L^2(\mathbb{R}^n)} .$$

We can then study the properties of the weighted space $L^2_w(\mathbb{R}^n)$; this transformed problem is simpler, as many of the problems involved are algebraic estimates of Fourier images.

Moreover, we can consider $W^{p,2}(\mathbb{R}^n)$ with fractional indices p. These lead to necessary and sufficient conditions for the trace problem: given $W^{m,2}(\Omega)$, find the space $W^{p,2}(\partial\Omega)$ in which $W^{m,2}(\Omega)$ is continuously imbedded. The inverse trace problem is, given $W^{p,2}(\Omega)$, find the maximal index m such that every element $u \in W^{p,2}(\partial\Omega)$

can be extended to $\overline{\Omega}$, $u^* \in W^{m,2}(\Omega)$, in such a way that

$$\|u^*\|_{W^{m,2}(\Omega)} \le c\,\|u\|_{W^{p,2}(\partial\Omega)} .$$

In this way, many results from the contemporary theory of elliptic (and other types of) equations and systems are obtained. We should mention that the trace theorems are formulated mostly for smooth manifolds, hence are not applicable to practical problems involving domains with corners.

As stated in Lemma 1.16.3, all Sobolev spaces are separable. The same holds for all the energy spaces we introduce.

2.2 Operator of Imbedding

Let Ω be a Jordan measurable compact set in $\mathbb{R}^n$. Any element of $C^{(k)}(\Omega)$ also belongs to $C(\Omega)$. This correspondence — between an element of the space $C^{(k)}(\Omega)$ and the same element in the space $C(\Omega)$ — is a linear operator. Although it resembles an identity operator, its domain and range differ and it is called the *imbedding operator* from $C^{(k)}(\Omega)$ to $C(\Omega)$. Clearly, $C^{(k)}(\Omega)$ is a proper subset of $C(\Omega)$; this situation is typical for an imbedding operator that gives the correspondence between the same elements but considered in different spaces. For a function u that belongs to both spaces, we have

$$\|u\|_{C(\Omega)} \le \|u\|_{C^{(k)}(\Omega)} ,$$

which shows that the imbedding operator is bounded or continuous (what can be said about its norm?). Moreover, by a consequence of Arzelà's theorem, it is compact.

We get a similar imbedding operator when considering the spaces ℓ^p and ℓ^q for $1 \le p < q$. As (1.4.3) states, any element $\mathbf{x}$ of ℓ^p belongs to ℓ^q and

$$\|\mathbf{x}\|_{\ell^q} \le c_1\,\|\mathbf{x}\|_{\ell^p} .$$

Here and below, the constants c_k are independent of the element of the spaces, so the imbedding operator from ℓ^p to ℓ^q is bounded.

We encounter another imbedding operator when considering the correspondence between the same equivalence classes in the spaces $L^p(\Omega)$ and $L^q(\Omega)$: if $1 \le p < q$, then any element u of $L^q(\Omega)$ belongs to $L^p(\Omega)$ and

$$\|u\|_{L^p(\Omega)} \le c_2\,\|u\|_{L^q(\Omega)} .$$

So the imbedding operator from $L^q(\Omega)$ to $L^p(\Omega)$ is also bounded. (Note that the relations between p and q are different in the spaces ℓ^r and $L^r(\Omega)$.)

The situation is slightly different when we consider the relation between the spaces $C(\Omega)$ and $L^p(\Omega)$. It is clear that any function u from $C(\Omega)$ has a finite norm in $L^p(\Omega)$. To use the imbedding idea, we identify the function u with the equivalence class from $L^p(\Omega)$ that contains the stationary sequence $(u, u, \ldots)$, and in this way consider u as an element of $L^p(\Omega)$. Because

$$\int_\Omega u\,d\Omega \le c_3 \|u\|_{C(\Omega)} ,$$

the imbedding operator from $C(\Omega)$ to $L^p(\Omega)$ is bounded for any $1 \le p < \infty$.

This practice of identification must be extended when we consider imbedding operators involving Sobolev spaces as treated in this book. In the imbedding from $W^{1,2}(\Omega)$ to $L^p(\Omega)$, the elements of $W^{1,2}(\Omega)$ — the equivalence classes of functions from $C^{(1)}(\Omega)$ — are different from the equivalence classes from $C(\Omega)$ that constitute $L^p(\Omega)$; however, each class from $C^{(1)}(\Omega)$ is contained by some class from $C(\Omega)$, and we identify these classes. In this sense, we say that $W^{1,2}(\Omega)$ is imbedded into $L^p(\Omega)$ and the imbedding operator is continuous (or even compact).

The imbedding of a Sobolev space into the space of continuous functions is even more complicated, although the general ideas are the same. We will consider this in more detail when dealing with the energy space for a plate. In this case we cannot directly identify an element of a Sobolev space, which is an equivalence class, with a continuous function. However, we observe that for a class of equivalent sequences in the norm of a Sobolev space, for any sequence that enters into the class, there is a limit element, a continuous function that does not differ for other representative sequences, and we identify the whole class with this function. The inequality relating the Sobolev norm of the class and the continuous norm of this function states that this identification or correspondence is a bounded operator whose linearity is obvious. Some other inequalities state that the operator is compact.

2.3 Some Energy Spaces

A Beam. Earlier we noted that the set S of all real-valued continuous functions $y(x)$ having continuous first and second derivatives on $[0, l]$ and satisfying the boundary conditions

$$y(0) = y'(0) = y(l) = y'(l) = 0 \tag{2.3.1}$$

is a metric space under the metric

$$d(y, z) = \left(\frac{1}{2}\int_0^l B(x)[y''(x) - z''(x)]^2\,dx\right)^{1/2} . \tag{2.3.2}$$

We called this an energy space for the clamped beam. We can introduce an inner product

$$(y, z) = \frac{1}{2}\int_0^l B(x)y''(x)z''(x)\,dx \tag{2.3.3}$$

and norm

$$\|y\| = \left(\frac{1}{2}\int_0^l B(x)[y''(x)]^2\,dx\right)^{1/2}$$

such that $d(y, z) = \|y - z\|$. But this space is not complete (it is clear that there are Cauchy sequences whose limits do not belong to $C^{(2)}(0, l)$; the reader should construct an example). To have a complete space, we must apply the completion theorem. The actual energy space denoted by E_B is the completion of S in the metric (2.3.2) (or, what amounts to the same thing, in the inner product (2.3.3)).

Let us consider some properties of the elements of E_B. An element $y(x) \in E_B$ is a set of Cauchy sequences equivalent in the metric (2.3.2). Let $\{y_n(x)\}$ be a representative sequence of $y(x)$. Then, according to (2.3.2),

$$\left(\frac{1}{2}\int_0^l B(x)[y_n''(x) - y_m''(x)]^2\,dx\right)^{1/2} \to 0 \text{ as } n, m \to \infty .$$

If we assume that

$$0 < m_1 \le B(x) \le m_2 ,$$

then the sequence of second derivatives $\{y_n''(x)\}$ is a Cauchy sequence in the norm of $L^2(0, l)$ because

$$m_1 \int_0^l [y_n''(x) - y_m''(x)]^2\,dx \le \int_0^l B(x)[y_n''(x) - y_m''(x)]^2\,dx .$$

Hence $\{y_n''(x)\}$ is a representative sequence of some element of $L^2(0, l)$, and we can regard $y \in E_b$ as having a second derivative that belongs to $L^2(0, l)$.

Now consider $\{y_n'(x)\}$. For any $y(x) \in S$ we get

$$y'(x) = \int_0^x y''(t)\,dt .$$

So for a representative $\{y_n(x)\}$ of a class $y(x) \in E_B$ we have

$$\begin{aligned}
|y_n'(x) - y_m'(x)| &\le \int_0^x |y_n''(t) - y_m''(t)|\,dt \le \int_0^l 1 \cdot |y_n''(t) - y_m''(t)|\,dt \\
&\le l^{1/2}\left(\int_0^l [y_n''(x) - y_m''(x)]^2\,dx\right)^{1/2} \\
&\le (l/m_1)^{1/2}\left(\int_0^l B(x)[y_n''(x) - y_m''(x)]^2\,dx\right)^{1/2} \\
&\to 0 \text{ as } n, m \to \infty .
\end{aligned} \tag{2.3.4}$$

It follows that $\{y_n'(x)\}$ converges uniformly on $[0, l]$; hence there exists a limit function $z(x)$ which is continuous on $[0, l]$. This function does not depend on the choice of representative sequence (Problem 2.3.1 below). The same holds for a sequence of functions $\{y_n(x)\}$: its limit is a function $y(x)$ continuous on $[0, l]$. Moreover,

$$y'(x) = z(x) .$$

To prove this, it is necessary to repeat the arguments of Sect. 1.11 on the differentiability of the elements of $C^{(k)}(\Omega)$, with due regard for (2.3.4). From (2.3.4) and the similar inequality for $\{y_n(x)\}$ we get

$$\max_{x\in[0,l]} (|y(x)| + |y'(x)|) \le m\left(\frac{1}{2}\int_0^l B(x)[y''(x)]^2\,dx\right)^{1/2} \tag{2.3.5}$$

for some constant m independent of $y(x) \in E_B$. So each element $y(x) \in E_B$ can be identified with an element $y(x) \in C^{(1)}(0,l)$ in such a way that (2.3.5) holds. This correspondence is an imbedding operator, and we interpret (2.3.5) as a statement that the imbedding operator from E_B to $C^{(1)}(0,l)$ is continuous (cf., Sect. 1.21). Henceforth we refer to the elements of E_B as if they were continuously differentiable functions, attaching the properties of the uniquely determined limit functions to the corresponding elements of E_B themselves.

Problem 2.3.1. Prove that the function $z(x)$, discussed above, is independent of the choice of representative sequence. □

We are interested in analyzing all terms that appear in the statement of the equilibrium problem for a body as a minimum potential energy problem. So we will consider the functional giving the work of external forces. For the beam, it is

$$A = \int_0^l F(x)y(x)\,dx\,.$$

This is well-defined on E_B if $F(x) \in L(\Omega)$; moreover, (2.3.5) shows that the work of external forces can contain terms of the form

$$\sum_k [F_k y(x_k) + M_k y'(x_k)]\,,$$

which can be interpreted as the work of point forces and point moments, respectively. This is a consequence of the continuity of the imbedding from E_B to $C^{(1)}(0,l)$.

Remark 2.3.1. Alternatively we can define E_B on a base set S_1 of smoother functions, in $C^{(4)}(0,l)$ say, satisfying (2.3.1). The result is the same, since S_1 is dense in S with respect to the norm of $C^{(2)}(0,l)$. However, sometimes such a definition is convenient. □

Remark 2.3.2. Readers familiar with the contemporary literature in this area may have noticed that Western authors usually deal with Sobolev spaces, studying the properties of operators corresponding to problems under consideration; we prefer to deal with energy spaces, studying first their properties and then those of the corresponding operators. Although these approaches lead to the same results, in our view the physics of a particular problem should play a principal role in the analysis — in this way the methodology seems simpler, clearer, and more natural. In papers devoted to the study of elastic bodies, we mainly find interest in the case of a clamped boundary. Sometimes this is done on the principle that it is better to deal solely with homogeneous Dirichlet boundary conditions, but often it is an unfortunate consequence of the use of the Sobolev spaces $H^k(\Omega)$. The theory of these spaces is well developed but is not amenable to the study of other boundary conditions. Success in the investigation of mechanics problems can be much more difficult without the benefit of the physical ideas brought out by the energy spaces. □

Remark 2.3.3. In defining the energy space of the beam, we left aside the question of smoothness of the stiffness function $B(x)$. From a mathematical standpoint this is risky since, in principle, $B(x)$ can be nonintegrable. But in the case of an actual physical beam, $B(x)$ can have no more than a finite number of discontinuities and must be differentiable everywhere else. For simplicity, we shall continue to make realistic assumptions concerning physical parameters such as stiffness and elastic constants; in particular, we shall suppose whatever degree of smoothness is required for our purposes. □

A Membrane (Clamped Edge). The subset of $C^{(1)}(\Omega)$ consisting of all functions satisfying

$$u(x,y)\big|_{\partial\Omega} = 0 \tag{2.3.6}$$

with the metric

$$d(u,v) = \left\{ \iint_\Omega \left[\left(\frac{\partial u}{\partial x} - \frac{\partial v}{\partial x} \right)^2 + \left(\frac{\partial u}{\partial y} - \frac{\partial v}{\partial y} \right)^2 \right] dx\,dy \right\}^{1/2} \tag{2.3.7}$$

is an incomplete metric space. If we introduce an inner product

$$(u,v) = \iint_\Omega \left(\frac{\partial u}{\partial x}\frac{\partial v}{\partial x} + \frac{\partial u}{\partial y}\frac{\partial v}{\partial y} \right) dx\,dy \tag{2.3.8}$$

consistent with (2.3.7), we get an inner product space. Its completion in the metric (2.3.7) is the energy space E_{MC} for the clamped membrane, a real Hilbert space.

What can we say about the elements $U(x,y) \in E_{MC}$? If $\{u_n(x,y)\}$ is a representative of $U(x,y)$, then

$$\left\{ \iint_\Omega \left[\left(\frac{\partial u_n}{\partial x} - \frac{\partial u_m}{\partial x} \right)^2 + \left(\frac{\partial u_n}{\partial y} - \frac{\partial u_m}{\partial y} \right)^2 \right] dx\,dy \right\}^{1/2} \to 0 \text{ as } n,m \to \infty\,.$$

and we see that the sequences of first derivatives $\{\partial u_n/\partial x\}$, $\{\partial u_n/\partial y\}$ are Cauchy sequences in the norm of $L^2(\Omega)$. What about $U(x,y)$ itself? If we extend each $u_n(x,y)$ by zero outside Ω, we can write

$$u_n(x,y) = \int_0^x \frac{\partial u_n(s,y)}{\partial s}\,ds$$

(assuming, without loss of generality, that Ω is confined to the band $0 \le x \le a$). Squaring both sides and integrating over Ω, we get

$$\iint_\Omega u_n^2(x,y)\,dx\,dy = \iint_\Omega \left[\int_0^x \frac{\partial u_n(s,y)}{\partial s}\,ds\right]^2 dx\,dy$$
$$\leq \iint_\Omega \left[\int_0^a 1\cdot\left|\frac{\partial u_n(s,y)}{\partial s}\right|\,ds\right]^2 dx\,dy$$
$$\leq \iint_\Omega \left[a\int_0^a \left(\frac{\partial u_n(s,y)}{\partial s}\right)^2 ds\right] dx\,dy$$
$$\leq a^2 \iint_\Omega \left(\frac{\partial u_n(x,y)}{\partial x}\right)^2 dx\,dy\,.$$

This means that if $\{\partial u_n/\partial x\}$ is a Cauchy sequence in the norm of $L^2(\Omega)$, then so is $\{u_n\}$. Hence we can consider elements $U(x,y) \in E_{MC}$ to be such that $U(x,y)$, $\partial U/\partial x$, and $\partial U/\partial y$ belong to $L^2(\Omega)$.

As a consequence of the last chain of inequalities, we get *Friedrichs' inequality*

$$\iint_\Omega U^2(x,y)\,dx\,dy \leq m \iint_\Omega \left[\left(\frac{\partial U}{\partial x}\right)^2 + \left(\frac{\partial U}{\partial y}\right)^2\right] dx\,dy$$

which holds for any $U(x,y) \in E_{MC}$ and a constant m independent of $U(x,y)$.

Membrane (Free Edge). Although it is natural to introduce the energy space using the energy metric (2.3.7), we cannot distinguish between two states $u_1(x,y)$ and $u_2(x,y)$ of the membrane with free edge if

$$u_2(x,y) - u_1(x,y) = c = \text{constant}\,.$$

This is the only form of "rigid" displacement possible for a membrane. We first show that no other rigid displacements (i.e., displacements associated with zero strain energy) are possible. The proof is a consequence of *Poincaré's inequality*

$$\iint_\Omega u^2\,dx\,dy \leq m\left\{\left(\iint_\Omega u\,dx\,dy\right)^2 + \iint_\Omega \left[\left(\frac{\partial u}{\partial x}\right)^2 + \left(\frac{\partial u}{\partial y}\right)^2\right] dx\,dy\right\} \tag{2.3.9}$$

for a function $u(x,y) \in C^{(1)}(\Omega)$. The constant m does not depend on $u(x,y)$.

Proof of Poincaré's inequality [9]. We first assume that Ω is the square $[0,a]\times[0,a]$ and write down the identity

$$u(x_2,y_2) - u(x_1,y_1) = \int_{x_1}^{x_2} \frac{\partial u(s,y_1)}{\partial s}\,ds + \int_{y_1}^{y_2} \frac{\partial u(x_2,t)}{\partial t}\,dt\,.$$

Squaring both sides and then integrating over the square, first with respect to the variables x_1 and y_1 and then with respect to x_2 and y_2, we get

$$\iint_\Omega \iint_\Omega \left[u^2(x_2,y_2) - 2u(x_2,y_2)u(x_1,y_1) + u^2(x_1,y_1)\right] dx_1\, dy_1\, dx_2\, dy_2$$

$$= \iint_\Omega \iint_\Omega \left[\int_{x_1}^{x_2} \frac{\partial u(s,y_1)}{\partial s}\, ds + \int_{y_1}^{y_2} \frac{\partial u(x_2,t)}{\partial t}\, dt\right]^2 dx_1\, dy_1\, dx_2\, dy_2$$

$$\leq \iint_\Omega \iint_\Omega \left[\int_0^a 1 \cdot \left|\frac{\partial u(s,y_1)}{\partial s}\right| ds + \int_0^a 1 \cdot \left|\frac{\partial u(x_2,t)}{\partial t}\right| dt\right]^2 dx_1\, dy_1\, dx_2\, dy_2$$

$$\leq 2a \iint_\Omega \iint_\Omega \left[\int_0^a \left(\frac{\partial u(s,y_1)}{\partial s}\right)^2 ds + \int_0^a \left(\frac{\partial u(x_2,t)}{\partial t}\right)^2 dt\right] dx_1\, dy_1\, dx_2\, dy_2$$

$$\leq 2a^4 \int_0^a \int_0^a \left[\left(\frac{\partial u}{\partial x}\right)^2 + \left(\frac{\partial u}{\partial y}\right)^2\right] dx\, dy\,.$$

Note that, along with the Schwarz inequality for integrals, we have used the elementary inequality $(a+b)^2 \leq 2(a^2+b^2)$ which follows from the fact that $(a-b)^2 \geq 0$. The beginning of this chain of inequalities is

$$a^2 \iint_\Omega u^2(x,y)\, dx\, dy - 2\left(\iint_\Omega u(x,y)\, dx\, dy\right)^2 + a^2 \iint_\Omega u^2(x,y)\, dx\, dy$$

so

$$2a^2 \iint_\Omega u^2\, dx\, dy \leq 2\left(\iint_\Omega u\, dx\, dy\right)^2 + 2a^4 \iint_\Omega \left[\left(\frac{\partial u}{\partial x}\right)^2 + \left(\frac{\partial u}{\partial y}\right)^2\right] dx\, dy$$

and we get (2.3.9) with $m = \max(a^2, 1/a^2)$. □

It can be shown that Poincaré's inequality holds on more general domains. A modification of (2.3.9) will appear in Sect. 2.12.

Let us return to the free membrane problem. Provided we consider only the membrane's state of stress, any two states are identical if they are described by functions $u_1(x,y)$ and $u_2(x,y)$ whose difference is constant. We gather all functions (such that the difference between any two is a constant) into a class denoted by $u_*(x,y)$. There is a unique representative of $u_*(x,y)$ denoted by $u_b(x,y)$ such that

$$\iint_\Omega u_b(x,y)\, dx\, dy = 0\,. \tag{2.3.10}$$

For this *balanced representative* (or *balanced function*), Poincaré's inequality becomes

$$\iint_\Omega u_b^2(x,y)\, dx\, dy \leq m \iint_\Omega \left[\left(\frac{\partial u_b}{\partial x}\right)^2 + \left(\frac{\partial u_b}{\partial y}\right)^2\right] dx\, dy\,. \tag{2.3.11}$$

As the right-hand side is zero for a "rigid" displacement, so is the left-hand side and it follows that the balanced representative associated with a rigid displacement must be zero. Hence $u(x,y) = c$ is the only permissible form for a rigid displacement of a membrane.

Because (2.3.11) has the same form as Friedrichs' inequality, we can repeat our former arguments to construct the energy space E_{MF} for a free membrane using the balanced representatives of the classes $u_*(x,y)$. In what follows we shall use this space E_{MF}, remembering that its elements satisfy (2.3.10).

The condition (2.3.10) is a geometrical constraint resulting from our mathematical technique. Solving the static free membrane problem, we must remember that the formulation of the equilibrium problem does not impose this constraint — the membrane can move as a "rigid body" in the direction normal to its own surface. But if we consider only deformations and the strain energy defined by the first partial derivatives of $u(x,y)$, the results must be independent of such motions. Consider then the functional of the work of external forces

$$A = \iint_{\Omega} F(x,y)U(x,y)\,dx\,dy\,.$$

If we use the space E_{MF}, then A makes sense if $F(x,y) \in L^2(\Omega)$ (guaranteed by (2.3.11) together with the Schwarz inequality in $L^2(\Omega)$). This is the only restriction on external forces for a clamped membrane. However, in the case of equilibrium for a free membrane, the functional A must be invariant under transformations of the form $u(x,y) \mapsto u(x,y) + c$ with any constant c. This requires

$$\iint_{\Omega} F(x,y)\,dx\,dy = 0\,. \tag{2.3.12}$$

Again, we consider the equilibrium problem where rigid motion, however, is possible. Since we did not introduce inertia forces, we have formally equated the mass of the membrane to zero. In this situation of zero mass, any forces with nonzero resultant would make the membrane as a whole move with infinite acceleration. Thus, (2.3.12) also precludes such physical nonsense.

Meanwhile, Sobolev's imbedding theorem permits us to incorporate forces ψ acting on the boundary S of Ω into the functional describing the work of external forces:

$$\iint_{\Omega} F(x,y)u(x,y)\,dx\,dy + \int_S \psi u\,ds\,. \tag{2.3.13}$$

In this case, the condition of invariance of the functional under constant motions yields the condition

$$\iint_{\Omega} F(x,y)\,dx\,dy + \int_S \psi\,ds = 0\,. \tag{2.3.14}$$

Note that in the formulation of the Neumann problem, this quantity ψ appears in the boundary condition:

$$\left.\frac{\partial u}{\partial n}\right|_S = \psi\,.$$

The reader may ask why, if the membrane can move as a rigid body, it cannot rotate freely in the manner of an ordinary free rigid body. The answer is that in this model, unlike the other models for elastic bodies in this book, rotation of the membrane as a rigid body alters the membrane strain energy. In the membrane model, the deflection $u(x, y)$ is the deflection that is imposed on the prestressed state of the membrane. The membrane equilibrium problem is an example of a problem for a prestressed body.

There is another way to formulate the equilibrium problem for a free membrane, based on a different method of introducing the energy space. Let us return to Poincaré's inequality (2.3.9). Denote

$$\|u\|_1 = \left\{ \left(\iint_\Omega u\,dx\,dy \right)^2 + D(u) \right\}^{1/2}$$

where

$$D(u) = \iint_\Omega \left[\left(\frac{\partial u}{\partial x} \right)^2 + \left(\frac{\partial u}{\partial y} \right)^2 \right] dx\,dy\,,$$

and recall that the norm in $W^{1,2}(\Omega)$ is

$$\|u\|_{W^{1,2}(\Omega)} = \left\{ \left(\iint_\Omega u^2\,dx\,dy \right)^2 + D(u) \right\}^{1/2}.$$

By (2.3.9), the norms $\|\cdot\|_1$ and $\|\cdot\|_{W^{1,2}(\Omega)}$ are equivalent on the set of continuously differentiable functions, and hence on $W^{1,2}(\Omega)$. Let us use the norm $\|\cdot\|_1$ on $W^{1,2}(\Omega)$. Now the space $W^{1,2}(\Omega)$ is the completion, with respect to the norm $\|\cdot\|_1$, of the functions continuously differentiable on $\overline{\Omega}$. Let us take an element $U(x, y) \in W^{1,2}(\Omega)$ and select from $U(x, y)$ an arbitrary representative Cauchy sequence $\{u_k(x, y)\}$. A smooth function $u_k(x, y)$ can be uniquely written in the form

$$u_k(x, y) = \tilde{u}_k(x, y) + a_k \quad \text{where} \quad \iint_\Omega \tilde{u}_k\,dx\,dy = 0\,.$$

Since

$$\|u_k - u_m\|_1^2 = \left(\iint_\Omega (a_k - a_m)\,dx\,dy \right)^2 + D(\tilde{u}_k - \tilde{u}_m) \to 0 \text{ as } k, m \to \infty\,,$$

we see that $\{a_k\}$ is a numerical Cauchy sequence that has a limit corresponding to U. This means that $\{a_k\}$ belongs to c, the space of numerical sequences each of which has a limit. It is easy to show that this limit does not depend on the choice of representative sequence $\{u_k(x, y)\}$, and we can regard it as a rigid displacement of the membrane. Now, because the membrane is not geometrically fixed and can be moved through any uniform displacement with no change in energy, we place all the elements of $W^{1,2}(\Omega)$ that characterize a strained state of the membrane into the same class such that for any two elements $U'(x, y)$ and $U''(x, y)$ of the class and any representative sequences $\{u_k'(x, y)\}$ and $\{u_k''(x, y)\}$ taken from them, $\{\tilde{u}_k'(x, y)\}$ and $\{\tilde{u}_k''(x, y)\}$ are equivalent in the norm (i.e., $D(\tilde{u}_k' - \tilde{u}_k'') \to 0$) and the difference sequence $\{a_k' - a_k''\}$ is in c. This construction of the classes is equivalent to the construction of the factor space $W^{1,2}(\Omega)/c$, which can also be called the energy space for the free membrane. The zero of this energy factor space is the set of the elements of $W^{1,2}(\Omega)$ each containing as a representative an element from c. Clearly if we wish to have the energy functional defined in this space, the necessary condition for its continuity is that its linear part — the work of external forces — must be zero over any constant displacement. The latter involves the self-balance condition (2.3.12). It is seen that the principal part of any class-element of the new energy space, defined by the sequences $\{\tilde{u}_k(x, y)\}$, coincides with the corresponding sequences for the elements of the space E_{MF}; moreover, this correspondence between the two energy spaces maintains equal norms in both spaces. So we can repeat the procedure to establish the existence-uniqueness theorem in the new energy space, using the proof in E_{MF}.

The restriction (2.3.12) (or (2.3.14)) is necessary for the functional of external forces to be uniquely defined for an element $U_*(x, y)$. We shall use the same notation E_{MF} for this type of energy space since there is a one-to-one correspondence, preserving distances and inner products, between the two types of energy space for the free membrane. Moreover, we shall always make clear which version we mean.

Those familiar with the theory of the Neumann problem for Laplace's equation should note that the necessary condition for solvability that arises in mathematical physics as a mathematical consequence, i.e.,

$$\int_S \psi \, ds = 0 \,,$$

is a particular case of (2.3.14) when $F(x, y) = 0$. This means that for solvability of the problem, the external forces acting on the membrane edge should be self-balanced.

Finally, we note that Poisson's equation governs not only membranes, but also situations in electricity, magnetism, hydrodynamics, mathematical biology, and other fields. So we can consider spaces such as E_M in various other sciences. It is clear that the results will be the same.

We will proceed to introduce other energy spaces in a similar manner: they will be completions of corresponding metric (inner product) spaces consisting of smooth functions satisfying certain boundary conditions. The problem is to determine properties of the elements of those completions. As a rule, metrics must contain all the strain energy terms (we now discuss only linear systems). For example, we can consider a membrane whose edge is elastically supported; then we must include the energy of elastic support in the expression for the energy metric.

Bending a Plate. Here we begin with the work of internal forces on variations of displacements

$$-(w_1, w_2) = -\iint_\Omega D^{\alpha\beta\gamma\delta} \rho_{\gamma\delta}(w_1)\, \rho_{\alpha\beta}(w_2)\, dx\, dy \qquad (2.3.15)$$

where $w_1(x, y)$ is the normal displacement of the plate midsurface Ω, $w_2(x, y)$ can be considered as its variation, $\rho_{\alpha\beta}(u)$ are components of the change-of-curvature tensor,

$$\rho_{11}(u) = \frac{\partial^2 u}{\partial x^2}\,, \qquad \rho_{12} = \frac{\partial^2 u}{\partial x \partial y}\,, \qquad \rho_{22} = \frac{\partial^2 u}{\partial y^2}\,,$$

$D^{\alpha\beta\gamma\delta}$ are components of the tensor of elastic constants of the plate such that

$$D^{\alpha\beta\gamma\delta} = D^{\gamma\delta\alpha\beta} = D^{\beta\alpha\gamma\delta} \tag{2.3.16}$$

and, for any symmetric tensor $\rho_{\alpha\beta}$ there exists a constant $m_0 > 0$ such that

$$D^{\alpha\beta\gamma\delta}\rho_{\gamma\delta}\,\rho_{\alpha\beta} \geq m_0 \sum_{\alpha,\beta=1}^{2} \rho_{\alpha\beta}^2 \,. \tag{2.3.17}$$

We suppose the $D^{\alpha\beta\gamma\delta}$ are constants, but piecewise continuity of these parameters would be sufficient.

For the theory of shells and plates, Greek indices will assume values from the set $\{1, 2\}$ while Latin indices will assume values from the set $\{1, 2, 3\}$. The repeated index convention for summation is also in force. For example, we have

$$a^{\alpha\beta}b_{\alpha\beta} \equiv \sum_{\alpha,\beta=1}^{2} a^{\alpha\beta}b_{\alpha\beta} \,.$$

We first consider a plate with clamped edge $\partial\Omega$:

$$w\big|_{\partial\Omega} = \frac{\partial w}{\partial n}\bigg|_{\partial\Omega} = 0 \,. \tag{2.3.18}$$

(Of course, the variation of w must satisfy (2.3.18) as well.) Let us show that on S_4, the subset of $C^{(4)}(\Omega)$ consisting of those functions which satisfy (2.3.18), the form (w_1, w_2) given in (2.3.15) is an inner product. We begin with the axiom P1:

$$\begin{aligned}(w, w) &= \iint_\Omega D^{\alpha\beta\gamma\delta}\rho_{\alpha\beta}(w)\,\rho_{\gamma\delta}(w)\,dx\,dy \geq m_0 \iint_\Omega \sum_{\alpha,\beta=1}^{2} \rho_{\alpha\beta}^2(w)\,dx\,dy \\ &= m_0 \iint_\Omega \left[\left(\frac{\partial^2 w}{\partial x^2}\right)^2 + 2\left(\frac{\partial^2 w}{\partial x \partial y}\right)^2 + \left(\frac{\partial^2 w}{\partial y^2}\right)^2\right] dx\,dy \geq 0 \,.\end{aligned}$$

If $w = 0$ then $(w, w) = 0$. If $(w, w) = 0$ then, on Ω,

$$\frac{\partial^2 w}{\partial x^2} = 0\,, \qquad \frac{\partial^2 w}{\partial x \partial y} = 0\,, \qquad \frac{\partial^2 w}{\partial y^2} = 0\,.$$

It follows that

$$w(x, y) = a_1 + a_2 x + a_3 y\,,$$

where the a_i are constants. By (2.3.18) then, $w(x, y) = 0$. Hence P1 is satisfied. Satisfaction of P2 follows from (2.3.16), and it is evident that P3 is also satisfied.

Thus S_4 with inner product (2.3.15) is an inner product space; its completion in the corresponding metric is the energy space E_{PC} for a clamped plate.

Let us consider some properties of the elements of E_{PC}. It was shown that

$$m_0 \iint_\Omega \left[\left(\frac{\partial^2 w}{\partial x^2} \right)^2 + 2 \left(\frac{\partial^2 w}{\partial x \partial y} \right)^2 + \left(\frac{\partial^2 w}{\partial y^2} \right)^2 \right]^2 dx\, dy$$
$$\leq \iint_\Omega D^{\alpha\beta\gamma\delta} \rho_{\gamma\delta}(w)\, \rho_{\alpha\beta}(w)\, dx\, dy \equiv (w, w)\,. \qquad (2.3.19)$$

From this and the Friedrichs inequality, written first for w and then for the first derivatives of $w \in S_4$ as well, we get

$$\iint_\Omega w^2\, dx\, dy \leq m_1 \iint_\Omega \left[\left(\frac{\partial w}{\partial x} \right)^2 + \left(\frac{\partial w}{\partial y} \right)^2 \right] dx\, dy$$
$$\leq m_2 \iint_\Omega \left[\left(\frac{\partial^2 w}{\partial x^2} \right)^2 + 2 \left(\frac{\partial^2 w}{\partial x \partial y} \right)^2 + \left(\frac{\partial^2 w}{\partial y^2} \right)^2 \right] dx\, dy$$
$$\leq m_3 \iint_\Omega D^{\alpha\beta\gamma\delta} \rho_{\gamma\delta}(w)\, \rho_{\alpha\beta}(w)\, dx\, dy \equiv m_3\, (w, w)\,. \qquad (2.3.20)$$

Hence if $\{w_n\} \subset S_4$ is a Cauchy sequence in E_{PC}, then the sequences

$$\{w_n\}, \quad \left\{ \frac{\partial w_n}{\partial x} \right\}, \quad \left\{ \frac{\partial w_n}{\partial y} \right\}, \quad \left\{ \frac{\partial^2 w_n}{\partial x^2} \right\}, \quad \left\{ \frac{\partial^2 w_n}{\partial x \partial y} \right\}, \quad \left\{ \frac{\partial^2 w_n}{\partial y^2} \right\},$$

are Cauchy sequences in $L^2(\Omega)$. So we can say that an element W of the completion E_{PC} is such that $W(x, y)$ and all its derivatives up to order two are in $L^2(\Omega)$.

We now investigate $W(x, y)$ further. Let $w \in S_4$ and $w(x, y) \equiv 0$ outside Ω. Suppose Ω lies in the domain $\{(x, y)\colon\ x > 0, y > 0\}$. Then the representation

$$w(x, y) = \int_0^x \int_0^y \frac{\partial^2 w(s, t)}{\partial s \partial t}\, ds\, dt$$

holds. Using Hölder's inequality and (2.3.20), we get

$$|w(x, y)| \leq \int_0^x \int_0^y \left| \frac{\partial^2 w(s, t)}{\partial s \partial t} \right| ds\, dt \leq \iint_\Omega 1 \cdot \left| \frac{\partial^2 w(s, t)}{\partial s \partial t} \right| ds\, dt$$
$$\leq (\operatorname{mes} \Omega)^{1/2} \left(\iint_\Omega \left| \frac{\partial^2 w(s, t)}{\partial s \partial t} \right|^2 ds\, dt \right)^{1/2} \leq m_4\, (w, w)^{1/2}. \qquad (2.3.21)$$

This means that if $\{w_n\} \subset S_4$ is a Cauchy sequence in the metric of E_{PC}, then it converges uniformly on Ω. Hence there exists a limit function

$$w_0(x, y) = \lim_{n \to \infty} w_n(x, y)$$

which is continuous on Ω; this function is identified, as above, with the corresponding element of E_{PC} and we shall say that E_{PC} is continuously imbedded into $C(\Omega)$.

The functional describing the work of external forces

$$A = \iint_\Omega F(x,y)\, W(x,y)\, dx\, dy$$

now makes sense if $F(x,y) \in L(\Omega)$; moreover, it can contain the work of point forces

$$\sum_k F(x_k, y_k)\, w_0(x_k, y_k)$$

and line forces

$$\int_\gamma F(x,y)\, w_0(x,y)\, ds$$

where γ is a line in Ω and $w_0(x,y)$ is the corresponding limit function for $W(x,y)$.

Remark 2.3.4. Modern books on partial differential equations often require that $F(x,y) \in H^{-2}(\Omega)$. This is a complete characterization of external forces — however, it is difficult for an engineer to verify this property. □

Now let us consider a plate with free edge. In this case, we also wish to use the inner product (2.3.15) to create an energy space. As in the case of a membrane with free edge, the axiom P1 is not fulfilled: we saw that from $(w,w) = 0$ it follows that

$$w(x,y) = a_1 + a_2 x + a_3 y\,. \tag{2.3.22}$$

This admissible motion of the plate as a rigid whole is called a rigid motion, but still differs from real "rigid" motions of the plate as a three-dimensional body.

Poincaré's inequality (2.3.9) implies that the zero element of the corresponding completion consists of functions of the form (2.3.22). Indeed, taking $w(x,y) \in C^{(4)}(\Omega)$ we write down Poincaré's inequality for $\partial w/\partial x$:

$$\iint_\Omega \left(\frac{\partial w}{\partial x}\right)^2 dx\, dy \le m \left\{ \left(\iint_\Omega \frac{\partial w}{\partial x}\, dx\, dy \right)^2 + \iint_\Omega \left[\left(\frac{\partial^2 w}{\partial x^2}\right)^2 + \left(\frac{\partial^2 w}{\partial x \partial y}\right)^2 \right] dx\, dy \right\},$$

and then the same inequality for $\partial w/\partial y$ with the roles of x and y interchanged. From these and (2.3.9) we get

$$\begin{aligned}
\iint_\Omega \left[w^2 + \left(\frac{\partial w}{\partial x}\right)^2 + \left(\frac{\partial w}{\partial y}\right)^2 \right] dx\, dy \le m_1 \Bigg\{ & \left(\iint_\Omega w\, dx\, dy \right)^2 \\
& + \left(\iint_\Omega \frac{\partial w}{\partial x}\, dx\, dy \right)^2 + \left(\iint_\Omega \frac{\partial w}{\partial y}\, dx\, dy \right)^2 \\
& + \iint_\Omega \left[\left(\frac{\partial^2 w}{\partial x^2}\right)^2 + 2\left(\frac{\partial^2 w}{\partial x \partial y}\right)^2 + \left(\frac{\partial^2 w}{\partial y^2}\right)^2 \right] dx\, dy \Bigg\}.
\end{aligned}$$

From (2.3.19) it follows that

$$\iint_{\Omega}\left[w^2+\left(\frac{\partial w}{\partial x}\right)^2+\left(\frac{\partial w}{\partial y}\right)^2\right]dx\,dy \le m_2\left\{\left(\iint_{\Omega} w\,dx\,dy\right)^2\right.$$
$$+\left(\iint_{\Omega}\frac{\partial w}{\partial x}\,dx\,dy\right)^2+\left(\iint_{\Omega}\frac{\partial w}{\partial y}\,dx\,dy\right)^2$$
$$\left.+\iint_{\Omega} D^{\alpha\beta\gamma\delta}\rho_{\gamma\delta}(w)\,\rho_{\alpha\beta}(w)\,dx\,dy\right\}. \tag{2.3.23}$$

For any function $w(x,y) \in C^{(4)}(\Omega)$, we can take suitable constants a_i and find a function $w_b(x,y)$ of the form

$$w_b(x,y) = w(x,y) + a_1 + a_2 x + a_3 y \tag{2.3.24}$$

such that

$$\iint_{\Omega} w_b\,dx\,dy = 0\,, \quad \iint_{\Omega}\frac{\partial w_b}{\partial x}\,dx\,dy = 0\,, \quad \iint_{\Omega}\frac{\partial w_b}{\partial y}\,dx\,dy = 0\,. \tag{2.3.25}$$

As for the membrane with free edge, we can now consider a subset S_{4b} of $C^{(4)}(\Omega)$ consisting of balanced functions satisfying (2.3.25). We construct an energy space E_{PF} for a plate with free edge as the completion of S_{4b} in the metric induced by the inner product (2.3.15).

From (2.3.25), (2.3.23), and (2.3.19), we see that an element $W(x,y) \in E_{PF}$ is such that $W(x,y)$ and all its "derivatives" up to order two are in $L^2(\Omega)$. We could show the existence of a limit function

$$w_0(x,y) = \lim_{n\to\infty} w_n \in C(\Omega)$$

for any Cauchy sequence $\{w_n\}$, but in this case the technique is more complicated and, in what follows, we have this result as a particular case of the Sobolev imbedding theorem.

Note that (2.3.25) can be replaced by

$$\iint_{\Omega} w(x,y)\,dx\,dy = 0\,, \quad \iint_{\Omega} x\,w(x,y)\,dx\,dy = 0\,, \quad \iint_{\Omega} y\,w(x,y)\,dx\,dy = 0\,,$$

since these also uniquely determine the a_i for a class of functions of the form (2.3.24). (This possibility follows from Sobolev's general result [33] on equivalent norms in Sobolev spaces.)

The system (2.3.25) represents constraints that are absent in nature. For a static problem, there must be a certain invariance of some objects under transformations of the form (2.3.24) with arbitrary constants a_k. In particular, the work of external forces should not depend on the a_k if the problem is stated correctly. This leads to the necessary conditions

$$\iint_\Omega F(x,y)\,dx\,dy = 0\,, \quad \iint_\Omega x\,F(x,y)\,dx\,dy = 0\,, \quad \iint_\Omega y\,F(x,y)\,dx\,dy = 0\,. \tag{2.3.26}$$

The mechanical sense of (2.3.26) is clear: the resultant force and moments must vanish. This is the condition for a self-balanced force system.

Problem 2.3.2. What is the form of (2.3.26) if the external forces contain point and line forces? □

An energy space for a free plate, as for the membrane with free edge, can be introduced in another way: namely, we begin with an element of $W^{2,2}(\Omega)$, selecting a representative sequence $\{\tilde{w}_k(x,y)+a_k+b_kx+c_ky\}$, where $\tilde{w}$ satisfies (2.3.25) and a_k, b_k, c_k are constants. It is easy to show that the numerical sequences $\{a_k\}$, $\{b_k\}$, $\{c_k\}$ are Cauchy and therefore belong to the space c. Next we combine into a class-element all the elements of $W^{2,2}(\Omega)$ whose differences are a linear polynomial $a + bx + cy$, and state that the energy space is the factor space of $W^{2,2}(\Omega)$ by the space which is the completion of the space of linear polynomials with respect to the norm

$$\|a + bx + cy\| = (a^2 + b^2 + c^2)^{1/2}.$$

In the factor space, the zero is the class of all Cauchy sequences whose differences are equivalent (in the norm of $W^{2,2}(\Omega)$) to a sequence $\{a_k + b_kx + c_ky\}$ with $\{a_k\}, \{b_k\}, \{c_k\} \in c$. The norm of W, an element in the factor space, is

$$\left(\iint_\Omega D^{\alpha\beta\gamma\delta}\rho_{\gamma\delta}(W)\,\rho_{\alpha\beta}(W)\,dx\,dy\right)^{1/2}.$$

The elements of the factor space are uniquely identified with the elements of the energy space E_{PF} whose elements satisfy (2.3.25). Moreover, the identification is isometric so, as for the membrane, we can use this factor space as the energy space and repeat the proof of the existence-uniqueness theorem in terms of the elements of the energy factor space. The self-balance condition for the external forces is a necessary condition to have the functional of the work of external forces be meaningful. The reader may also consider mixed boundary conditions: how must the treatment be modified if the plate is clamped only along a segment $AB \subset \Omega$ so that

$$w(x,y)\Big|_{AB} = 0\,,$$

with the rest of the boundary free of geometrical constraints?

Linear Elasticity. We return to the problem of linear elasticity, considered in Sect. 1.5. Let us introduce a functional describing the work of internal forces on variations $\mathbf{v}(\mathbf{x})$ of the displacement field $\mathbf{u}(\mathbf{x})$:

$$-(\mathbf{u},\mathbf{v}) = -\iiint_\Omega c^{ijkl}\epsilon_{kl}(\mathbf{u})\,\epsilon_{ij}(\mathbf{v})\,d\Omega\,. \tag{2.3.27}$$

For the notation, see (1.5.7)–(1.5.9). We recall that the strain energy $\mathcal{E}_4(\mathbf{u})$ of an elastic body occupying volume Ω is related to the introduced inner product as follows:

$$(\mathbf{u}, \mathbf{u}) = \|\mathbf{u}\|^2 = 2\mathcal{E}_4(\mathbf{u}) .$$

The elastic moduli c^{ijkl} may be piecewise continuous functions satisfying (1.5.8) and (1.5.9), which guarantee that all inner product axioms are satisfied by $(\mathbf{u}, \mathbf{v})$ for vector functions $\mathbf{u}, \mathbf{v}$ continuously differentiable on Ω, except P1: from $(\mathbf{u}, \mathbf{u}) = 0$ it follows that $\mathbf{u} = \mathbf{a} + \mathbf{b} \times \mathbf{x}$. Note that $(\mathbf{u}, \mathbf{v})$ is consistent with the metric (1.5.10).

Let us consider boundary conditions prescribed by

$$\mathbf{u}(\mathbf{x})\big|_{\partial\Omega} = \mathbf{0} . \tag{2.3.28}$$

If we use the form (2.3.27) on the set S_3 of vector-functions $\mathbf{u}(\mathbf{x})$ satisfying (2.3.28) and such that each of their components is of class $C^{(2)}(\Omega)$, then $(\mathbf{u}, \mathbf{v})$ is an inner product and S_3 with this inner product becomes an inner product space. Its completion E_{EC} in the corresponding metric (or norm) is the energy space of an elastic body with clamped boundary. To describe the properties of the elements of E_{EC}, we establish *Korn's inequality*.

Lemma 2.3.1 (Korn). For a vector function $\mathbf{u}(\mathbf{x}) \in S_3$, we have

$$\iiint_\Omega \left[|\mathbf{u}|^2 + \sum_{i,j=1}^{3} \left(\frac{\partial u_i}{\partial x_j} \right)^2 \right] d\Omega \le m \iiint_\Omega c^{ijkl} \epsilon_{kl}(\mathbf{u})\, \epsilon_{ij}(\mathbf{u})\, d\Omega$$

for some constant m which does not depend on $\mathbf{u}(\mathbf{x})$.

Proof. By (1.5.9) and Friedrichs' inequality, it is sufficient to show that

$$\iiint_\Omega \sum_{i,j=1}^{3} \left(\frac{\partial u_i}{\partial x_j} \right)^2 d\Omega \le m_1 \iiint_\Omega \sum_{\substack{i,j=1 \\ i\le j}}^{3} \epsilon_{ij}^2(\mathbf{u})\, d\Omega .$$

Consider the term on the right:

$$\begin{aligned} A &\equiv \iiint_\Omega \sum_{\substack{i,j=1 \\ i\le j}}^{3} \epsilon_{ij}^2(\mathbf{u})\, d\Omega = \frac{1}{4} \iiint_\Omega \sum_{\substack{i,j=1 \\ i\le j}}^{3} \left(\frac{\partial u_i}{\partial x_j} + \frac{\partial u_j}{\partial x_i} \right)^2 d\Omega \\ &= \iiint_\Omega \left\{ \sum_{i=1}^{3} \left(\frac{\partial u_i}{\partial x_i} \right)^2 + \frac{1}{4} \sum_{\substack{i,j=1 \\ i<j}}^{3} \left[\left(\frac{\partial u_i}{\partial x_j} \right)^2 + \left(\frac{\partial u_j}{\partial x_i} \right)^2 + 2 \frac{\partial u_i}{\partial x_j} \frac{\partial u_j}{\partial x_i} \right] \right\} d\Omega . \end{aligned}$$

Integrating by parts (twice) the term

$$B \equiv \frac{1}{2} \iiint_\Omega \sum_{\substack{i,j=1 \\ i<j}}^{3} \frac{\partial u_i}{\partial x_j} \frac{\partial u_j}{\partial x_i}\, d\Omega = \frac{1}{2} \iiint_\Omega \sum_{\substack{i,j=1 \\ i<j}}^{3} \frac{\partial u_i}{\partial x_i} \frac{\partial u_j}{\partial x_j}\, d\Omega$$

and using the elementary inequality $|ab| \le (a^2 + b^2)/2$, we get

$$|B| \le \frac{1}{4} \iiint_\Omega \sum_{\substack{i,j=1 \\ i<j}}^{3} \left[\left(\frac{\partial u_i}{\partial x_i} \right)^2 + \left(\frac{\partial u_j}{\partial x_j} \right)^2 \right] d\Omega = \frac{1}{2} \iiint_\Omega \sum_{i=1}^{3} \left(\frac{\partial u_i}{\partial x_i} \right)^2 d\Omega \, .$$

Therefore

$$A \ge \frac{1}{4} \iiint_\Omega \sum_{i,j=1}^{3} \left(\frac{\partial u_i}{\partial x_j} \right)^2 d\Omega \, ,$$

which completes the proof. □

By Korn's inequality, each component of an element $U \in E_{EC}$ belongs to E_{MC}, i.e., the u_i and their first derivatives belong to $L^2(\Omega)$.

Note that the construction of an energy space is the same if the boundary condition (2.3.28) is given only on some part $\partial\Omega_1$ of the boundary of Ω:

$$\mathbf{u}(\mathbf{x})\Big|_{\partial\Omega_1} = \mathbf{0} \, .$$

Korn's inequality also holds, but its proof is more complicated (see, for example, [26, 11]).

If we consider an elastic body with free boundary, we encounter issues similar to those for a membrane or plate with free edge: we must circumvent the difficulty with the zero element of the energy space. The restrictions

$$\iiint_\Omega \mathbf{u} \, d\Omega = \mathbf{0} \, , \qquad \iiint_\Omega \mathbf{x} \times \mathbf{u}(\mathbf{x}) \, d\Omega = \mathbf{0} \, , \tag{2.3.29}$$

provide that the rigid motion $\mathbf{u} = \mathbf{a} + \mathbf{b} \times \mathbf{x}$ becomes zero, and that Korn's inequality remains valid for smooth vector functions satisfying (2.3.29). So by completion, we get an energy space E_{EF} with known properties: all Cartesian components of vectors pertain to the space $W^{1,2}(\Omega)$.

As for a free membrane, we can also organize an energy space of classes — a factor space — in which the zero element is the set of all elements whose differences between any representative sequences in the norm of $(W^{1,2}(\Omega))^3$ are equivalent to a sequence of the form $\{\mathbf{a}_k + \mathbf{b}_k \times \mathbf{x}\}$ such that the Cartesian components of the vectors $\{\mathbf{a}_k\}$ and $\{\mathbf{b}_k\}$ constitute some elements of the space c.

2.4 Generalized Solutions in Mechanics

We now discuss how to introduce generalized solutions in mechanics. We begin with Poisson's equation

$$-\Delta u(x, y) = F(x, y) \, , \qquad (x, y) \in \Omega \, , \tag{2.4.1}$$

where Ω is a bounded open domain in $\mathbb{R}^2$. The Dirichlet problem consists of this equation supplemented by the boundary condition

$$u\big|_{\partial\Omega} = 0 . \tag{2.4.2}$$

Let $u(x, y)$ be its classical solution; i.e., let $u \in C^{(2)}(\overline{\Omega})$ satisfy (2.4.1) and (2.4.2). Let $\varphi(x, y)$ be a function with compact support in Ω. Again, this means that $\varphi \in C^{(\infty)}(\overline{\Omega})$ and the closure of the set $M = \{(x, y) \in \Omega : \varphi(x, y) \neq 0\}$ lies in Ω.

Multiplying both sides of (2.4.1) by $\varphi(x, y)$ and integrating over Ω, we get

$$-\iint_\Omega \varphi(x, y)\,\Delta u(x, y)\,dx\,dy = \iint_\Omega F(x, y)\,\varphi(x, y)\,dx\,dy . \tag{2.4.3}$$

If this equality holds for every infinitely differentiable function $\varphi(x, y)$ with compact support in Ω, and if $u \in C^{(2)}(\overline{\Omega})$ and satisfies (2.4.2), then, as is well known from the classical calculus of variations, $u(x, y)$ is the unique classical solution to the Dirichlet problem.

But using (2.4.3), we can pose this Dirichlet problem directly without using the differential equation (2.4.1); namely, $u(x, y)$ is a solution to the Dirichlet problem if, obeying (2.4.2), it satisfies (2.4.3) for every $\varphi(x, y)$ that is infinitely differentiable with compact support in Ω. If $F(x, y)$ belongs to $L^p(\Omega)$ then we can take, as it seems, $u(x, y)$ having second derivatives in the space $L^p(\Omega)$; such a $u(x, y)$ is not a classical solution, and it is natural to call it a generalized solution.

We can go further by applying integration by parts to the left-hand side of (2.4.3) as follows:

$$\iint_\Omega \left(\frac{\partial u}{\partial x}\frac{\partial \varphi}{\partial x} + \frac{\partial u}{\partial y}\frac{\partial \varphi}{\partial y} \right) dx\,dy = \iint_\Omega F(x, y)\,\varphi(x, y)\,dx\,dy . \tag{2.4.4}$$

In such a case we may impose weaker restrictions on a solution $u(x, y)$ and call it the generalized solution if it belongs to E_{MC}, the energy space for a clamped membrane. Equation (2.4.4) defines this solution if it holds for every $\varphi(x, y)$ that has a compact support in Ω. Note the disparity in requirements on $u(x, y)$ and $\varphi(x, y)$.

Further integration by parts on the left-hand side of (2.4.4) yields

$$-\iint_\Omega u(x, y)\,\Delta\varphi(x, y)\,dx\,dy = \iint_\Omega F(x, y)\,\varphi(x, y)\,dx\,dy . \tag{2.4.5}$$

Now we can formally consider solutions from the space $L(\Omega)$ and this is a new class of generalized solutions.

This way leads to the theory of distributions originated by Schwartz [32]. He extended the notion of generalized solution to a class of linear continuous functionals, or *distributions*, defined on the set $\mathcal{D}(\Omega)$ of all functions infinitely differentiable in Ω and with compact support in Ω. For this it is necessary to introduce the convergence and other structures of continuity in $\mathcal{D}(\Omega)$. Unfortunately $\mathcal{D}(\Omega)$ is not a normed space (see, for example, Yosida [44] — it is a locally convex topological space) and its presentation would exceed our scope. This theory justifies, in particular, the use of the so-called δ-function, which was introduced in quantum mechanics via the equality

$$\int_{-\infty}^{\infty} \delta(x-a) f(x)\, dx = f(a) \,, \tag{2.4.6}$$

valid for every continuous function $f(x)$. Physicists considered $\delta(x)$ to be a function vanishing everywhere except at $x = 0$, where its value is infinity. Any known theory of integration gave zero for the value of the integral on the left-hand side of (2.4.6), and the theory of distributions explained how to understand such strange functions. It is interesting to note that the δ-function was well known in classical mechanics, too; if we consider $\delta(x - a)$ as a unit point force applied at $x = a$, then the integral on the left-hand side of (2.4.6) is the work of this force on the displacement $f(a)$, which is indeed $f(a)$.

So we have several generalized statements of the Dirichlet problem, but which one is most natural from the viewpoint of mechanics?

From mechanics it is known that a solution to the problem is a minimizer of the total potential energy functional

$$I(u) = \iint_{\Omega} \left[\left(\frac{\partial u}{\partial x}\right)^2 + \left(\frac{\partial u}{\partial y}\right)^2 \right] dx\, dy - 2 \iint_{\Omega} F u\, dx\, dy \,. \tag{2.4.7}$$

According to the calculus of variations, a minimizer of $I(u)$ on the subset of $C^{(2)}(\overline{\Omega})$ consisting of all functions satisfying (2.4.2), if it exists, is a classical solution to the Dirichlet problem. But we also can consider $I(u)$ on the energy space E_{MC} if $F(x, y) \in L^p(\Omega)$ for $p > 1$. Indeed, the first term in $I(u)$ is well-defined in E_{MC} and can be written in the form $\|u\|^2$; the second,

$$\Phi(u) = - \iint_{\Omega} F(x, y)\, u(x, y)\, dx\, dy \,,$$

is a linear functional with respect to $u(x, y)$. It is also continuous in E_{MC}; by virtue of Hölder's inequality with exponents p and $q = p/(p - 1)$, we have

$$\left| \iint_{\Omega} F u\, dx\, dy \right| \le \left(\iint_{\Omega} |F|^p\, dx\, dy \right)^{1/p} \left(\iint_{\Omega} |u|^q\, dx\, dy \right)^{1/q}$$
$$\le m_1 \|F\|_{L^p(\Omega)} \|u\|_{W^{1,2}(\Omega)} \le m_2 \|u\|_{E_{MC}} \,.$$

To show this, we have used the imbedding Theorem 2.1.2 and the Friedrichs inequality. By Theorem 1.21.1, $\Phi(u)$ is continuous in E_{MC}, and therefore so is $I(u)$.

Thus $I(u)$ is of the form

$$I(u) = \|u\|^2 + 2\Phi(u) \,. \tag{2.4.8}$$

Let $u_0 \in E_{MC}$ be a minimizer of $I(u)$, i.e.,

$$I(u_0) \le I(u) \quad \text{for all } u \in E_{MC} \,. \tag{2.4.9}$$

We try a method from the classical calculus of variations. Take $u = u_0 + \epsilon v$ where v is an arbitrary element of E_{MC}. Then

$$\begin{aligned} I(u) &= I(u_0 + \epsilon v) = \|u_0 + \epsilon v\|^2 + 2\Phi(u_0 + \epsilon v) \\ &= (u_0 + \epsilon v, u_0 + \epsilon v) + 2\Phi(u_0 + \epsilon v) \\ &= \|u_0\|^2 + 2\epsilon(u_0, v) + \epsilon^2 \|v\|^2 + 2\Phi(u_0) + 2\epsilon\Phi(v) \\ &= \|u_0\|^2 + 2\Phi(u_0) + 2\epsilon[(u_0, v) + \Phi(v)] + \epsilon^2 \|v\|^2 . \end{aligned}$$

From (2.4.9), we get

$$2\epsilon[(u_0, v) + \Phi(v)] + \epsilon^2 \|v\|^2 \geq 0 .$$

Since ϵ is an arbitrary real number (in particular it can take either sign, and the first term, if nonzero, dominates for ϵ small in magnitude), it follows that

$$(u_0, v) + \Phi(v) = 0 . \qquad (2.4.10)$$

In other words,

$$\iint_\Omega \left(\frac{\partial u_0}{\partial x} \frac{\partial v}{\partial x} + \frac{\partial u_0}{\partial y} \frac{\partial v}{\partial y} \right) dx\, dy - \iint_\Omega F(x, y)\, v(x, y)\, dx\, dy = 0 . \qquad (2.4.11)$$

This equality holds for every $v \in E_{MC}$ and defines the minimizer $u_0 \in E_{MC}$. Note that (2.4.11) has the same form as (2.4.4).

So we have introduced a notion of generalized solution that has an explicit mechanical background.

Definition 2.4.1. An element $u \in E_{MC}$ is called the *generalized solution* to the Dirichlet problem if u satisfies (2.4.11) for any $v \in E_{MC}$.

We can also obtain (2.4.11) from the principle of virtual displacements (work). This asserts that in the state of equilibrium, the sum of the work of internal forces (which is now the variation of the strain energy with a negative sign) and the work of external forces is zero on all virtual (admissible) displacements.

In the case under consideration, both approaches to introducing generalized energy solutions are equivalent. In general, however, this is not so, and the virtual work principle has wider applicability. If $F(x, y)$ is a nonconservative load depending on $u(x, y)$, we cannot use the principle of minimum total energy; however, (2.4.11) remains valid since it has the mathematical form of the virtual work principle. In what follows, we often use this principle to pose problems in equation form.

Since the part of the presentation from (2.4.8) up to (2.4.10) is general and does not depend on the specific form (2.4.11) of the functional $I(u)$, we can formulate

Theorem 2.4.1. Let u_0 be a minimizer of a functional $I(u) = \|u\|^2 + 2\Phi(u)$ given in an inner product (Hilbert) space H, where the functional $\Phi(u)$ is linear and continuous. Then u_0 satisfies (2.4.10) for every $v \in H$.

Equation (2.4.10) is a necessary condition for minimization of the functional $I(u)$, analogous to the condition that the first derivative of an ordinary function must vanish at a point of minimum.

We can obtain (2.4.10) formally by evaluating

$$\frac{d}{d\epsilon} I(u_0 + \epsilon v)\Big|_{\epsilon=0} = 0 . \tag{2.4.12}$$

This is valid for the following reason. Given u_0 and v, the functional $I(u_0 + \epsilon v)$ is an ordinary function of the numerical variable ϵ, and assumes a minimum value at $\epsilon = 0$. The left-hand side of (2.4.12) can be interpreted as a partial derivative at $u = u_0$ in the direction v, and is called the *Gâteaux derivative* of $I(u)$ at $u = u_0$ in the direction of v. We shall return to this issue later.

The Dirichlet problem for a clamped membrane is a touchstone for all static problems in continuum mechanics. In a similar way we can introduce a natural notion of generalized solution for other problems under consideration. As we said, each of them can be represented as a minimization problem for a total potential energy functional of the form (2.4.10) in an energy space. For example, equation (2.4.11), a particular form of (2.4.10) for a clamped membrane, is the same for a free membrane — we need only replace E_{MC} by E_{MF}. The quantity $\Phi(u)$, to be a continuous linear functional in E_{MF}, must be supplemented with self-balance condition (2.3.12) for the load.

Let us concretize equation (2.4.10) for each of the other problems we have under consideration.

Plate. The definition of generalized solution $w_0 \in E_P$ is given by the equation

$$\iint_\Omega D^{\alpha\beta\gamma\delta} \rho_{\gamma\delta}(w_0)\, \rho_{\alpha\beta}(w)\, dx\, dy - \iint_\Omega F(x,y)\, w(x,y)\, dx\, dy$$
$$- \sum_{k=1}^{m} F_k w(x_k, y_k) - \int_\gamma f(s)\, w(x,y)\, ds = 0 \tag{2.4.13}$$

(see the notation of Sect. 2.3) which must hold for every $w \in E_P$. The equation is the same for any kind of homogeneous boundary conditions (i.e., for usual ones) but the energy space will change from one set of boundary conditions to another. If a plate can move as a rigid whole, the requirement that

$$F(x,y) \in L(\Omega)\,, \qquad f(s) \in L(\gamma)\,,$$

for Φ to be a continuous linear functional, must be supplemented with self-balance conditions for the load:

$$\iint_\Omega F(x,y)\, w_i(x,y)\, dx\, dy + \sum_{k=1}^{m} F_k w_i(x_k, y_k) + \int_\gamma f(s)\, w_i(x,y)\, ds = 0 \tag{2.4.14}$$

for $i = 1, 2, 3$, where $w_1(x, y) = 1$, $w_2(x, y) = x$, and $w_3(x, y) = y$. This condition is necessary if we use the space where the set of rigid plate motions is the zero of the space. If from each element of the energy space we select a representative using (2.3.25), then on the energy space of all representatives with norm $\|\cdot\|_P$ we can prove existence and uniqueness of the energy solution without self-balance condition (2.4.14). But for solvability of the initial problem for a free plate, we still should require (2.4.14) to hold as constraints (2.3.25) are absent in the problem statement.

Note that for each concrete problem we must specify the energy space. The same is true for the following problem.

Linear Elasticity. The generalized solution $\mathbf{u}$ is defined by the integro-differential equation

$$\iiint_\Omega c^{ijkl} \epsilon_{kl}(\mathbf{u})\, \epsilon_{ij}(\mathbf{v})\, d\Omega - \iiint_\Omega \mathbf{F}(x, y, z) \cdot \mathbf{v}(x, y, z)\, d\Omega - \iint_\Gamma \mathbf{f}(x, y, z) \cdot \mathbf{v}(x, y, z)\, dS = 0\,, \tag{2.4.15}$$

where $\mathbf{F}$ and $\mathbf{f}$ are forces distributed over Ω and over some surface $\Gamma \subset \Omega$, respectively. If we consider the Dirichlet (or first) problem of elasticity, which is

$$\mathbf{u}(\mathbf{x})\Big|_{\partial\Omega} = \mathbf{0}$$

where $\partial\Omega$ is the boundary of Ω, the solution $\mathbf{u}$ should belong to the space E_{EC} and equation (2.4.15) must hold for every virtual displacement $\mathbf{v} \in E_{EC}$. Note that in this case,

$$\iint_{\partial\Omega} \mathbf{f}(x, y, z) \cdot \mathbf{v}(x, y, z)\, dS = 0\,.$$

The load, thanks to Theorem 2.1.4 and Korn's inequality, is of the class

$$F_i(x, y, z) \in L^{6/5}(\Omega)\,, \quad f_i(x, y, z) \in L^{4/3}(\Gamma) \qquad (i = 1, 2, 3)\,,$$

where $F_i(x, y, z)$ and f_i are Cartesian components of $\mathbf{F}$ and $\mathbf{f}$, respectively, and Γ is a piecewise smooth surface in $\overline{\Omega}$. This provides continuity of $\Phi(w)$.

In the second problem of elasticity there are given forces distributed over the boundary:

$$c^{ijkl} \epsilon_{kl}(\mathbf{x})\, n_j(\mathbf{x})\Big|_{\partial\Omega} = f_i(\mathbf{x})\,,$$

where the n_j are Cartesian components of the unit exterior normal to $\partial\Omega$. This can be written in tensor notation as

$$\boldsymbol{\sigma}(\mathbf{x}) \cdot \mathbf{n}\Big|_{\partial\Omega} = \mathbf{f}(\mathbf{x})$$

where the components of stress tensor $\boldsymbol{\sigma}$ are related to the components of the strain tensor by $\sigma_{ij} = c^{ijkl} \epsilon_{kl}$.

As for equilibrium problems for free membranes and plates, in the case of a free elastic body we must require that the load be self-balanced:

$$\iiint_{\Omega} \mathbf{F}(\mathbf{x})\, d\Omega + \iint_{\Gamma} \mathbf{f}(\mathbf{x})\, dS = \mathbf{0}\,,$$

$$\iiint_{\Omega} \mathbf{x} \times \mathbf{F}(\mathbf{x})\, d\Omega + \iint_{\Gamma} \mathbf{x} \times \mathbf{f}(\mathbf{x})\, dS = \mathbf{0}\,. \tag{2.4.16}$$

The solution is required to be in E_{EF}.

We have argued that it is legitimate to introduce the generalized solution in such a way. Of course, full legitimacy will be assured when we prove that this solution exists and is unique in the corresponding space.

We emphasize once more that the definition of generalized solution arose in a natural way from the variational principle of mechanics.

2.5 Existence of Energy Solutions to Some Mechanics Problems

In Sect. 2.4 we introduced generalized solutions for several mechanics problems and reduced those problems to a solution of the abstract equation

$$(u, v) + \Phi(v) = 0 \tag{2.5.1}$$

in an energy (Hilbert) space. We obtained some restrictions on the forces to provide continuity of the linear functional $\Phi(v)$ in the energy space. The following theorem guarantees solvability of those mechanics problems in a generalized sense.

Theorem 2.5.1. Assume $\Phi(v)$ is a continuous linear functional given on a Hilbert space H. Then there is a unique element $u \in H$ that satisfies (2.5.1) for every $v \in H$.

Proof. By the Riesz representation theorem there is a unique $u_0 \in H$ such that the continuous linear functional $\Phi(v)$ is represented in the form $\Phi(v) = (v, u_0) \equiv (u_0, v)$. Hence (2.5.1) takes the form

$$(u, v) + (u_0, v) = 0\,. \tag{2.5.2}$$

We need to find $u \in H$ that satisfies (2.5.2) for every $v \in H$. Rewriting it in the form

$$(u + u_0, v) = 0\,,$$

we see that its unique solution is $u = -u_0$. □

This theorem answers the question of solvability, in the generalized sense, of the problems treated in Sect. 2.4. To demonstrate this, we rewrite Theorem 2.5.1 in concrete terms for a pair of problems.

Theorem 2.5.2. Assume $F(x, y) \in L(\Omega)$ and $f(x, y) \in L(\gamma)$ where $\Omega \subset \mathbb{R}^2$ is compact and γ is a piecewise smooth curve in Ω. The equilibrium problem for a plate with clamped edge has a unique generalized solution: there is a unique $w_0 \in E_{PC}$ which satisfies (2.4.13) for all $w \in E_{PC}$.

Changes for a plate which is free of clamping are evident: we must add the self-balance condition (2.4.14) for forces and replace the space E_{PC} by E_{PF}.

Theorem 2.5.3. Assume all Cartesian components of the volume forces $\mathbf{F}(x,y,z)$ are in $L^{6/5}(\Omega)$ and those of the surface forces $\mathbf{f}(x,y,z)$ are in $L^{4/3}(S)$, where Ω is compact in $\mathbb{R}^3$ and S is a piecewise smooth surface in Ω. Then the problem of equilibrium of an elastic body occupying Ω, with clamped boundary, has a unique generalized solution $\mathbf{u} \in E_{EC}$; namely, $\mathbf{u}(x,y,z)$ satisfies (2.4.15) for every $\mathbf{v} \in E_{EC}$.

In both theorems, the load restrictions provide continuity of the corresponding functionals Φ, the work of external forces.

Problem 2.5.1. Formulate existence theorems for the other mechanics problems discussed in Sect. 2.4. □

2.6 Operator Formulation of an Eigenvalue Problem

We have seen how to use the Riesz representation theorem to prove the existence and uniqueness of a generalized solution. Now let us consider another application of the Riesz representation theorem: how to cast a problem as an operator equation.

The eigenvalue equation for a membrane has the form

$$\frac{\partial^2 u}{\partial x^2} + \frac{\partial^2 u}{\partial y^2} = -\lambda u\,. \tag{2.6.1}$$

Similar to the equilibrium problem for a membrane, we can introduce a generalized solution to the eigenvalue problem for a clamped membrane by the integro-differential equation

$$\iint_\Omega \left(\frac{\partial u}{\partial x}\frac{\partial v}{\partial x} + \frac{\partial u}{\partial y}\frac{\partial v}{\partial y}\right) dx\,dy = \lambda \iint_\Omega uv\,dx\,dy\,. \tag{2.6.2}$$

The eigenvalue problem is to find a nontrivial element $u \in E_{MC}$ and a corresponding number λ such that u satisfies (2.6.2) for every $v \in E_{MC}$.

First we reformulate this problem as an operator equation

$$u = \lambda K u \tag{2.6.3}$$

in the space E_{MC}. For this, consider the term

$$F(v) = \iint_\Omega uv\,dx\,dy$$

as a functional in E_{MC}, with respect to v, when u is a fixed element of E_{MC}. It is seen that $F(v)$ is a linear functional. By the Schwarz inequality

$$|F(v)| = \left| \iint_\Omega uv\,dx\,dy \right| \le \left(\iint_\Omega u^2\,dx\,dy \right)^{1/2} \left(\iint_\Omega v^2\,dx\,dy \right)^{1/2}$$

hence by the Friedrichs inequality

$$|F(v)| \le m\,\|u\|\,\|v\| = m_1\|v\| \tag{2.6.4}$$

(hereafter the norm $\|\cdot\|$ and the inner product $(\cdot,\cdot)$ are taken in E_{MC}). So $F(v)$ is a continuous linear functional acting in the Hilbert space E_{MC}. By the Riesz representation theorem, $F(v)$ has the unique representation

$$F(v) \equiv \iint_\Omega uv\,dx\,dy = (v, f) = (f, v)\,. \tag{2.6.5}$$

What have we shown? For every $u \in E_{MC}$, by this representation, there is a unique element $f \in E_{MC}$. Hence the correspondence

$$u \mapsto f$$

is an operator $f = K(u)$ from E_{MC} to E_{MC}.

Let us display some properties of this operator. First we show that it is linear. Let

$$f_1 = K(u_1) \quad \text{and} \quad f_2 = K(u_2)\,.$$

Then

$$\iint_\Omega (\lambda_1 u_1 + \lambda_2 u_2) v\,dx\,dy = (K(\lambda_1 u_1 + \lambda_2 u_2), v)$$

while on the other hand,

$$\begin{aligned}\iint_\Omega (\lambda_1 u_1 + \lambda_2 u_2) v\,dx\,dy &= \lambda_1 \iint_\Omega u_1 v\,dx\,dy + \lambda_2 \iint_\Omega u_2 v\,dx\,dy \\ &= \lambda_1 (K(u_1), v) + \lambda_2 (K(u_2), v) \\ &= (\lambda_1 K(u_1) + \lambda_2 K(u_2), v)\,.\end{aligned}$$

Combining these we have

$$(K(\lambda_1 u_1 + \lambda_2 u_2), v) = (\lambda_1 K(u_1) + \lambda_2 K(u_2), v)\,,$$

hence

$$K(\lambda_1 u_1 + \lambda_2 u_2) = \lambda_1 K(u_1) + \lambda_2 K(u_2)$$

because $v \in E_{MC}$ is arbitrary. Therefore linearity is proven.

Now let us rewrite (2.6.4) in terms of this representation:

$$|(K(u), v)| \le m\,\|u\|\,\|v\|\,.$$

Take $v = K(u)$; then

$$\|K(u)\|^2 \le m\,\|u\|\,\|K(u)\|$$

and it follows that

$$\|K(u)\| \le m\,\|u\| \,. \tag{2.6.6}$$

Hence K is a continuous operator in E_{MC}.

Equation (2.6.2) can now be written in the form

$$(u, v) = \lambda(K(u), v) \,.$$

Since v is an arbitrary element of E_{MC}, this equation is equivalent to the operator equation

$$u = \lambda K(u)$$

with a continuous linear operator K.

By (2.6.6), we get

$$\|\lambda K(u) - \lambda K(v)\| = |\lambda|\,\|K(u - v)\| \le m\,|\lambda|\,\|u - v\| \,.$$

If

$$m\,|\lambda| < 1 \,,$$

then λK is a contraction operator in E_{MC} and, by the contraction mapping principle, there is a unique fixed point of λK which clearly is $u = 0$. So the set $|\lambda| < 1/m$ does not contain real eigenvalues of the problem. Further, we shall see (and this is well known in mechanics) that eigenvalues in this problem must be real. The fact that the set $|\lambda| < 1/m$ does not contain real eigenvalues, and so any eigenvalues of the problem, has a clear mechanical sense: the lowest eigenfrequency of oscillation of a bounded clamped membrane is strictly positive. Note that from (2.6.2), when $v = u$ it follows that an eigenvalue must be positive.

In a similar way, we can introduce eigenvalue problems for plates and elastic bodies. Here we can obtain corresponding equations of the form (2.6.3) with continuous linear operators and can also show that the corresponding lowest eigenvalues are strictly positive. All this we leave to the reader; later we shall consider eigenvalue problems in more detail.

In what follows, we shall see that, using the Riesz representation theorem, one can also introduce operators and operator equations for nonlinear problems of mechanics. One of them is presented in the next section.

2.7 Problem of Elastico-Plasticity; Small Deformations

Following the lines of a paper by I.I. Vorovich and Yu.P. Krasovskij [40] that was published in a sketchy form, we consider a variant of the theory of elastico-plasticity (Il'yushin [16]), and justify the *method of elastic solutions* for corresponding boundary value problems.

The system of partial differential equations describing the behavior of an elastic-plastic body occupying a bounded volume Ω is

$$\left(\frac{\nu}{\nu-2}-\frac{\omega}{3}\right)\frac{\partial\theta}{\partial x_k}+(1-\omega)\Delta u_k$$
$$-\frac{2}{3}e_I\frac{d\omega}{de_I}\sum_{s,t=1}^{3}\epsilon^*_{ks}\sum_{l=1}^{3}\epsilon^*_{lt}\frac{\partial^2 u_l}{\partial x_s\partial x_t}+\frac{F_k}{G}=0\quad(k=1,2,3)\,, \tag{2.7.1}$$

where ν is Poisson's ratio, G is the shear modulus, $\mathbf{F}=(F_1,F_2,F_3)$ are the volume forces, and $\omega(e_i)$ is a function of the variable e_I, the intensity of the strain tensor which defines plastic properties of the material with hardening:

$$e_I=\frac{\sqrt{2}}{3}\left[(\epsilon_{11}-\epsilon_{22})^2+(\epsilon_{11}-\epsilon_{33})^2+(\epsilon_{22}-\epsilon_{33})^2+6(\epsilon_{12}^2+\epsilon_{13}^2+\epsilon_{23}^2)\right]^{1/2}.$$

The function $\omega(e_I)$ must satisfy

$$0\le\omega(e_I)\le\omega(e_I)+e_I\frac{d\omega(e_I)}{de_I}\le\lambda<1\,. \tag{2.7.2}$$

Other bits of notation are

$$\theta\equiv\theta(\mathbf{u})=\epsilon_{11}(\mathbf{u})+\epsilon_{22}(\mathbf{u})+\epsilon_{33}(\mathbf{u})\,,$$

and

$$\epsilon_{ij}=\frac{1}{2}\left(\frac{\partial u_i}{\partial x_j}+\frac{\partial u_j}{\partial x_i}\right),\qquad \epsilon^*_{ks}=\begin{cases}\left(\dfrac{\partial u_k}{\partial x_s}-\dfrac{\theta}{3}\right)\dfrac{\sqrt{2}}{e_I}, & k=s\,,\\[2ex] \left(\dfrac{\partial u_k}{\partial x_s}+\dfrac{\partial u_s}{\partial x_k}\right)\dfrac{1}{\sqrt{2}e_I}, & k\ne s\,.\end{cases}$$

If $\omega(e_I)\equiv 0$ we get the equations of linear elasticity for an isotropic homogeneous body. By analogy with elasticity problems, to pose a boundary value problem for (2.7.1) we must supplement the equations with boundary conditions. We consider a mixed boundary value problem: a part S_0 of the boundary $\partial\Omega$ of a body occupying the domain Ω is fixed,

$$\mathbf{u}\big|_{S_0}=\mathbf{0}\,, \tag{2.7.3}$$

and the remainder $S_1=\partial\Omega\setminus S_0$ is subjected to surface forces $\mathbf{f}(\mathbf{x})$ (see [16]):

$$\boldsymbol{\sigma}\cdot\mathbf{n}\big|_{S_1}=\mathbf{f}\,, \tag{2.7.4}$$

where $\boldsymbol{\sigma}$ is the stress tensor and $\mathbf{n}$ is the external unit normal to S_1.

When $\omega(e_I)$ is small (as it is if e_I is small) we have a nonlinear boundary value problem which is, in a certain way, a perturbation of a corresponding boundary value problem of linear elasticity. It leads to the idea of using an iterative procedure, the method of elastic solutions, to solve the former. This procedure looks like that of the contraction mapping principle if we can make the problem take the corresponding operator form. Then it remains to show that the operator of the problem is a contraction. Now we begin to carry out the program.

Let us introduce the notation

$$\begin{aligned}\langle \mathbf{u}, \mathbf{v} \rangle = \tfrac{2}{9}\{&[\epsilon_{11}(\mathbf{u}) - \epsilon_{22}(\mathbf{u})][\epsilon_{11}(\mathbf{v}) - \epsilon_{22}(\mathbf{v})] \\ &+ [\epsilon_{11}(\mathbf{u}) - \epsilon_{33}(\mathbf{u})][\epsilon_{11}(\mathbf{v}) - \epsilon_{33}(\mathbf{v})] \\ &+ [\epsilon_{22}(\mathbf{u}) - \epsilon_{33}(\mathbf{u})][\epsilon_{22}(\mathbf{v}) - \epsilon_{33}(\mathbf{v})] \\ &+ 6[\epsilon_{12}(\mathbf{u})\epsilon_{12}(\mathbf{v}) + \epsilon_{13}(\mathbf{u})\epsilon_{13}(\mathbf{v}) + \epsilon_{23}(\mathbf{u})\epsilon_{23}(\mathbf{v})]\} \,. \end{aligned} \tag{2.7.5}$$

If we consider the terms on the right-hand side of (2.7.5) as coordinates of vectors $\mathbf{a} = (a_1, \ldots, a_6)$, $\mathbf{b} = (b_1, \ldots, b_6)$,

$$a_i = c_i(\mathbf{u}) \,, \qquad b_i = c_i(\mathbf{v}) \qquad (i = 1, \ldots, 6) \,,$$

$$\begin{aligned} c_1(\mathbf{w}) &= \tfrac{\sqrt{2}}{3}[\epsilon_{11}(\mathbf{w}) - \epsilon_{22}(\mathbf{w})] \,, & c_2(\mathbf{w}) &= \tfrac{\sqrt{2}}{3}[\epsilon_{11}(\mathbf{w}) - \epsilon_{33}(\mathbf{w})] \,, \\ c_3(\mathbf{w}) &= \tfrac{\sqrt{2}}{3}[\epsilon_{22}(\mathbf{w}) - \epsilon_{33}(\mathbf{w})] \,, & c_4(\mathbf{w}) &= \tfrac{2}{\sqrt{3}}\epsilon_{12}(\mathbf{w}) \,, \\ c_5(\mathbf{w}) &= \tfrac{2}{\sqrt{3}}\epsilon_{13}(\mathbf{w}) \,, & c_6(\mathbf{w}) &= \tfrac{2}{\sqrt{3}}\epsilon_{23}(\mathbf{w}) \,, \end{aligned}$$

then the form $\langle \mathbf{u}, \mathbf{v} \rangle$ is a scalar product between $\mathbf{a}$ and $\mathbf{b}$ in $\mathbb{R}^6$:

$$\langle \mathbf{u}, \mathbf{v} \rangle = \sum_{i=1}^{6} a_i b_i \,.$$

Besides,

$$\langle \mathbf{u}, \mathbf{u} \rangle = \sum_{i=1}^{6} c_i^2(\mathbf{u}) = e_I^2(\mathbf{u}) \tag{2.7.6}$$

and by the Schwarz inequality we get

$$|\langle \mathbf{u}, \mathbf{v} \rangle| = \left| \sum_{i=1}^{6} c_i(\mathbf{u}) c_i(\mathbf{v}) \right| \le e_I(\mathbf{u}) e_I(\mathbf{v}) \,. \tag{2.7.7}$$

On the set C_2 of vector functions satisfying the boundary condition (2.7.3) and such that each of their components is of class $C^{(2)}(\Omega)$, we introduce an inner product

$$(\mathbf{u}, \mathbf{v}) = \iiint_\Omega \left(\tfrac{3}{2} G \langle \mathbf{u}, \mathbf{v} \rangle + \tfrac{1}{2} K \,\theta(\mathbf{u})\, \theta(\mathbf{v}) \right) d\Omega \,. \tag{2.7.8}$$

This coincides with a special case of the inner product (2.3.27) in the linear theory of elasticity. So the completion of C_2 in the metric corresponding to (2.7.8) is the energy space of linear elasticity E_{EM} (M for "mixed") if we suppose that the condition (2.7.3) provides $\mathbf{u} = 0$ if

$$\|\mathbf{u}\|^2 = \iiint_\Omega \left(\tfrac{3}{2}Ge_I^2(\mathbf{u}) + \tfrac{1}{2}K\theta^2(\mathbf{u})\right) d\Omega = 0\,.$$

The norm of E_{EM} is equivalent to one of $W^{1,2}(\Omega)\times W^{1,2}(\Omega)\times W^{1,2}(\Omega)$ (see Sect. 2.3 and Fichera [11]). (By $H_1 \times H_2$ we denote the *Cartesian product* of Hilbert spaces H_1 and H_2, the elements of which are pairs (x, y) for $x \in H_1$ and $y \in H_2$. The scalar product in $H_1 \times H_2$ is defined by the expression $(x_1, x_2)_1 + (y_1, y_2)_2$ where $x_1, x_2 \in H_1$ and $y_1, y_2 \in H_2$.)

By the principle of virtual displacements, the integro-differential equation of equilibrium of an elastico-plastic body is

$$(\mathbf{u}, \mathbf{v}) - \tfrac{3}{2}G \iiint_\Omega \omega(e_I(\mathbf{u}))\langle \mathbf{u}, \mathbf{v}\rangle \, d\Omega - \sum_{i=1}^{3} \iiint_\Omega F_i v_i \, d\Omega - \sum_{i=1}^{3} \iint_{S_1} f_i v_i \, dS = 0\,. \tag{2.7.9}$$

This equation can be obtained using the equations (2.7.1) and the boundary conditions (2.7.3)–(2.7.4). Conversely, using the technique of the classical calculus of variations we can get (2.7.1) and the natural boundary conditions (2.7.4). Thus, in a certain way, (2.7.9) is equivalent to the above statement of the problem. So we can introduce

Definition 2.7.1. A vector function $\mathbf{u} \in E_{EM}$ is called the generalized solution of the problem of elastico-plasticity with boundary conditions (2.7.3)–(2.7.4) if it satisfies (2.7.9) for every $\mathbf{v} \in E_{EM}$.

For correctness of this definition we must impose some restrictions on external forces. It is evident that they coincide with those for linear elasticity. So we assume that

$$F_i(x_1, x_2, x_3) \in L^{6/5}(\Omega)\,, \qquad f_i(x_1, x_2, x_3) \in L^{4/3}(S_1)\,. \tag{2.7.10}$$

Consider the form

$$B[\mathbf{u}, \mathbf{v}] = \tfrac{3}{2}G \iiint_\Omega \omega(e_I(\mathbf{u}))\langle \mathbf{u}, \mathbf{v}\rangle \, d\Omega + \sum_{i=1}^{3} \iiint_\Omega F_i v_i \, d\Omega + \sum_{i=1}^{3} \iint_{S_1} f_i v_i \, dS$$

as a functional in E_{EM} with respect to $\mathbf{v}(x_1, x_2, x_3)$ when $\mathbf{u}(x_1, x_2, x_3) \in E_{EM}$ is fixed. As in linear elasticity, the load terms, thanks to (2.7.10), are continuous linear functionals with respect to $\mathbf{v} \in E_{EM}$. In accordance with (2.7.5) and (2.7.2), we get

$$\left|\tfrac{3}{2}G \iiint_\Omega \omega(e_I(\mathbf{u}))\langle \mathbf{u}, \mathbf{v}\rangle \, d\Omega\right| \le \lambda \tfrac{3}{2}G \iiint_\Omega |\langle \mathbf{u}, \mathbf{v}\rangle| \, d\Omega \le \lambda \, \|\mathbf{u}\| \, \|\mathbf{v}\|\,,$$

so this part of the functional is also continuous.

Therefore we can apply the Riesz representation theorem to $B[\mathbf{u}, \mathbf{v}]$ and obtain

$$B[\mathbf{u}, \mathbf{v}] = (\mathbf{v}, \mathbf{g}) \equiv (\mathbf{g}, \mathbf{v})\,.$$

This representation uniquely defines a correspondence

$$\mathbf{u} \mapsto \mathbf{g}$$

where $\mathbf{u}, \mathbf{g} \in E_{EM}$. We obtain an operator A acting in E_{EM} by the equality

$$\mathbf{g} = A(\mathbf{u}) .$$

Equation (2.7.9) is now equivalent to

$$(\mathbf{u}, \mathbf{v}) - (A(\mathbf{u}), \mathbf{v}) = 0 \tag{2.7.11}$$

or, since $\mathbf{v} \in E_{EM}$ is arbitrary,

$$\mathbf{u} = A(\mathbf{u}) . \tag{2.7.12}$$

The operator A is nonlinear. We shall show that it is a contraction operator. For this, take arbitrary elements $\mathbf{u}, \mathbf{v}, \mathbf{w} \in E_{EM}$ and consider

$$(A(\mathbf{u}) - A(\mathbf{v}), \mathbf{w}) = \tfrac{3}{2}G \iiint_{\Omega} [\omega(e_I(\mathbf{u}))\langle \mathbf{u}, \mathbf{w}\rangle - \omega(e_I(\mathbf{v}))\langle \mathbf{v}, \mathbf{w}\rangle] \, d\Omega . \tag{2.7.13}$$

First, let $\mathbf{u}, \mathbf{v}, \mathbf{w}$ be in C_2. At every point of Ω, by (2.7.7), we can estimate the integrand from (2.7.13) as follows. We have

$$\begin{aligned} \text{Int} &\equiv \left|\omega(e_I(\mathbf{u}))\,\langle \mathbf{u}, \mathbf{w}\rangle - \omega(e_I(\mathbf{v}))\,\langle \mathbf{v}, \mathbf{w}\rangle\right| \\ &= \left|\omega(e_I(\mathbf{u})) \sum_{i=1}^{6} c_i(\mathbf{u})\, c_i(\mathbf{w}) - \omega(e_I(\mathbf{v})) \sum_{i=1}^{6} c_i(\mathbf{v})\, c_i(\mathbf{w})\right| . \end{aligned}$$

Let us introduce a real-valued function $f(t)$ of a real variable t by the relation

$$f(t) = \sum_{i=1}^{6} \omega(e_I(t\mathbf{u} + (1-t)\mathbf{v}))c_i(t\mathbf{u} + (1-t)\mathbf{v})c_i(\mathbf{w}) .$$

It is seen that

$$\text{Int} = |f(1) - f(0)| .$$

As $f(t)$ is continuously differentiable, the classical mean value theorem gives

$$f(1) - f(0) = f'(z)(1-0) = f'(z) \text{ for some } z \in [0, 1] ,$$

or, in the above terms, we get

$$\text{Int} = \left| \frac{d}{dt} \left\{ \sum_{i=1}^{6} \omega(e_I(t\mathbf{u} + (1-t)\mathbf{v}))\, c_i(t\mathbf{u} + (1-t)\mathbf{v})\, c_i(\mathbf{w}) \right\}_{t=z} \right|$$

$$= \left| \left\{ \frac{d\omega(e_I(t\mathbf{u} + (1-t)\mathbf{v}))}{de_I} \frac{de_I(t\mathbf{u} + (1-t)\mathbf{v})}{dt} \cdot \right. \right.$$

$$\left. \left. \cdot \sum_{i=1}^{6} c_i(t\mathbf{u} + (1-t)\mathbf{v})\, c_i(\mathbf{w}) + \omega \sum_{i=1}^{6} c_i(\mathbf{u} - \mathbf{v})\, c_i(\mathbf{w}) \right\}_{t=z} \right| .$$

(Here we have used the linearity of $c_i(\mathbf{u})$ in $\mathbf{u}$ and, thus, in t.) Let us consider the term

$$T = \sum_{i=1}^{6} \frac{de_I(t\mathbf{u} + (1-t)\mathbf{v})}{dt}\, c_i(t\mathbf{u} + (1-t)\mathbf{v})\, c_i(\mathbf{w})$$

$$= \sum_{i=1}^{6} \frac{d}{dt} \Big(\sum_{j=1}^{6} c_j^2(t\mathbf{u} + (1-t)\mathbf{v}) \Big)^{1/2} c_i(t\mathbf{u} + (1-t)\mathbf{v})\, c_i(\mathbf{w})$$

$$= \sum_{i=1}^{6} \frac{2 \sum_{j=1}^{6} c_j(t\mathbf{u} + (1-t)\mathbf{v})\, c_j(\mathbf{u} - \mathbf{v})}{2 \Big(\sum_{j=1}^{6} c_j^2(t\mathbf{u} + (1-t)\mathbf{v}) \Big)^{1/2}}\, c_i(t\mathbf{u} + (1-t)\mathbf{v})\, c_i(\mathbf{w}) .$$

Applying the Schwarz inequality, we obtain

$$|T| \le \sum_{i=1}^{6} \frac{\Big(\sum_{j=1}^{6} c_j^2(t\mathbf{u} + (1-t)\mathbf{v}) \Big)^{1/2} \Big(\sum_{j=1}^{6} c_j^2(\mathbf{u} - \mathbf{v}) \Big)^{1/2}}{\Big(\sum_{j=1}^{6} c_j^2(t\mathbf{u} + (1-t)\mathbf{v}) \Big)^{1/2}} \cdot |c_i(t\mathbf{u} + (1-t)\mathbf{v})|\, |c_i(\mathbf{w})|$$

so that by (2.7.6)

$$|T| \le \Big(\sum_{j=1}^{6} c_j^2(\mathbf{u} - \mathbf{v}) \Big)^{1/2} \sum_{i=1}^{6} |c_i(t\mathbf{u} + (1-t)\mathbf{v})|\, |c_i(\mathbf{w})|$$

$$\le e_I(\mathbf{u} - \mathbf{v}) \Big(\sum_{i=1}^{6} c_i^2(t\mathbf{u} + (1-t)\mathbf{v}) \Big)^{1/2} \Big(\sum_{i=1}^{6} c_i^2(\mathbf{w}) \Big)^{1/2}$$

$$= e_I(\mathbf{u} - \mathbf{v})\, e_I(t\mathbf{u} + (1-t)\mathbf{v})\, e_I(\mathbf{w}) .$$

Similarly,

$$\Big|\sum_{i=1}^{6} c_i(\mathbf{u}-\mathbf{v})\, c_i(\mathbf{w})\Big| \le \Big(\sum_{i=1}^{6} c_i^2(\mathbf{u}-\mathbf{v})\Big)^{1/2}\Big(\sum_{i=1}^{6} c_i^2(\mathbf{w})\Big)^{1/2} = e_I(\mathbf{u}-\mathbf{v})\, e_I(\mathbf{w}) \,.$$

Combining all these, we get

$$\begin{aligned}\text{Int} &\le \left\{\frac{d\omega(e_I(t\mathbf{u}+(1-t)\mathbf{v}))}{de_I} e_I(t\mathbf{u}+(1-t)\mathbf{v}))\, e_I(\mathbf{u}-\mathbf{v})\, e_I(\mathbf{w})\right. \\ &\quad \left.\left. + \, \omega(e_I(t\mathbf{u}+(1-t)\mathbf{v}))\, e_I(\mathbf{u}-\mathbf{v})\, e_I(\mathbf{w})\right\}\right|_{t=z} \\ &= \left\{\omega(e_I(t\mathbf{u}+(1-t)\mathbf{v})) + \frac{d\omega(e_I(t\mathbf{u}+(1-t)\mathbf{v}))}{de_I}\,\cdot\right. \\ &\quad \left.\left. \cdot\, e_I(t\mathbf{u}+(1-t)\mathbf{v}))\right\}\right|_{t=z} e_I(\mathbf{u}-\mathbf{v})\, e_I(\mathbf{w}) \,.\end{aligned}$$

By the condition (2.7.2), we have

$$\text{Int} \le \lambda e_I(\mathbf{u}-\mathbf{v})\, e_I(\mathbf{w}) \tag{2.7.14}$$

at every point of Ω.

Returning to (2.7.13) we have, using (2.7.14),

$$|(A(\mathbf{u})-A(\mathbf{v}),\mathbf{w})| \le \lambda \iiint_\Omega \tfrac{3}{2} G\, e_I(\mathbf{u}-\mathbf{v})\, e_I(\mathbf{w})\, d\Omega \,.$$

In accordance with the norm of E_{EM} it follows that

$$|(A(\mathbf{u})-A(\mathbf{v}),\mathbf{w})| \le \lambda \,\|\mathbf{u}-\mathbf{v}\|\, \|\mathbf{w}\|$$

or, putting $\mathbf{w} = A(\mathbf{u}) - A(\mathbf{v})$, we get

$$\|A(\mathbf{u})-A(\mathbf{v})\| \le \lambda \,\|\mathbf{u}-\mathbf{v}\| \,, \qquad \lambda = \text{constant} < 1 \,. \tag{2.7.15}$$

Being obtained for $\mathbf{u},\mathbf{v},\mathbf{w} \in C_2$, this inequality holds for all $\mathbf{u},\mathbf{v},\mathbf{w} \in E_{EM}$ since in this inequality we can pass to the limit for corresponding Cauchy sequences in E_{EM}.

Inequality (2.7.15) states that A is a contraction operator in E_{EM}; hence, by the contraction mapping principle, (2.7.12) has a unique solution that can be found using the iterative procedure

$$\mathbf{u}_{k+1} = A(\mathbf{u}_k) \qquad (k = 0, 1, 2, \ldots) \,.$$

This procedure begins with an arbitrary element $\mathbf{u}_0 \in E_{EM}$; when $\mathbf{u}_0 = \mathbf{0}$, it is called the method of elastic solutions since at each step we must solve a problem of linear elasticity with some given load terms. From a practical standpoint, the method works best when the constant λ is small.

So we can formulate

Theorem 2.7.1. Assume S_0 is a piecewise smooth surface of nonzero area and that conditions (2.7.2) and (2.7.10) hold. Then a mixed boundary value problem of elastico-plasticity has a unique generalized solution in the sense of Definition 2.7.1; the iterative procedure (2.7.15) defines a sequence of successive approximations $\mathbf{u}_k \in E_{EM}$ that converges to the solution $\mathbf{u} \in E_{EM}$ and

$$\|\mathbf{u}_k - \mathbf{u}\| \leq \frac{\lambda^k}{1-\lambda} \|\mathbf{u}_0 - \mathbf{u}_1\| . \qquad (2.7.16)$$

It is clear that we cannot apply this theorem when, say, $S_1 = \partial\Omega$. In such a case, we must add the self-balance conditions (2.4.16). These guarantee that we can repeat the above method for a free elastic-plastic body, and so we can formulate

Theorem 2.7.2. Assume that all the requirements of Theorem 2.7.1 and the self-balance conditions (2.4.16) are met. Then there is a unique generalized solution of the boundary value problem for a bounded elastic-plastic body, and it can be found by an iterative procedure of the form (2.7.15).

Problem 2.7.1. Is an estimate of the type (2.7.16) valid in Theorem 2.7.2? □

We recommend that the reader prove Theorem 2.7.2 in detail, in order to gain experience with the technique.

Remark 2.7.1. We should call attention to the way in which we obtained the main inequality of this section: it was proved for smooth functions and then extended to the general case. This is a standard technique in the treatment of nonlinear problems of mechanics. □

2.8 Bases and Complete Systems; Fourier Series

If a linear space Y has finite dimension n, then there is a set $\{g_1, \ldots, g_n\}$ of n linearly independent elements, called a *basis* of Y, such that every $y \in Y$ has a unique representation

$$y = \sum_{k=1}^{n} \alpha_k g_k$$

where the α_k are scalars. We now consider an infinite dimensional normed space X.

Definition 2.8.1. A system of elements $\{e_k\}$ is a (*countable*) *basis* of X if any element $x \in X$ has a unique representation

$$x = \sum_{k=1}^{\infty} \alpha_k e_k$$

where the α_k are scalars.

It is clear that a basis $\{e_k\}$ is linearly independent since the equation

$$0 = \sum_{k=1}^{\infty} \alpha_k e_k$$

has the unique solution $\alpha_k = 0$ for each k.

A normed space with a countable basis is separable: a countable set of all linear combinations $\sum_{k=1}^{n} c_k e_k$ (n arbitrary) with rational coefficients c_k is dense in the space.

Problem 2.8.1. Prove this. □

We are familiar with some systems of functions which could be bases in certain spaces: for example,

$$\{g_k\} = \left\{ \frac{1}{\sqrt{2\pi}} e^{ikx} \right\} \tag{2.8.1}$$

in $L^2(0, 2\pi)$. Later, we confirm this example.

Now we consider the system of monomials $\{x^k\}$ $(k = 1, 2, \ldots)$ in $C(0, 1)$. If it is a basis, then any function $f(x) \in C(0, 1)$ could be represented in the form

$$f(x) = \sum_{k=0}^{\infty} \alpha_k x^k ,$$

where the series converges uniformly on $[0, 1]$. This means the function is analytic, but we know there are continuous functions on $[0, 1]$ that are not analytic. Hence the system $\{x^k\}$ is not a basis. On the other hand, the Weierstrass theorem states that this system possesses properties similar to those of a basis. To generalize this similarity, we introduce

Definition 2.8.2. A countable system $\{g_k\}$ of elements in a normed space X is *complete* (or *total*) in X if for any $x \in X$ and any positive number ε there is a finite linear combination $\sum_{i=1}^{n(\varepsilon)} \alpha_i g_i$ such that

$$\left\| x - \sum_{i=1}^{n(\varepsilon)} \alpha_i g_i \right\| < \varepsilon .$$

By Definition 2.8.2 and the Weierstrass theorem, the system of monomials $\{x^k\}$ is complete in $C(0, 1)$. Because $C(0, 1)$ is dense in $L^p(0, 1)$ for $p \geq 1$, this system is also complete in $L^p(0, 1)$.

Problem 2.8.2. Which systems are complete in $L^p(\Omega)$ or $W^{k,p}(\Omega)$? □

If a normed space has a countable complete system, then the space is separable. The reader should be able to name a countable dense set to verify this.

Problem 2.8.3. Name such a set. □

The problem of existence of a basis in a certain normed space is difficult, but there is a special case where it is fully solved: a separable Hilbert space. The reader will find here the theory of Fourier series largely repeated in abstract terms. We begin with

Definition 2.8.3. A system $\{x_k\}$ of elements of a Hilbert space H is *orthonormal* if for all integers m, n,

$$(x_m, x_n) = \delta_{mn}$$

where δ_{mn} is the Kronecker delta symbol.

We know that, at least for $\mathbb{R}^n$, there are some advantages in using an orthonormal system of vectors as a basis.

Suppose we have an arbitrary linearly independent system of elements of a Hilbert space H, say $\{f_1, \ldots, f_n\}$, and let H_n be the subspace of H spanned by this system. We would like to use the system to construct an orthonormal system $\{g_1, \ldots, g_n\}$ that is also a basis of H_n. This can be accomplished by the *Gram–Schmidt procedure*:

(1) The first element of the new system is $g_1 = f_1/\|f_1\|$, $\|g_1\| = 1$.

(2) Take $e_2 = f_2 - (f_2, g_1)g_1$; then $(e_2, g_1) = (f_2, g_1) - (f_2, g_1)\|g_1\|^2 = 0$, so the second element is $g_2 = e_2/\|e_2\|$.

(3) Take $e_3 = f_3 - (f_3, g_1)g_1 - (f_3, g_2)g_2$; then $(e_3, g_1) = 0$ and $(e_3, g_2) = 0$. Since $e_3 \neq 0$, we get the third element as $g_3 = e_3/\|e_3\|$.

$\vdots$

(i) Let $e_i = f_i - (f_i, g_1)g_1 - \cdots - (f_i, g_{i-1})g_{i-1}$. It is seen that $(e_i, g_k) = 0$ for $k = 1, \ldots, i-1$, hence we set $g_i = e_i/\|e_i\|$.

This process can be continued ad infinitum since all $e_k \neq 0$ (why?). So we obtain an orthonormalized system $\{g_1, g_2, g_3, \ldots\}$. The process is, however, found to be unstable for numerical computation.

As is known from linear algebra, a system $\{f_1, \ldots, f_n\}$ is linearly independent in an inner product space if and only if the *Gram determinant*

$$\begin{vmatrix} (f_1, f_1) & \cdots & (f_1, f_n) \\ \vdots & \ddots & \vdots \\ (f_n, f_1) & \cdots & (f_n, f_n) \end{vmatrix}$$

is nonzero. For an orthonormal system of elements the Gram determinant, being the determinant of the identity matrix, equals $+1$; hence an orthonormal system is linearly independent.

Problem 2.8.4. Provide a more direct proof that an orthonormal system is linearly independent. □

Let $\{g_k\}$ $(k = 1, 2, \ldots)$ be an orthonormal system in a complex Hilbert space H. For an element $f \in H$, the numbers α_k defined by $\alpha_k = (f, g_k)$ are called the *Fourier coefficients* of f. Now we prove

Theorem 2.8.1. A complete orthonormal system $\{g_k\}$ in a Hilbert space H is a basis of H; any $f \in H$ has the unique *Fourier series* representation

$$f = \sum_{k=1}^{\infty} \alpha_k g_k \tag{2.8.2}$$

where $\alpha_k = (f, g_k)$ are the Fourier coefficients of f.

Proof. First we consider the problem of the best approximation of an element $f \in H$ by elements of a subspace H_n spanned by $g_1, \ldots, g_n$. In Sect. 1.19 we showed that this problem has a unique solution. Now we show that it is $\sum_{k=1}^{n} \alpha_k g_k$. Indeed, consider an arbitrary linear combination $\sum_{k=1}^{n} c_k g_k$. Then

$$\begin{aligned}
\left\| f - \sum_{k=1}^{n} c_k g_k \right\|^2 &= \left(f - \sum_{k=1}^{n} c_k g_k, f - \sum_{k=1}^{n} c_k g_k \right) \\
&= \|f\|^2 - \left(f, \sum_{k=1}^{n} c_k g_k \right) - \left(\sum_{k=1}^{n} c_k g_k, f \right) + \left\| \sum_{k=1}^{n} c_k g_k \right\|^2 \\
&= \|f\|^2 - \sum_{k=1}^{n} \overline{c_k} \alpha_k - \sum_{k=1}^{n} c_k \overline{\alpha_k} + \sum_{k=1}^{n} c_k \overline{c_k} \\
&= \|f\|^2 - \sum_{k=1}^{n} |\alpha_k|^2 + \sum_{k=1}^{n} |c_k - \alpha_k|^2 .
\end{aligned}$$

Because the right-hand side takes its minimum value when $c_k = \alpha_k$, we have

$$\left\| f - \sum_{k=1}^{n} \alpha_k g_k \right\|^2 = \min_{c_1, \ldots, c_n} \left\| f - \sum_{k=1}^{n} c_k g_k \right\|^2 = \|f\|^2 - \sum_{k=1}^{n} |\alpha_k|^2 \geq 0 ; \tag{2.8.3}$$

moreover, we obtain *Bessel's inequality*

$$\sum_{k=1}^{n} |(f, g_k)|^2 \leq \|f\|^2 . \tag{2.8.4}$$

Denote by

$$f_n = \sum_{k=1}^{n} \alpha_k g_k \tag{2.8.5}$$

the nth partial sum of the Fourier series for f. Let us show that $\{f_n\}$ is a Cauchy sequence in H. By Bessel's inequality,

$$\sum_{k=1}^{n} |\alpha_k|^2 \leq \|f\|^2 ;$$

hence

$$\|f_n - f_{n+m}\|^2 = \left\| \sum_{k=n+1}^{n+m} \alpha_k g_k \right\|^2 = \sum_{k=n+1}^{n+m} |\alpha_k|^2 \to 0 \text{ as } n \to \infty .$$

Now we show that $\{f_n\}$ converges to f. Indeed, by completeness of the system $\{g_k\}$ in H, for any $\varepsilon > 0$ we can find a number N and coefficients $c_k(\varepsilon)$ such that

$$\left\| f - \sum_{k=1}^{N} c_k(\varepsilon) g_k \right\|^2 < \varepsilon .$$

By (2.8.3),

$$\|f - f_N\|^2 = \left\| f - \sum_{k=1}^{N} \alpha_k g_k \right\|^2 \le \left\| f - \sum_{k=1}^{N} c_k(\varepsilon) g_k \right\|^2 < \varepsilon ,$$

so the sequence $\{f_N\}$ converges to f and thus

$$f = \lim_{n\to\infty} f_n . \tag{2.8.6}$$

This completes the proof. □

From (2.8.6) we can obtain *Parseval's equality*

$$\sum_{k=1}^{\infty} |(f, g_k)|^2 = \|f\|^2 , \tag{2.8.7}$$

which holds whenever $\{g_k\}$ is a complete orthonormal system in H. Indeed, by (2.8.3),

$$0 = \lim_{n\to\infty} \left\| f - \sum_{k=1}^{n} \alpha_k g_k \right\|^2 = \lim_{n\to\infty} \left(\|f\|^2 - \sum_{k=1}^{n} |\alpha_k|^2 \right) .$$

Now we introduce

Definition 2.8.4. A system $\{e_k\}$ $(k = 1, 2, \ldots)$ in a Hilbert space H is *closed in H* if from the system of equations

$$(f, e_k) = 0 \text{ for all } k = 1, 2, 3, \ldots$$

it follows that $f = 0$.

It is clear that a complete orthonormal system of elements is closed in H.

Problem 2.8.5. Provide a detailed proof. □

The converse statement holds as well. We formulate

Theorem 2.8.2. Let $\{g_k\}$ be an orthonormal system of elements in a Hilbert space H. This system is complete in H if and only if it is closed in H.

Proof. We need to demonstrate only that a closed orthonormal system in H is complete. Proving Theorem 2.8.1, we established that for any element $f \in H$ the sequence of partial Fourier sums (2.8.5) is a Cauchy sequence. By completeness of H, there exists $f^* = \lim_{n\to\infty} f_n$ that belongs to H. To complete the proof we need to show that $f = f^*$. We have

$$(f - f^*, g_m) = \lim_{n\to\infty}\left(f - \sum_{k=1}^{n} \alpha_k g_k, g_m\right) = \alpha_m - \alpha_m = 0\,.$$

By Definition 2.8.4, it follows that $f = f^*$, hence $\{g_k\}$ is complete. □

It is normally simpler to check whether a system is closed than to check whether it is complete. At the beginning of this section we established that any system of linearly independent elements in H can be transformed into an orthonormal system equivalent to the original system in a certain way. So we draw the following conclusion.

Theorem 2.8.3. A complete system $\{g_k\}$ in H is closed in H; conversely, a system closed in H is complete in H.

Problem 2.8.6. Write out a detailed proof. □

As stated above, the existence of a countable basis in a Hilbert space provides its separability. The converse statement is also valid. We formulate that as

Theorem 2.8.4. A Hilbert space H has a countable orthonormal basis if and only if H is separable.

The proof follows immediately from the previous theorem. Indeed, in H select a countable set of elements that is dense everywhere in H. Using the Gram–Schmidt procedure, produce an orthonormal system of elements from this set (removing any linearly dependent elements). Since the initial system is dense it is complete and thus, as a result of the Gram–Schmidt procedure, we get an orthonormal basis of the space.

Remember that all of the energy spaces we introduced above are separable. Hence each of them has a countable orthonormal basis (nonunique, of course). If a Hilbert space is not separable, by Bessel's inequality and Lemma 1.16.5, it follows that for any element x of a nonseparable space the set of nonzero coefficients of Fourier α_k is countable. Repeating the above considerations we can get that (2.8.2) is valid in this case as well.

In conclusion, we consider whether the system (2.8.1) is a basis of the complex space $L^2(0, 2\pi)$. From standard calculus it is known that the system is orthonormal in $L^2(0, 2\pi)$ (the reader, however, can check this). Weierstrass's theorem on the approximation of a function continuous on $[0, 2\pi]$ can be formulated as the statement that the set of trigonometric polynomials, i.e., finite sums of the form $\sum_k \alpha_k e^{ikx}$, is dense in the complex space $C(0, 2\pi)$. But the set of functions $C(0, 2\pi)$ is the base for construction of $L^2(0, 2\pi)$, hence the finite sums $\sum_k \alpha_k e^{ikx}$ are dense in $L^2(0, 2\pi)$. This shows that (2.8.1) is an orthonormal basis of $L^2(0, 2\pi)$.

2.9 Weak Convergence in a Hilbert Space

We know that in $\mathbb{R}^n$, the convergence of a sequence of vectors is equivalent to coordinate-wise convergence.

In a Hilbert space H, the Fourier coefficients (f, g_k) of an element $f \in H$ play the role of the coordinates of f. Suppose $\{g_k\}$ is an orthonormal basis of H. What can we say about convergence of a sequence $\{f_n\}$ if, for every fixed k, the numerical sequence $\{(f_n, g_k)\}$ is convergent?

Let us consider $\{g_n\}$ as a sequence. It is seen that for every k,

$$\lim_{n\to\infty} (g_n, g_k) = 0 ,$$

hence we have coordinate-wise convergence of $\{g_n\}$ to zero. But the sequence $\{g_n\}$ is not convergent, since

$$\|g_n - g_m\| = \sqrt{2} \text{ for } n \neq m .$$

Therefore, coordinate-wise convergence in a Hilbert space is not equivalent to the usual form of convergence in the space. We define a new type of convergence in a Hilbert space.

Definition 2.9.1. Let H be a Hilbert space. A sequence $\{x_k\} \subset H$ is *weakly convergent* to $x_0 \in H$ if for every continuous linear functional F in H,

$$\lim_{k\to\infty} F(x_k) = F(x_0) .$$

If every numerical sequence $\{F(x_k)\}$ is a Cauchy sequence, then $\{x_k\}$ is a *weak Cauchy sequence*.

To distinguish between weak convergence and convergence as defined on p. 29, we shall refer to the latter as *strong convergence*. We retain the notation $x_k \to x$ for strong convergence and adopt $x_k \rightharpoonup x$ for weak convergence.

Definition 2.9.1 is given in a form which (with suitable modifications) is valid in a metric space. But in a Hilbert space any continuous linear functional, by the Riesz representation theorem, takes the form $F(x) = (x, f)$ where f is an element of H. So Definition 2.9.1 may be rewritten as follows:

Definition 2.9.2. Let H be a Hilbert space. A sequence $\{x_n\} \subset H$ is weakly convergent to $x_0 \in H$ if for every element $f \in H$ we have

$$\lim_{n\to\infty} (x_n, f) = (x_0, f) .$$

If every numerical sequence $\{(x_n, f)\}$ is a Cauchy sequence, then $\{x_k\}$ is a weak Cauchy sequence.

We have seen that some weak Cauchy sequences in H are not strong Cauchy sequences. But a strong Cauchy sequence is always a weak Cauchy sequence, by virtue of the continuity of the linear functionals in the definition.

We formulate a simple sufficient condition for strong convergence of a weakly convergent sequence:

Theorem 2.9.1. Suppose that $x_k \rightharpoonup x_0$, where x_k, x_0 belong to a Hilbert space H. Then $x_k \to x_0$ if $\|x_k\| \to \|x_0\|$.

Proof. Consider $\|x_k - x_0\|^2$. We get

$$\|x_k - x_0\|^2 = (x_k - x_0, x_k - x_0) = \|x_k\|^2 - (x_0, x_k) - (x_k, x_0) + \|x_0\|^2 .$$

By Definition 2.9.2 we have

$$\lim_{k\to\infty} [(x_0, x_k) + (x_k, x_0)] = 2\,\|x_0\|^2 ,$$

hence $\|x_k - x_0\|^2 \to 0$. □

We shall see later that for some numerical methods it is easier to first establish weak convergence of approximate solutions and then strong convergence, than to establish strong convergence directly. The last theorem allows us to justify a method successively, beginning with a simple approximate result and then passing to the needed one. That is why weak convergence is a major preoccupation in this presentation.

Theorem 2.9.2. In a Hilbert space, every weak Cauchy sequence $\{x_n\}$ is bounded.

Proof. We will prove the theorem for a complex Hilbert space; the proof is valid for a real space as well. Suppose to the contrary that there is a weak Cauchy sequence $\{x_n\}$ which is not bounded in H. So let $\|x_n\| \to \infty$ as $n \to \infty$. We will show that this yields a contradiction.

First we consider the set U of all numbers of the form (x_n, y), where y belongs to a closed ball $\overline{B}(y_0, \varepsilon)$ with arbitrary $\varepsilon > 0$ and center $y_0 \in H$, which are momentarily fixed. We claim that U is unbounded from above. Indeed, elements of the form $y_n = y_0 + \varepsilon x_n/(2\,\|x_n\|)$ belong to $\overline{B}(y_0, \varepsilon)$ since

$$\|y_n - y_0\| = \left\| \frac{\varepsilon x_n}{2\,\|x_n\|} \right\| = \frac{\varepsilon}{2} .$$

As $\{x_n\}$ is a weak Cauchy sequence, the numerical sequence $\{(x_n, y_0)\}$ converges and therefore is bounded. Since $\|x_n\| \to \infty$ we get

$$|(x_n, y_n)| = \left|(x_n, y_0) + \frac{\varepsilon}{2\,\|x_n\|}(x_n, x_n)\right| = \left|(x_n, y_0) + \frac{\varepsilon}{2}\,\|x_n\|\right| \to \infty$$

as $n \to \infty$. We see that U is unbounded for any fixed y_0.

Now we show that unboundedness of any set U for any y_0 yields a contradiction. Take the ball $\overline{B}(y_0, \varepsilon_1)$ with $\varepsilon_1 = 1$ and $y_0 = 0$. Because U is unbounded from above, we can take any $y_1 \in \overline{B}(y_0, \varepsilon_1)$ and then find x_{n_1} such that

$$|(x_{n_1}, y_1)| > 1 . \tag{2.9.1}$$

By continuity of the inner product in both its variables, we can find a closed ball $\overline{B}(y_1, \varepsilon_2)$ such that $\overline{B}(y_1, \varepsilon_2) \subset \overline{B}(y_0, \varepsilon_1)$ and such that (2.9.1) holds not only for y_1 but for all $y \in \overline{B}(y_1, \varepsilon_2)$:

$$|(x_{n_1}, y)| > 1 \quad \text{for all } y \in \overline{B}(y_1, \varepsilon_2) .$$

Then, in the ball $\overline{B}(y_1, \varepsilon_2)$, we similarly take an interior point y_2 and find x_{n_2}, with $n_2 > n_1$, such that

$$|(x_{n_2}, y_2)| > 2 ,$$

and, after this, a closed ball $\overline{B}(y_2, \varepsilon_3)$ such that $\overline{B}(y_2, \varepsilon_3) \subset \overline{B}(y_1, \varepsilon_2)$ and

$$|(x_{n_2}, y)| > 2 \quad \text{for all } y \in \overline{B}(y_2, \varepsilon_3) .$$

Repeating this procedure ad infinitum, we produce a sequence of closed balls $\overline{B}(y_k, \varepsilon_{k+1})$ such that $\overline{B}(y_0, \varepsilon_1) \supset \overline{B}(y_1, \varepsilon_2) \supset \overline{B}(y_2, \varepsilon_3) \supset \cdots$, and corresponding terms x_{n_k}, $n_{k+1} > n_k$, of the sequence $\{x_n\}$ such that

$$|(x_{n_k}, y)| > k \quad \text{for all } y \in \overline{B}(y_k, \varepsilon_{k+1}) .$$

Since H is a Hilbert space there is at least one element y^* which belongs to every $\overline{B}(y_k, \varepsilon_{k+1})$, so

$$|(x_{n_k}, y^*)| > k .$$

Thus we find a continuous linear functional $F^*(x) = (x, y^*)$ for which the numerical sequence $\{F^*(x_{n_k})\}$ is not a Cauchy sequence. This contradicts the definition of weak convergence of $\{x_k\}$. □

This proof yields another important result:

Lemma 2.9.1. Assume $\{x_k\}$ is an unbounded sequence in H, i.e., $\|x_k\| \to \infty$. Then there exists $y^* \in H$ and a subsequence $\{x_{n_k}\}$ such that $(x_{n_k}, y^*) \to \infty$ as $k \to \infty$.

Proof. Let $z_n = x_n/\|x_n\|$. For any y with unit norm, the numerical sequence (z_n, y) is bounded and thus we can select a convergent subsequence from it. If there exists such a unit element y^* and a subsequence $\{z_{n_k}\}$ for which $(z_{n_k}, y^*) \to a \neq 0$, then the statement of the lemma is valid for the subsequence $\{x_{n_k}\}$ and y^* if $a > 0$; if $a < 0$, then y^* must be changed to $-y^*$. Indeed, if $a > 0$ then $(x_{n_k}, y^*) = (z_{n_k}, y^*)\,\|x_{n_k}\| \to \infty$.

Now we suppose that we cannot find such an element y^* and a subsequence $\{z_{n_k}\}$ for which $(z_{n_k}, y^*) \to a \neq 0$. So $(z_n, y) \to 0$ for any $y \in H$. By the Riesz representation theorem, this means that $\{z_n\}$ converges weakly to zero. We will prove that the statement of Lemma 2.9.1 holds for the latter class of sequences as well. For this we repeat two steps of the proof of Theorem 2.9.2.

First we show that for any center y_0 and radius ε, the numerical set (x_n, y) with y running over $\overline{B}(y_0, \varepsilon)$ is unbounded. Indeed, taking the sequence $y_n = y_0 + \varepsilon/(2\,\|x_n\|)\, x_n$ we get an element from $\overline{B}(y_0, \varepsilon)$. Next,

$$(x_n, y_n) = (x_n, y_0) + \frac{\varepsilon}{2\,\|x_n\|}(x_n, x_n) = \left((z_n, y_0) + \frac{\varepsilon}{2}\right)\|x_n\| .$$

Since ε is finite and $(z_n, y_0) \to 0$ as $n \to \infty$, we have $(x_n, y_n) \to \infty$.

Another step of the proof of Theorem 2.9.2, establishing the existence of a subsequence $\{x_{n_k}\}$ and an element y^* such that $(x_{n_k}, y^*) \to \infty$, requires only that $\|x_n\| \to \infty$ and that for any $\varepsilon > 0$ the set (x_n, y) is unbounded when y runs over $\overline{B}(y_0, \varepsilon)$, which was just proved. Thus we immediately state the validity of Lemma 2.9.1 for all the unbounded sequences. □

This is used in proving the *principle of uniform boundedness*, which we have established in a more general form (Theorem 1.23.2).

Theorem 2.9.3. Let $\{F_k(x)\}$ $(k = 1, 2, \ldots)$ be a family of continuous linear functionals defined on a Hilbert space H. If $\sup_k |F_k(x)| < \infty$, then $\sup_k \|F_k\| < \infty$.

Proof. By the Riesz representation theorem, each of the functionals $F_k(x)$ has the form

$$F_k(x) = (x, f_k) \,, \quad \text{where} \quad f_k \in H \,, \quad \|f_k\| = \|F_k\| \,.$$

So the condition of the theorem can be rewritten as

$$\sup_k |(x, f_k)| < \infty \,. \tag{2.9.2}$$

By Lemma 2.9.1, the assumption that $\sup_k \|f_k\| = \infty$ implies the existence of $x_0 \in H$ and $\{f_{k_n}\}$ such that

$$|(x_0, f_{k_n})| \to \infty \ \text{ as } k \to \infty \,.$$

This contradicts (2.9.2). □

Corollary 2.9.1. Let $\{F_k(x)\}$ be a sequence of continuous linear functionals given on H, such that for every $x \in H$ the numerical sequence $\{F_k(x)\}$ is a Cauchy sequence. Then there is a continuous linear functional $F(x)$ on H such that

$$F(x) = \lim_{k\to\infty} F_k(x) \ \text{ for all } x \in H \tag{2.9.3}$$

and

$$\|F\| \le \liminf_{k\to\infty} \|F_k\| < \infty \,. \tag{2.9.4}$$

Proof. The limit on the right-hand side of (2.9.3), existing by the condition, defines a functional $F(x)$ which is clearly linear. Since the condition of Theorem 2.9.3 is met, we have $\sup_k \|F_k\| < \infty$; from

$$|F(x)| = \lim_{k\to\infty} |F_k(x)| \le \sup_k \|F_k\| \, \|x\|$$

it follows that $F(x)$ is continuous. Moreover (recall Problem 1.23.1),

$$|F(x)| = \lim_{k\to\infty} |F_k(x)| \le \liminf_{k\to\infty} \|F_k\| \, \|x\| \,,$$

i.e., (2.9.4) is proved also. □

The following theorem gives an equivalent but more convenient definition of weak convergence.

Theorem 2.9.4. A sequence $\{x_n\}$ is weakly Cauchy in a Hilbert space H if and only if the following pair of conditions holds:

(i) $\{x_n\}$ is bounded in H, i.e., there is a constant M such that $\|x_n\| \le M$;

(ii) for any $f_\alpha \in H$ from a system $\{f_\alpha\}$ which is complete in H, the numerical sequence (x_n, f_α) is a Cauchy sequence.

Proof. Necessity of the conditions follows from the definition of weak convergence and Theorem 2.9.2.

Now we prove sufficiency. Suppose the conditions (i) and (ii) hold. Take an arbitrary continuous linear functional defined, by the Riesz representation theorem, by an element $f \in H$ and consider the numerical sequence

$$d_{nm} = (x_n, f) - (x_m, f) .$$

As the system $\{f_\alpha\}$ is complete, there is a linear combination

$$f_\varepsilon = \sum_{k=1}^{N} c_k f_k$$

such that

$$\|f - f_\varepsilon\| < \varepsilon/3M .$$

Then

$$\begin{aligned} |d_{nm}| &= |(x_n - x_m, f)| \\ &= |(x_n - x_m, f_\varepsilon + f - f_\varepsilon)| \\ &\le |(x_n - x_m, f_\varepsilon)| + |(x_n - x_m, f - f_\varepsilon)| \\ &\le \sum_{k=1}^{N} |c_k|\, |(x_n - x_m, f_k)| + (\|x_n\| + \|x_m\|)\, \|f - f_\varepsilon\| . \end{aligned}$$

Since, by (ii), the sequences $\{(x_n, f_k)\}$ $(k = 1, \ldots, N)$ are Cauchy sequences, we can find a number R such that

$$\sum_{k=1}^{N} |c_k|\, |(x_n - x_m, f_k)| < \varepsilon/3 \quad \text{for all } m, n > R$$

hence

$$|d_{nm}| \le \varepsilon/3 + 2M\varepsilon/(3M) = \varepsilon \quad \text{for } m, n > R .$$

This means that $\{(x_n, f)\}$ is a weak Cauchy sequence. □

Problem 2.9.1. Show that a sequence $\{x_n\}$ is weakly convergent to x_0 in H if and only if the following pair of conditions holds:

(i) $\{x_n\}$ is bounded in H;

(ii) for any f_α from a system $\{f_\alpha\}$, $f_\alpha \in H$, which is complete in H, we have $\lim_{n\to\infty}(x_n, f_\alpha) = (x_0, f_\alpha)$. □

Because weak convergence differs from strong convergence, we are led to consider *weak completeness* of a Hilbert space.

Theorem 2.9.5. Any weak Cauchy sequence $\{x_n\}$ in a Hilbert space H converges weakly to an element of this space.

In other words, a Hilbert space H is also weakly complete.

Proof. For any fixed $y \in H$ we define $F(y) = \lim_{n\to\infty}(y, x_n)$. The functional $F(y)$, whose linearity is evident, is defined on the whole of H. From the inequality

$$|(y, x_n)| \le M \|y\|$$

where M is a constant such that $\|x_n\| \le M$, it follows that

$$|F(y)| \le M \|y\| \quad \text{and} \quad \|F\| \le M .$$

Therefore $F(y)$ is a continuous linear functional which, by the Riesz representation theorem, can be written in the form

$$F(y) = (y, f) , \qquad f \in H , \quad \|f\| = \|F\| \le M .$$

But this means that f is a weak limit of $\{x_n\}$. □

From this proof also follows

Lemma 2.9.2. If a sequence $\{x_n\} \subset H$ converges weakly to x_0 in H and $\|x_n\| \le M$ for all n, then $\|x_0\| \le M$.

Problem 2.9.2. Provide the details. □

This states that a closed ball about zero is weakly closed. Any closed subspace of a Hilbert space is also weakly closed. We also formulate Mazur's theorem that any closed convex set in a Hilbert space is weakly closed. The interested reader can find a proof in Yosida [44].

Theorem 2.9.6 (Mazur). Assume that a sequence $\{x_n\}$ in a Hilbert space H converges weakly to $x_0 \in H$. Then there is a subsequence $\{x_{n_k}\}$ of $\{x_n\}$ such that the sequence of arithmetic means $\frac{1}{N}\sum_{k=1}^{N} x_{n_k}$ converges strongly to x_0.

Problem 2.9.3. Show that a weakly closed set is closed. □

Let us consider the problem of *weak compactness* of a set in a Hilbert space. We have seen that a ball in an infinite dimensional Hilbert space is not strongly compact. But for weak compactness, an analog of the Bolzano–Weierstrass theorem holds as follows:

Theorem 2.9.7. A bounded sequence $\{x_n\}$ in a separable Hilbert space contains a weak Cauchy subsequence.

In other words, a bounded set in a Hilbert space is *weakly precompact.*

Proof. In a separable Hilbert space there is an orthonormal basis $\{g_n\}$. By Theorem 2.9.4 it suffices to show that there is a subsequence $\{x_{n_k}\}$ such that, for each fixed g_m, the numerical sequence $\{(x_{n_k}, g_m)\}$ is a Cauchy sequence.

The bounded numerical sequence $\{(x_n, g_1)\}$ contains a convergent subsequence $\{(x_{n_1}, g_1)\}$. Considering the numerical sequence $\{(x_{n_1}, g_2)\}$, for the same reason we can choose a convergent sequence $\{(x_{n_2}, g_2)\}$. Continuing this process, on the kth step we obtain a convergent numerical subsequence $\{(x_{n_k}, g_k)\}$.

Choosing now the elements x_{n_n}, we obtain a sequence $\{x_{n_n}\}$ such that for any fixed g_m the numerical sequence $\{(x_{n_n}, g_m)\}$ is a Cauchy sequence. That is, $\{x_{n_n}\}$ is a weak Cauchy sequence. □

This theorem has important applications. In justifying certain numerical methods we can sometimes prove boundedness of the set of approximate solutions in a Hilbert (as a rule, energy) space, and hence obtain a subsequence of approximations that converges weakly to an element; then we can show that this element is a solution.

Let us apply this procedure to the approximation problem, namely, we want to minimize a functional

$$F(x) = \|x - x_0\|^2 \tag{2.9.5}$$

over a real Hilbert space when x_0 is a fixed element of H, $x_0 \notin M$, and x is an arbitrary element of a closed subspace $M \subset H$.

In Sect. 1.19 we established the existence of a minimizer of $F(x)$. We now treat this problem once more, as though this existence were unknown to us.

This very simple problem (at least in theory) exhibits the following typical steps, which are common for the justification of approximate solutions to many boundary value problems:

1. the formulation of an approximation problem and the demonstration of its solvability;
2. a global *a priori* estimate of the approximate solutions that does not depend on the step of approximation;
3. the demonstration of convergence of the approximate solutions to a solution of the initial problem, and a study of the nature of convergence.

Thus we begin to study our problem with *Step 1*, the formulation of the approximation problem.

We try to solve the problem approximately, using the *Ritz method*. Assume $\{g_k\}$ is a complete system in M such that any of its finite subsystems is linearly independent. Consider M_n spanned by $(g_1, \ldots, g_n)$ and find an element which minimizes $F(x)$ on M_n. A solution of this problem, denoted by x_n, is the nth *Ritz approximation* of the solution.

A real-valued function $f(t) = F(x_n + tg_k)$ of the real variable t takes its minimal value at $t = 0$ and, thanks to differentiability of $f(t)$,

$$\left.\frac{df(t)}{dt}\right|_{t=0} = 0 .$$

This yields

$$0 = \frac{d}{dt}\|x_n - x_0 + tg_k\|^2\Big|_{t=0} = \frac{d}{dt}(x_n - x_0 + tg_k, x_n - x_0 + tg_k)\Big|_{t=0} = 2(x_n - x_0, g_k) ,$$

so $x_n - x_0$ is orthogonal to each g_k $(k = 1, \ldots, n)$.

Using the representation

$$x_n = \sum_{k=1}^{n} c_{kn} g_k ,$$

we get a linear system of algebraic equations called the Ritz system of nth approximation:

$$\sum_{k=1}^{n} c_{kn}(g_k, g_m) = (x_0, g_m) \qquad (m = 1, \ldots, n) . \tag{2.9.6}$$

The determinant of this system is the Gram determinant of a linearly independent system $(g_1, \ldots, g_m)$ that is not equal to zero. So the system (2.9.6) has a unique solution $(\hat{c}_{1n}, \ldots, \hat{c}_{kn})$.

Step 2. Now we will find a global estimate of the approximate solutions that does not depend on n. Although, in this case, we know that the approximate solution exists, we can get the estimate without this knowledge. Hence it is called an *a priori* estimate.

We begin with the definition of x_n:

$$\|x_n - x_0\|^2 \le \|x - x_0\|^2 \ \text{ for all } x \in M_n .$$

As $x = 0 \in M_n$, it follows that

$$\|x_n - x_0\|^2 \le \|x_0\|^2 ,$$

from which

$$\|x_n\|^2 \le 2\,\|x_n\|\,\|x_0\| ,$$

hence

$$\|x_n\| \le 2\,\|x_0\| . \tag{2.9.7}$$

This is the required estimate.

Remark 2.9.1. It is possible to get a sharper estimate than (2.9.7); however, for this problem it is only necessary to establish the existence of a bound. □

Step 3. Our last goal is to show that the sequence of approximations converges to a solution of the problem. First we demonstrate that this convergence is weak, and then that it is strong.

By (2.9.7), the sequence $\{x_n\}$ is bounded and, thanks to Theorem 2.9.7, contains a weakly convergent subsequence $\{x_{n_k}\}$ whose weak limit x^* belongs to M (remember that a closed subspace is weakly closed).

For any fixed g_m, we can pass to the limit as $k \to \infty$ in the equality

$$(x_{n_k} - x_0, g_m) = 0$$

and get

$$(x^* - x_0, g_m) = 0$$

because (x, g_m) is a continuous linear functional in $x \in H$.

Now consider $(x^* - x_0, h)$ where $h \in M$ is arbitrary but fixed. By completeness of the system $g_1, g_2, g_3, \ldots$ in M, given $\varepsilon > 0$ we can find a finite linear combination

$$h_\varepsilon = \sum_{k=1}^{N} c_k g_k$$

such that

$$\|h - h_\varepsilon\| \le \varepsilon/(3\,\|x_0\|)\,.$$

Then

$$\begin{aligned}
|(x^* - x_0, h)| &= |(x^* - x_0, h - h_\varepsilon + h_\varepsilon)| \\
&\le |(x^* - x_0, h - h_\varepsilon)| + |(x^* - x_0, h_\varepsilon)| \\
&= |(x^* - x_0, h - h_\varepsilon)| \\
&\le \|x^* - x_0\|\,\|h - h_\varepsilon\| \\
&\le (\|x^*\| + \|x_0\|)\,\|h - h_\varepsilon\| \\
&\le (2\,\|x_0\| + \|x_0\|)\,\varepsilon/(3\,\|x_0\|) = \varepsilon\,.
\end{aligned}$$

Therefore, for any $h \in M$ we get

$$(x^* - x_0, h) = 0\,. \tag{2.9.8}$$

Finally, considering values of (2.9.5) on elements of the form $x = x^* + h$ when $h \in M$, we obtain, by (2.9.8),

$$\begin{aligned}
F(x^* + h) &= (x^* - x_0 + h, x^* - x_0 + h) \\
&= \|x^* - x_0\|^2 + 2(x^* - x_0, h) + \|h\|^2 \\
&= \|x^* - x_0\|^2 + \|h\|^2 \ge \|x^* - x_0\|^2 = F(x^*)\,.
\end{aligned}$$

It follows that x^* is a solution of the problem, and existence of solution has been proved.

Now we can show that the approximation sequence converges strongly to a solution of the problem. By Theorem 1.19.3, a minimizer of $F(x)$ is unique; this gives us weak convergence of the sequence $\{x_n\}$ on the whole. Indeed, suppose to the contrary that $\{x_n\}$ does not converge weakly to x^*. Then there is an element $f \in H$ such that

$$(x_n, f) \nrightarrow (x^*, f) . \tag{2.9.9}$$

By boundedness of the numerical set $\{(x_n, f)\}$, the statement (2.9.9) implies that there is a subsequence $\{x_{n_k}\}$ such that there exists

$$\lim_{k\to\infty} (x_{n_k}, f) \neq (x^*, f) . \tag{2.9.10}$$

Problem 2.9.4. Prove (2.9.10). □

But, for the subsequence $\{x_{n_k}\}$, we can repeat the above considerations and find that $\{x_{n_k}\}$ contains a subsequence which converges weakly to a solution of the problem. Since the solution is unique, this contradicts (2.9.10). Finally, multiplying both sides of (2.9.6) by the Ritz coefficient $\hat{c}_{mn}$ and summing over m, we get

$$(x_n, x_n) = (x_0, x_n) .$$

We can pass to the limit as $n \to \infty$, obtaining

$$\lim_{n\to\infty} (x_n, x_n) = \lim_{n\to\infty} (x_0, x_n) = (x_0, x^*) .$$

By (2.9.8) with $h = x^*$ we have

$$(x_0, x^*) = (x^*, x^*) ,$$

so

$$\lim_{n\to\infty} \|x_n\|^2 = \|x^*\|^2 .$$

Therefore, by Theorem 2.9.1, the sequence $\{x_n\}$ converges strongly to x^*.

So we have demonstrated, via the Ritz method, a general way of justifying the solution of a minimal problem and the Ritz method itself. The method is common to a wide variety of problems, some nonlinear. In the latter case, many difficulties center on Steps 2 or 3, depending on the problem. The problem under discussion can also be interpreted another way, and this is of so much importance that we devote a separate section to it.

2.10 The Ritz and Bubnov–Galerkin Methods in Linear Problems

We reconsider the problem of minimizing the quadratic functional (2.4.8) in a Hilbert space, namely,

$$I(x) = \|x\|^2 + 2\Phi(x) \to \min_{x\in H} . \tag{2.10.1}$$

Assuming $\Phi(x)$ is a continuous linear functional, the Riesz representation theorem yields

$$\Phi(x) = (x, -x_0)$$

where $x_0 \in H$ is uniquely defined by $\Phi(x)$. Then

$$I(x) = \|x\|^2 - 2(x, x_0) = \|x - x_0\|^2 - \|x_0\|^2 .$$

Since $\|x_0\|^2$ is fixed, the problem (2.4.1) is equivalent to

$$F(x) = \|x - x_0\|^2 \to \min_{x \in H} .$$

This problem has the unique (and obvious) solution $x = x_0$. Of much interest is the fact that it coincides with the problem of the previous section if $M = H$. So application of the Ritz method in this problem is justified. Let us recall those results in terms of the new problem.

Let $\{g_k\}$ be a complete system in H, every finite subsystem of which is linearly independent, and let the nth Ritz approximation to a minimizer be

$$x_n = \sum_{k=1}^{n} c_{kn} g_k .$$

The system giving the nth approximation of the Ritz method is

$$\sum_{k=1}^{n} c_{kn}(g_k, g_m) = -\Phi(g_m) \qquad (m = 1, \ldots, n) . \tag{2.10.2}$$

Let us collect the results in

Theorem 2.10.1. The following statements hold.

(i) For each $n \geq 1$, the system (2.10.2) of nth approximation of the Ritz method has the unique solution $c_{1n}, \ldots, c_{nn}$.

(ii) The sequence $\{x_n\}$ of Ritz approximations defined by (2.10.2) converges strongly to the minimizer of the quadratic functional $\|x\|^2 + 2\Phi(x)$, where $\Phi(x)$ is a continuous linear functional on H.

It is interesting to note that if $\{g_k\}$ is an orthonormal basis of H, then (2.10.2) gives the Fourier coefficients of the solution in H.

Concerning Bubnov's method, we only mention that it appeared when A.S. Bubnov, reviewing an article by S. Timoshenko, noted that the Ritz equations can be obtained by multiplying by g_m, a function of a complete system, the differential equation of equilibrium in which u was replaced by

$$u_n = \sum_{k=1}^{n} c_{kn} g_k ,$$

integrating the latter over the region, and then integrating by parts. In our terms this is

$$(u_n, g_m) = -\Phi(g_m) \qquad (m = 1, \ldots, n) .$$

Since this system indeed coincides with (2.10.2), Theorem 2.10.1 also justifies Bubnov's method.

Galerkin was the first to propose multiplying by f_m, a function of another system, for better approximation of the residual. The corresponding system is, in our notation,

$$(u_n, f_m) = -\Phi(f_m) \qquad (m = 1, \ldots, n) .$$

Discussion of this modification of the method can be found in Mikhlin [27].

Finally, we note that the finite element method for solution of mechanics problems is a particular case of the Bubnov–Galerkin method, hence it is also justified for the problems we consider.

2.11 Curvilinear Coordinates, Nonhomogeneous Boundary Conditions

We have considered some problems of mechanics using the Cartesian coordinate system. Almost all of the textbooks present the theory of the same problems in Cartesian frames; the few exceptions are the textbooks on the theory of shells and curvilinear beams, where it is impossible to consider the problems in Cartesian frames. However, in practice other coordinate systems occur frequently. The question arises whether it is necessary to investigate the boundary value problems for other coordinates, or whether it is enough to reformulate the results for Cartesian systems. For the generalized statements of mechanics problems in energy spaces, the answer is simple: it is possible to reformulate the results, and a key tool is a simple change of the coordinates. This change allows us to reformulate the imbedding theorems in energy spaces, to establish the requirements for admitting classes of loads, etc. We note that it is a hard problem to obtain similar results independently, without the use of coordinate transformations, if the coordinate frame has singular points.

Let us consider a simple example of a circular membrane with fixed edge (Dirichlet problem). In Cartesian coordinates we have the Sobolev imbedding theorem

$$\left(\iint_\Omega |u(x)|^p \, dx\, dy \right)^{1/p} \le m \left\{ \iint_\Omega \left[\left(\frac{\partial u}{\partial x} \right)^2 + \left(\frac{\partial u}{\partial y} \right)^2 \right] dx\, dy \right\}^{1/2} \qquad (2.11.1)$$

for $p \ge 1$, which is valid for any $u \in \dot{W}^{1,2}(\Omega) \equiv E_{MC}$ satisfying the boundary condition

$$u\big|_{\partial\Omega} = 0 . \qquad (2.11.2)$$

Taking a function $u \in C^{(1)}(\Omega)$ satisfying (2.11.2), in both integrals of (2.11.1) we pass to the polar coordinate system:

$$\left(\int_0^R \int_0^{2\pi} |u|^p r\, d\phi\, dr\right)^{1/p} \leq m \left\{\int_0^R \int_0^{2\pi} \left[\left(\frac{\partial u}{\partial r}\right)^2 + \frac{1}{r^2}\left(\frac{\partial u}{\partial \phi}\right)^2\right] r\, d\phi\, dr\right\}^{1/2} \quad (2.11.3)$$

where (r, ϕ) are the polar coordinates in a disk of radius R. Passing to the limit along a Cauchy sequence of E_{MC} in the inequality (2.11.1), which is valid in Cartesian coordinates, shows us that it remains valid in the form (2.11.3) in polar coordinates. Inequality (2.11.3) is an imbedding theorem in the energy space of the circular membrane in terms of polar coordinates. The expression

$$\|u\| = \left\{\int_0^R \int_0^{2\pi} \left[\left(\frac{\partial u}{\partial r}\right)^2 + \frac{1}{r^2}\left(\frac{\partial u}{\partial \phi}\right)^2\right] r\, d\phi\, dr\right\}^{1/2} \quad (2.11.4)$$

is the norm in this coordinate system and

$$(u, v) = \int_0^R \int_0^{2\pi} \left(\frac{\partial u}{\partial r}\frac{\partial v}{\partial r} + \frac{1}{r^2}\frac{\partial u}{\partial \phi}\frac{\partial v}{\partial \phi}\right) r\, d\phi\, dr$$

is the corresponding inner product.

The requirement imposed on forces for existence of a generalized solution has the form

$$\int_0^R \int_0^{2\pi} |F|^q r\, d\phi\, dr < \infty \qquad (q > 1)\,.$$

We have a natural form of the norm in the energy space (which is determined by the energy itself) using curvilinear coordinates, as well as a form of the imbedding theorem (i.e., properties of elements of the energy space and natural requirements on forces for the problem to be uniquely solvable).

Then we note that we can replace formally the Cartesian system by any other system of coordinates which is admissible for smooth functions, and also change formally any variables in any expression which makes sense in the energy space considered in Cartesian coordinates.

Finally, note that a norm like (2.11.4) is usually called a *weighted norm* because of the presence of weight factors, here connected with powers of r. There is an abstract theory of such weighted Sobolev spaces, not being so elementary as in the space we have considered.

For more complicated problems such as elasticity problems, we can use the same method of introducing curvilinear coordinates; here we can change not only the independent variables (x_1, x_2, x_3), but also unknown components of vectors of displacements and prescribed forces, to the new coordinate system. We leave it to the reader to write down an equation determining a generalized solution, the forms of norm and scalar product, and restrictions for forces as well as imbedding inequalities, in other curvilinear coordinate systems such as cylindrical and spherical.

Now let us consider two questions connected with nonhomogeneous boundary value problems in mechanics. The first is to identify the whole class of admissible external forces for which an energy solution exists. We know that the condition for existence of a solution is that the functional of external forces

$$\int_\Omega F(\mathbf{x})v(\mathbf{x})\,d\Omega \tag{2.11.5}$$

(say, in the membrane problem) is continuous and linear with respect to $v(\mathbf{x})$ on an energy space. We shall show how this condition can be expressed in terms of so-called spaces with negative norms, a notion due to P.D. Lax [17].

The functional (2.11.5) can be considered as the scalar product of $F(\mathbf{x})$ by $v(\mathbf{x})$ in $L^2(\Omega)$. But $v(\mathbf{x})$ belongs to an energy space E whose norm, for simplicity, is assumed to be such that $\|v\|_E = 0$ implies $v = 0$. We know that $v \in L^2(\Omega)$ if $v \in E$; moreover, E is dense in $L^2(\Omega)$. For any $F(\mathbf{x}) \in L^2(\Omega)$, we can introduce a new norm

$$\|F\|_E = \sup_{\|v\|_E \le 1} \left| \int_\Omega F(\mathbf{x})v(\mathbf{x})\,d\Omega \right| .$$

It is clear that $L^2(\Omega)$ with this norm is not complete (since all $v \in L^p(\Omega)$ for any $p > 2$, $p < \infty$). The completion of $L^2(\Omega)$ in the norm $\|\cdot\|_E$ is called the space with negative norm, denoted E^-. In Lax [17] (and in other books, for example, Yosida [44]) it is shown that the set of all continuous linear functionals on E can be identified with E^- since E is dense in $L^2(\Omega)$.

So the condition $F(\mathbf{x}) \in E^-$ is necessary and sufficient for the work functional (2.11.5) to be continuous with respect to $v(\mathbf{x})$ on E.

In Lax [17], such a construction was introduced for a Sobolev space $\dot{W}^{k,2}(\Omega)$; the corresponding space with negative norm was denoted by $W^{-k,2}(\Omega)$. An equivalent approach to the introduction of $W^{-k,2}(\Omega)$ involves use of the Fourier transformation in Sobolev spaces (cf., Yosida [44]).

The notion of the space with negative norm is useful for studying problems, but it is not too informative when we want to know whether certain forces are of a needed class; here sufficient conditions are more convenient.

Secondly, we discuss how to handle nonhomogeneous boundary conditions (of Dirichlet type). Consider, for example, the problem

$$-\Delta v = F , \tag{2.11.6}$$

$$v\big|_{\partial\Omega} = \varphi . \tag{2.11.7}$$

We can try the classical approach, finding a function $v_0(\mathbf{x})$ that satisfies (2.11.7), i.e.,

$$v_0\big|_{\partial\Omega} = \varphi .$$

Now we are seeking $v(\mathbf{x})$ in the form $v = u + v_0$, where $u(\mathbf{x})$ satisfies the homogeneous boundary condition

$$u\big|_{\partial\Omega} = 0 . \tag{2.11.8}$$

An integro-differential equation of equilibrium of the membrane is

$$\iint_\Omega \left(\frac{\partial u}{\partial x}\frac{\partial \psi}{\partial x} + \frac{\partial u}{\partial y}\frac{\partial \psi}{\partial y} \right) d\Omega + \iint_\Omega \left(\frac{\partial v_0}{\partial x}\frac{\partial \psi}{\partial x} + \frac{\partial v_0}{\partial y}\frac{\partial \psi}{\partial y} \right) d\Omega = \iint_\Omega F\psi \, d\Omega \quad (2.11.9)$$

wherein virtual displacements must also satisfy (2.11.8):

$$\psi|_{\partial\Omega} = 0 \,.$$

We recognize the term

$$\iint_\Omega \left(\frac{\partial v_0}{\partial x}\frac{\partial \psi}{\partial x} + \frac{\partial v_0}{\partial y}\frac{\partial \psi}{\partial y} \right) d\Omega$$

as a continuous linear functional on E_{MC} if $\partial v_0/\partial x$ and $\partial v_0/\partial y$ belong to $L^2(\Omega)$. In that case there is a generalized solution to the problem, i.e., $u \in E_{MC}$ satisfying (2.11.9) for any $\psi \in E_{MC}$.

We have supposed that there exists an element of $W^{1,2}(\Omega)$ satisfying (2.11.7). In more detailed textbooks on the theory of partial differential equations, one may find the conditions for a function φ given on the boundary that are sufficient for the existence of v_0. Corresponding theorems for v_0 from Sobolev spaces are called trace theorems. The trace theorems assume the boundary is sufficiently smooth. The case of a piecewise smooth boundary, frequently encountered in practice, has not been completely studied yet. The problem of the traces of functions is beyond the scope of this book.

A final remark is in order. In mathematics we normally deal with dimensionless quantities, and we have followed that practice here. However, variables with dimensional units can be used without difficulty, provided we check carefully for units in all inequalities and equations, and introduce additional factors as needed. In particular, the constants in imbedding theorems normally carry dimensional units, hence these constants change if the units are changed.

2.12 Bramble–Hilbert Lemma and Its Applications

This lemma is widely used to establish the convergence rate for the finite element method (see, for example, Ciarlet [7]). It gives a bound for a functional with special properties in a Sobolev space. We remark that the lemma can be viewed as a simple consequence of the theorem on equivalent norming of $W^{l,p}(\Omega)$ in Sobolev [33].

Recall Poincaré's inequality (2.3.9),

$$\iint_S u^2 \, dS \le m \left\{ \left(\iint_S u \, dS \right)^2 + \iint_S \left[\left(\frac{\partial u}{\partial x} \right)^2 + \left(\frac{\partial u}{\partial y} \right)^2 \right] dS \right\}, \quad (2.12.1)$$

which was derived when S was the square $[0,a] \times [0,a]$.

The proof of (2.3.9) is easily extended to the case of an n-dimensional cube. We now discuss how to extend it to a compact set Ω that is star-shaped with respect to a square S; that is, any ray starting in S intersects the boundary of Ω exactly once. We shall establish the following estimate, which is also called *Poincaré's inequality*:

$$\iint_\Omega u^2\,d\Omega \le m_1\Big(\iint_S u\,dS\Big)^2 + m_2\iint_\Omega\left[\left(\frac{\partial u}{\partial x}\right)^2 + \left(\frac{\partial u}{\partial y}\right)^2\right]d\Omega\,. \tag{2.12.2}$$

Let us rewrite this in a system of polar coordinates (r,ϕ) having origin at the center of S. Let $\partial\Omega$ be given by the equation $r = R(\phi) \ge a/2$, $R(\phi) < R_0$. Then (2.12.2) has the form

$$\int_0^{2\pi}\int_0^{R(\phi)} u^2 r\,dr\,d\phi \le m_1\Big(\iint_S u\,dS\Big)^2 + m_2\int_0^{2\pi}\int_0^{R(\phi)}\left[\left(\frac{\partial u}{\partial r}\right)^2 + \frac{1}{r^2}\left(\frac{\partial u}{\partial \phi}\right)^2\right] r\,dr\,d\phi\,.$$

Because of (2.12.1), it follows that it is sufficient to get the estimate

$$\int_0^{2\pi}\int_{a/2}^{R(\phi)} u^2 r\,dr\,d\phi \le m_3\int_0^{2\pi}\int_{a/4}^{a/2} u^2 r\,dr\,d\phi + m_4\int_0^{2\pi}\int_{a/4}^{R(\phi)} r\left(\frac{\partial u}{\partial r}\right)^2 dr\,d\phi \tag{2.12.3}$$

with constants independent of $u \in C^{(1)}(\Omega)$ ($C^{(1)}(\Omega)$ is introduced in Cartesian coordinates!). We now proceed to prove this.

The starting point is the representation

$$u(r_2,\phi) = u(r_1,\phi) + \int_{r_1}^{r_2}\frac{\partial u(r,\phi)}{\partial r}\,dr\,, \qquad \begin{array}{l} a/4 \le r_1 \le a/2\,, \\ a/4 \le r_2 \le R_0\,, \end{array}$$

from which, by squaring both sides and applying elementary transformations, we get

$$\begin{aligned} u^2(r_2,\phi) &\le 2u^2(r_1,\phi) + 2\left[\int_{r_1}^{r_2}\frac{1}{\sqrt{r}}\left(\sqrt{r}\,\frac{\partial u}{\partial r}\right)dr\right]^2 \\ &\le 2u^2(r_1,\phi) + 2\int_{r_1}^{r_2}\frac{dr}{r}\int_{r_1}^{r_2} r\left(\frac{\partial u}{\partial r}\right)^2 dr \\ &\le 2u^2(r_1,\phi) + m_5\int_{a/4}^{R(\phi)} r\left(\frac{\partial u}{\partial r}\right)^2 dr\,, \quad m_5 = 2\ln\frac{4R_0}{a}\,. \end{aligned}$$

Multiplying this chain of inequalities by $r_1 r_2$ and then integrating it first with respect to r_2 from $a/2$ to $R(\phi)$ and then with respect to r_1 from $a/4$ to $a/2$, we have

$$\begin{aligned} \int_{a/4}^{a/2} r_1\int_{a/2}^{R(\phi)} u^2(r_2,\phi) r_2\,dr_2\,dr_1 &\le 2\int_{a/4}^{a/2} u^2(r_1,\phi) r_1\,dr_1\int_{a/2}^{R(\phi)} r_2\,dr_2 \\ &\quad + m_5\int_{a/4}^{a/2}\int_{a/2}^{R(\phi)} r_1 r_2\,dr_1\,dr_2\int_{a/4}^{R(\phi)} r\left(\frac{\partial u}{\partial r}\right)^2 dr \end{aligned}$$

or

$$\tfrac{3}{32}a^2\int_{a/2}^{R(\phi)} u^2(r,\phi)r\,dr \le R_0^2\int_{a/4}^{a/2} u^2(r,\phi)r\,dr + \tfrac{3}{64}a^2R_0^2m_5\int_{a/4}^{R(\phi)} r\left(\frac{\partial u}{\partial r}\right)^2 dr\,.$$

Finally, integrating this with respect to ϕ over $[0, 2\pi]$ and multiplying it by $32/(3a^2)$, we establish (2.12.3) and hence (2.12.2).

We can similarly extend Poincaré's inequality to the case of a multiconnected domain Ω which is a union of star-shaped domains, and to the case of an n-dimensional domain Ω with $n > 2$. The latter extension is

$$\int_\Omega u^2\,d\Omega \le m_1\Big(\int_C u\,d\Omega\Big)^2 + m_2\sum_{i=1}^{n}\int_\Omega\Big(\frac{\partial u}{\partial x_i}\Big)^2 d\Omega\,, \tag{2.12.4}$$

where $C \subset \Omega$ is a hypercube in $\mathbb{R}^n$.

We can apply the inequality (2.12.4) to any derivative $D^\alpha u$, $|\alpha| < k$. Combining these estimates successively, we derive the inequality needed to prove the Bramble–Hilbert lemma:

$$\|u\|^2_{W^{k,2}(\Omega)} \le m_3\sum_{0\le|\alpha|<k}\Big(\int_C D^\alpha u\,d\Omega\Big)^2 + m_4\sum_{|\alpha|=k}\int_\Omega |D^\alpha u|^2\,d\Omega\,. \tag{2.12.5}$$

This estimate permits us to introduce another form of equivalent norm in $W^{k,2}(\Omega)$. (Question to the reader: Which one?) Note that the estimate was obtained for functions of $C^{(k)}(\Omega)$, but the now standard procedure of completion provides that it is valid for any $u \in W^{k,2}(\Omega)$.

Lemma 2.12.1 (Bramble–Hilbert [5]). Assume $F(u)$ is a continuous linear functional on $W^{k,2}(\Omega)$ such that for any polynomial $P_r(\mathbf{x})$ of order less than k,

$$F(P_r(\mathbf{x})) = 0\,. \tag{2.12.6}$$

Then there is a constant m^* depending only on Ω such that

$$|F(u)| \le m^*\,\|F\|_{W^{k,2}(\Omega)}\Big(\sum_{|\alpha|=k}\int_\Omega |D^\alpha u|^2\,d\Omega\Big)^{1/2}. \tag{2.12.7}$$

Proof. From (2.12.5) and continuity of $F(u)$ on $W^{k,2}(\Omega)$, it follows that

$$|F(u)| \le m\,\|F\|_{W^{k,2}(\Omega)}\left[\sum_{0\le|\alpha|<k}\Big(\int_C D^\alpha u\,d\Omega\Big)^2 + \sum_{|\alpha|=k}\int_\Omega |D^\alpha u|^2\,d\Omega\right]^{1/2}. \tag{2.12.8}$$

By (2.12.6),

$$F(u(\mathbf{x}) + P_{k-1}(\mathbf{x})) = F(u(\mathbf{x}))$$

where $P_{k-1}(\mathbf{x})$ is an arbitrary polynomial of order $k-1$. Fixing $u(\mathbf{x}) \in W^{k,2}(\Omega)$, we can always choose a polynomial $P^*_{k-1}(\mathbf{x})$ such that

$$\int_C D^\alpha(u(\mathbf{x}) + P^*_{k-1}(\mathbf{x}))\, d\Omega = 0 \quad \text{for all } 0 \le |\alpha| \le k-1 .$$

Substituting $u(\mathbf{x}) + P^*_{k-1}(\mathbf{x})$ into (2.12.8), we get (2.12.7) since

$$D^\alpha P^*_{k-1}(\mathbf{x}) = 0 \quad \text{for } |\alpha| = k .$$

This completes the proof. □

Let us consider some simple applications of this lemma. Assume that we find numerically, by Simpson's rule,

$$\int_0^1 u(x)\, dx \quad \text{for } u(x) \in W^{2,2}(0,1) .$$

What is a bound on the error? First we find the error in one step of the trapezoidal rule:

$$F_k(u) = \int_{x_k}^{x_k+h} u(x)\, dx - \frac{h}{2}[u(x_k + h) + u(x_k)] .$$

It is clear that $F_k(u)$ is a linear and continuous functional in $W^{2,2}(0,1)$. Making the change of variable $x = x_k + hz$ in the integral, we get

$$|F_k(u)| = h \left| \int_0^1 u(x_k + hz)\, dz - \tfrac{1}{2}[u(x_k) + u(x_k + h)] \right| \le 2h \max_{z\in[0,1]} |u(x_k + zh)| . \tag{2.12.9}$$

By the elementary inequality (Problem 2.12.1 below)

$$\max_{x\in[0,1]} |f(x)| \le \sqrt{2} \left[\int_0^1 \left(f^2(x) + [f'(x)]^2 \right) dx \right]^{1/2} \le \sqrt{2}\, \|f\|_{W^{2,2}(0,1)} , \tag{2.12.10}$$

relation (2.12.9) gives

$$|F_k(u)| \le 2\sqrt{2}h\, \|u(x_k + hz)\|_{W^{2,2}(0,1)} .$$

Since $F_k(a+bx) = 0$ for any constants a, b we can apply the Bramble–Hilbert lemma and obtain

$$|F_k(u)| \le 2\sqrt{2}hm \left(\int_0^1 [u''(x_k + hz)]^2\, dz \right)^{1/2} = m_1 h^{5/2} \left(\int_{x_k}^{x_k+h} [u''(x)]^2\, dx \right)^{1/2} .$$

This is the needed error bound for one step of integration.

Consider now the bound on total error when $[0,1]$ is subdivided into N equal parts

$$F(u) = \int_0^1 u(x)\, dx - \frac{h}{2} \sum_{k=0}^{N-1} [u(x_k) + u(x_{k+1})] , \qquad x_k = kh .$$

This is linear and continuous in $W^{2,2}(0,1)$, and

$$f(u) = \sum_{k=0}^{N-1} F_k(u) \, .$$

We get

$$|F(u)| = \left| \sum_{k=0}^{N-1} F_k(u) \right| \le \sum_{k=0}^{N-1} |F_k(u)| \le m_1 h^{5/2} \sum_{k=0}^{N-1} \left(\int_{x_k}^{x_k+h} [u''(x)]^2 \, dx \right)^{1/2}$$

$$\le m_1 h^{5/2} \sqrt{N} \left(\sum_{k=0}^{N-1} \int_{x_k}^{x_k+h} [u''(x)]^2 \, dx \right)^{1/2} .$$

Thus the needed bound on the error of the trapezoidal rule is

$$|F(u)| \le m_1 h^2 \left(\int_0^1 [u''(x)]^2 \, dx \right)^{1/2} .$$

No improvements in the order of the error result if we take functions smoother than those from $W^{2,2}(0,1)$. But if $v \in W^{1,2}(0,1)$, the bound is worse:

$$|F(v)| \le m_2 h \left(\int_0^1 [v'(x)]^2 \, dx \right)^{1/2} .$$

Problem 2.12.1. Prove (2.12.10). □

Another example of the application of Lemma 2.12.1 is given by

Problem 2.12.2. Show that the local error of approximation of the first derivatives of a function $u(x_1, x_2) \in W^{3,2}(\Omega)$, $\Omega \subset \mathbb{R}^2$, by symmetric differences, is

$$l(u) = \left| \frac{\partial u(0,0)}{\partial x_1} - \frac{u(h_1,0) - u(-h_1,0)}{2h_1} \right| + \left| \frac{\partial u(0,0)}{\partial x_2} - \frac{u(0,h_2) - u(0,-h_2)}{2h_2} \right|$$

$$\le \frac{M(h_1^2 + h_2^2)}{\sqrt{h_1 h_2}} \|u\|_{W^{3,2}(\Omega)}$$

if $0 < c_1 < h_1/h_2 < c_2 < \infty$. □

Chapter 3
Some Spectral Problems of Mechanics

3.1 Introduction

We obtain a spectral problem by formally considering a solution u of the form

$$u(\mathbf{x}, t) = e^{i\omega t} v(\mathbf{x})$$

for a dynamic equation

$$B(u(\mathbf{x}, t)) = \rho \frac{\partial^2 u(\mathbf{x}, t)}{\partial t^2}$$

where B is a differential operator having coefficients independent of t, defined by the model of an elastic body, and ρ is the density of the body. The spectral equation is then

$$B(u(\mathbf{x}, t)) = -\rho\omega^2 u(\mathbf{x}, t) . \tag{3.1.1}$$

This equation should be supplemented with boundary conditions corresponding to those for the original dynamics problem. As in the traditional eigenvalue problems of linear algebra, we seek the set of values ω_k such that for $\omega = \omega_k$ there is a nontrivial solution of (3.1.1) that satisfies homogeneous boundary equations. Then ω_k is called an *eigenfrequency* of the problem and the corresponding solution is an *eigenfunction* or *eigenvector*.

In this chapter we establish some properties of the spectral problems of linear mechanics. In particular, we show that for various models of bounded elastic bodies, including the membrane, all the ω_k^2 are positive (there can be a few $\omega_k = 0$ that correspond to the motions of a body as a free rigid body) and the spectrum is discrete. We also establish completeness of the corresponding sets of eigenvectors, a property necessary for application of the separation of variables method to boundary value problems.

3.2 Adjoint Operator

This notion will be introduced for operators acting in a Hilbert space, although it can also be applied in other settings, as when we consider problems of linear continuum mechanics.

Let H be a Hilbert space and A a continuous linear operator acting in H. Consider the inner product (Ax, y) as a functional with respect to the variable $x \in H$ when $y \in H$ is arbitrary but fixed. This functional, thanks to the linearity of A, is linear and bounded because

$$|(Ax, y)| \leq \|Ax\| \, \|y\| \leq (\|A\| \, \|y\|) \, \|x\| \, .$$

By the Riesz representation theorem, it can be represented in the form

$$(Ax, y) = (x, z)$$

where the element z is uniquely defined by y and A. So the correspondence $y \mapsto z$ can be viewed as an operator A^*, $z = A^* y$, and we call A^* the *adjoint* of A.

Let us consider some properties of A^*.

Lemma 3.2.1. The adjoint A^* is a linear operator.

Proof. By definition we get

$$(Ax, y_1) = (x, A^* y_1) \, , \qquad (Ax, y_2) = (x, A^* y_2) \, ,$$

and

$$(Ax, \alpha_1 y_1 + \alpha_2 y_2) = (x, A^*(\alpha_1 y_1 + \alpha_2 y_2)) \, .$$

But

$$(Ax, \alpha_1 y_1 + \alpha_2 y_2) = \overline{\alpha_1}(Ax, y_1) + \overline{\alpha_2}(Ax, y_2) \, ,$$

so

$$(x, A^*(\alpha_1 y_1 + \alpha_2 y_2)) = \overline{\alpha_1}(x, A^* y_1) + \overline{\alpha_2}(x, A^* y_2) = (x, \alpha_1 A^* y_1) + (x, \alpha_2 A^* y_2) \, .$$

Since x is an arbitrary element of H, we have

$$A^*(\alpha_1 y_1 + \alpha_2 y_2) = \alpha_1 A^* y_1 + \alpha_2 A^* y_2 \, .$$

This completes the proof. □

Lemma 3.2.2. We have

(i) $(A + B)^* = A^* + B^*$,

(ii) $(AB)^* = B^* A^*$.

Proof. Property (i) is evident. Comparing the equalities

$$((AB)x, y) = (x, (AB)^* y)$$

and

$$(A(Bx), y) = (Bx, A^*y) = (x, B^*(A^*y)) ,$$

we prove (ii). □

Lemma 3.2.3. If A is a continuous linear operator, then A^* is continuous; moreover, we have $\|A^*\| = \|A\|$.

Proof. By definition of A^*,

$$(Ax, y) = (x, A^*y) . \qquad (3.2.1)$$

Putting $x = A^*y$ and using the Schwarz inequality, we get

$$\|A^*y\|^2 = (A(A^*y), y) \le \|A(A^*y)\| \, \|y\| \le \|A\| \, \|A^*y\| \, \|y\| .$$

Canceling $\|A^*y\|$ from both sides of the inequality, we have

$$\|A^*y\| \le \|A\| \, \|y\| ,$$

which means that A^* is bounded and $\|A^*\| \le \|A\|$. Next in (3.2.1) we put $y = Ax$. Similarly we obtain

$$\|Ax\|^2 = (x, A^*(Ax)) \le \|x\| \, \|A^*\| \, \|Ax\|$$

and it follows that $\|A\| \le \|A^*\|$. So $\|A^*\| = \|A\|$. □

Lemma 3.2.4. $(A^*)^* = A$.

Proof. Since A^* is continuous, then by definition

$$(A^*x, y) = (x, (A^*)^*y) .$$

On the other hand

$$(A^*x, y) = \overline{(y, A^*x)} = \overline{(Ay, x)} = (x, Ay) ,$$

so we get

$$(x, (A^*)^*y) = (x, Ay) .$$

Since x and y are arbitrary elements of H, we conclude that $(A^*)^* = A$. □

We have introduced the adjoint operator for a continuous linear operator in a Hilbert space. If A is unbounded, we can try to introduce A^* by the same equality

$$(Ax, y) = (x, A^*y) , \qquad x \in D(A)$$

which defines a value of A^*y uniquely if $D(A)$ is dense in H. What are the properties of A^* in this case? The interested reader can consult Yosida [44]. The notion of adjoint can also be introduced for operators acting in Banach spaces (see, e.g., [44]). In what follows, we will not use the notion of the adjoint to an unbounded operator.

We consider some examples.

1. *A matrix operator in* ℓ^2. This was considered in Sect. 1.32:

$$(Ax)_i = \sum_{j=1}^{\infty} a_{ij} x_j \,.$$

By (1.32.9), its norm in ℓ^2 is bounded by

$$\|A\| \le \Big(\sum_{i=1}^{\infty} \sum_{j=1}^{\infty} |a_{ij}|^2 \Big)^{1/2} .$$

The adjoint of A is defined as follows:

$$(Ax, y) = \sum_{i=1}^{\infty} \sum_{j=1}^{\infty} a_{ij} x_j \overline{y_i} = \sum_{j=1}^{\infty} x_j \overline{\Big(\sum_{i=1}^{\infty} \overline{a_{ij}} y_i \Big)} = (x, A^* y) \,,$$

so

$$(A^* y)_j = \sum_{i=1}^{\infty} \overline{a_{ij}}\, y_i \,.$$

Here the subscript j denotes the jth component of the vector A^*y. Sometimes it happens that $a_{ij} = \overline{a_{ji}}$ so that $A = A^*$.

Definition 3.2.1. We say that A is *self-adjoint* if $A^* = A$.

2. *An integral operator.* We consider an integral operator of the form

$$(Bf)(x) = \int_0^1 K(x, y) f(y)\, dy$$

in $L^2(0, 1)$. If we suppose that

$$K(x, y) \in L^2([0, 1] \times [0, 1]) \,, \tag{3.2.2}$$

then B is bounded in $L^2(0, 1)$. Indeed

$$\|Bf\|_{L^2(0,1)} = \Big(\int_0^1 \Big| \int_0^1 K(x, y) f(y)\, dy \Big|^2 dx \Big)^{1/2} .$$

Using the Schwarz inequality, we get

$$\begin{aligned} \|Bf\|_{L^2(0,1)} &\le \Big[\int_0^1 \Big(\int_0^1 |K(x, y)|^2\, dy \int_0^1 |f(y)|^2\, dy \Big) dx \Big]^{1/2} \\ &= \Big(\int_0^1 \int_0^1 |K(x, y)|^2\, dy\, dx \Big)^{1/2} \|f\|_{L^2(0,1)} \,. \end{aligned}$$

Thus

$$\|B\| \le \left(\int_0^1 \int_0^1 |K(x, y)|^2 \, dy \, dx \right)^{1/2} .$$

Let us introduce B^* under the assumption (3.2.2). From the equality

$$(Bf, g) = \int_0^1 \int_0^1 K(x, y) f(y) \, dy \, \overline{g(x)} \, dx = \int_0^1 f(y) \overline{\int_0^1 \overline{K(x, y)} g(x) \, dx} \, dy = (f, B^* g)$$

it follows that

$$(B^* g)(y) = \int_0^1 \overline{K(x, y)} \, g(x) \, dx .$$

Therefore B^* is also an integral operator. If $K(x, y) = \overline{K(y, x)}$, then B is self-adjoint.

3. *Stability of a thin plate.* In the theory of stability of an elastic nonlinear plate under tension, the linearized equation defining critical load points can be written in the generalized form

$$(w, \varphi)_{E_P} - \mu C(w, \varphi) = 0 , \tag{3.2.3}$$

$$C(w, \varphi) = \iint_\Omega \left[T_x \frac{\partial w}{\partial x} \frac{\partial \varphi}{\partial x} + T_{xy} \left(\frac{\partial w}{\partial y} \frac{\partial \varphi}{\partial x} + \frac{\partial w}{\partial x} \frac{\partial \varphi}{\partial y} \right) + T_y \frac{\partial w}{\partial y} \frac{\partial \varphi}{\partial y} \right] dx \, dy$$

where $(w, \varphi)_{E_P}$ is introduced by (2.3.15), $w(x, y) \in E_{PC}$ is the transverse displacement, and $\varphi \in E_{PC}$ is an admissible displacement. A particular spectral problem is to find the minimal μ such that there is a nontrivial function $w \in E_{PC}$ satisfying (3.2.3) for every $\varphi \in E_{PC}$. So this is the problem of finding the least eigenvalue. This minimal μ defines the critical load that is important for applications, and we will consider the problem later. Now we only note that it is another type of spectral problem that does not follow from a dynamic problem.

For definiteness, we consider the problem when T_x, T_{xy}, and T_y are considered as given functions from $L^2(\Omega)$. Later we will use the condition (3.12.2) that the plate is under tension, which provides for positiveness of all the eigenvalues. The following results of this section do not depend on (3.12.2).

Let us transform this problem into operator form. For this, consider $C(w, \varphi)$. Applying Hölder's inequality to a term

$$R(w, \varphi) = \iint_\Omega T_x \frac{\partial w}{\partial x} \frac{\partial \varphi}{\partial x} \, dx \, dy ,$$

we get

$$|R(w, \varphi)| \le \left(\iint_\Omega T_x^2 \, dx \, dy \right)^{1/2} \left(\iint_\Omega \left(\frac{\partial w}{\partial x} \right)^4 dx \, dy \right)^{1/4} \left(\iint_\Omega \left(\frac{\partial \varphi}{\partial x} \right)^4 dx \, dy \right)^{1/4} .$$

Remembering the imbedding theorem in E_{PC}, we obtain

$$|R(w, \varphi)| \le m \, \|w\|_{E_P} \, \|\varphi\|_{E_P} .$$

In a similar way, we can bound other terms in $C(w, \varphi)$ and so

$$|C(w, \varphi)| \leq m_1 \, \|w\|_{E_P} \, \|\varphi\|_{E_P} \,. \tag{3.2.4}$$

The linearity of $C(w, \varphi)$ with respect to both w and φ is evident.

Let $w \in E_{PC}$ be fixed. Then, thanks to (3.2.4), $C(w, \varphi)$ is a continuous linear functional with respect to $\varphi \in E_{PC}$ and thus, by the Riesz representation theorem, can be represented in the form

$$C(w, \varphi) = (\varphi, v)_{E_P} \,.$$

Since for every $w \in E_{PC}$ there is a unique element $v \in E_{PC}$ we have obtained a correspondence $w \mapsto v$ that is an operator G, $v = Gw$, whose linearity is evident. By (3.2.4) we have

$$|C(w, \varphi)| = |(\varphi, Gw)_{E_P}| \leq m_1 \, \|w\|_{E_P} \, \|\varphi\|_{E_P} \,.$$

Putting $\varphi = Gw$, we get

$$(Gw, Gw)_{E_P} \leq m_1 \, \|w\|_{E_P} \, \|Gw\|_{E_P}$$

or

$$\|Gw\|_{E_P} \leq m_1 \, \|w\|_{E_P} \,.$$

So the operator G is continuous.

Since the form $C(w, \varphi)$ is symmetric in w and φ, we get

$$(\varphi, Gw)_{E_P} = C(w, \varphi) = C(\varphi, w) = (w, G\varphi) \text{ for all } w, \varphi \in E_{PC} \,.$$

This means that G is self-adjoint. Later we will use this to establish some results on critical loads for the plate.

4. *Differentiation operator.* Let us consider an example of an unbounded operator. This is the operator

$$D_t = i \frac{d}{dt}$$

acting in $L^2(0, 1)$ whose domain consists of functions $\dot{W}^{1,2}(0, 1)$, i.e., functions $f(x)$ that, along with their derivatives, belong to $L^2(0, 1)$, and that satisfy

$$f(0) = f(1) = 0 \,.$$

As above, we shall find $(D_t)^*$:

$$(D_t f, g) = \int_0^1 i \frac{df(t)}{dt} \overline{g(t)} \, dt = \int_0^1 f(t) \overline{i \frac{dg(t)}{dt}} \, dt = (f, D_t^* g) \,.$$

This formula is valid if $g \in C^{(1)}(0, 1)$. Passage to the limit shows that it remains valid if $g \in W^{1,2}(0, 1)$.

Thus

$$D_t^* = i\frac{d}{dt}\ ;$$

i.e., D_t^* has the same form as D_t but its domain includes $W^{1,2}(0,1)$. So $D(D_t^*)$ is wider than $D(D_t)$, and thus $D_t^* \neq D_t$. In this case the operator is said to be *symmetric*.

We now obtain some simple but useful lemmas.

Lemma 3.2.5. If a linear operator A is strongly continuous on a Hilbert space then it is also weakly continuous; that is, it takes any weakly convergent sequence into a weakly convergent sequence.

Proof. Let $x_n \rightharpoonup x_0$ in H. An arbitrary continuous linear functional $F(x)$ takes the form $F(x) = (x, f)$, $f \in H$, and hence we must show that for any $f \in H$

$$(Ax_n - Ax_0, f) \to 0 \quad \text{as } n \to \infty\ .$$

But $(Ax, f) = (x, A^*f)$, and so

$$(Ax_n - Ax_0, f) = (x_n - x_0, A^*f) \to 0 \quad \text{as } n \to \infty$$

since $A^*f \in H$ and $\{x_n\}$ converges weakly to x_0. □

This is a particular case of the following.

Lemma 3.2.6. A continuous linear operator A acting from a normed space X into a normed space Y is also weakly continuous so if $x_n \rightharpoonup x_0$ as $n \to \infty$ then $Ax_n \rightharpoonup Ax_0$.

The proof follows from the fact that for any continuous linear functional $F(y)$ given on Y, the functional $\Phi(x) = F(Ax)$ is also continuous and linear, but on X.

Problem 3.2.1. Supply the details. □

Lemma 3.2.7. Assume A is a continuous linear operator acting in a Hilbert space H. Let $x_n \rightharpoonup x_0$ and $y_n \to y_0$ in H. Then

$$(Ax_n, y_n) \to (Ax_0, y_0) \quad \text{as } n \to \infty\ .$$

Proof. Because $(Ax, y) = (x, A^*y)$, where A^* is continuous, we get

$$R_n = (Ax_n, y_n) - (Ax_0, y_0) = (x_n, A^*y_n) - (x_0, A^*y_0)\ .$$

Transforming this, we have

$$\begin{aligned} R_n &= (x_n, A^*y_n) - (x_0, A^*y_0) + (x_n, A^*y_0) - (x_n, A^*y_0) \\ &= (x_n, A^*(y_n - y_0)) + (x_n - x_0, A^*y_0)\ ; \end{aligned}$$

but $(x_n - x_0, A^*y_0) \to 0$ because $x_n \rightharpoonup x_0$ in H, and

$$|(x_n, A^*(y_n - y_0))| \leq \|x_n\|\,\|A^*\|\,\|y_n - y_0\| \to 0$$

since $\{x_n\}$ is bounded as a weakly convergent sequence. □

Similarly, one may obtain an extension of Lemma 3.2.7:

Problem 3.2.2. Assume A and B are continuous linear operators acting in a Hilbert space H. Let $x_n \rightharpoonup x_0$ and $y_n \to y_0$ in H. Show that $(Ax_n, By_n) \to (Ax_0, By_0)$ as $n \to \infty$. □

Finally, we propose a simple but important

Problem 3.2.3. Let an operator K be defined via the Riesz representation theorem (cf., Sect. 2.6) from the equality

$$(Ku, \varphi)_E = \int_\Omega \rho(\mathbf{x})\, u(\mathbf{x})\, \varphi(\mathbf{x})\, d\Omega \tag{3.2.5}$$

in an energy space E when $\rho(\mathbf{x})$ is a bounded piecewise continuous function (density) on a compact set Ω. Show that K is a self-adjoint continuous linear operator in all of the energy spaces E_M, E_P, and E_E introduced earlier. (For the body with free boundary, only the spaces of balanced functions should be considered!) □

For a self-adjoint operator, the norm can be defined in another way.

Theorem 3.2.1. If A is a self-adjoint continuous linear operator given on a Hilbert space H, then

$$\|A\| = \sup_{\|x\|\le 1} |(Ax, x)|\,. \tag{3.2.6}$$

Proof. We denote $\sup_{\|x\|\le 1} |(Ax, x)| = \gamma$. Using the Schwarz inequality, we get

$$\gamma \le \sup_{\|x\|\le 1} \{\|Ax\|\, \|x\|\} \le \sup_{\|x\|\le 1} \{\|A\|\, \|x\|^2\} = \|A\|\,.$$

We now show the reverse inequality. By definition of γ, we have

$$|(Ax, x)| \le \gamma \|x\|^2. \tag{3.2.7}$$

Setting $x_1 = y + \lambda z$ and $x_2 = y - \lambda z$ where λ is a real number and $y, z \in H$, we have

$$C \equiv |(Ax_1, x_1) - (Ax_2, x_2)| = |2\lambda|\, |(Ay, z) + (Az, y)| = |2\lambda|\, |(Ay, z) + (z, Ay)|\,.$$

On the other hand

$$C \le |(Ax_1, x_1)| + |(Ax_2, x_2)| \le \gamma(\|x_1\|^2 + \|x_2\|^2) = 2\gamma(\|y\|^2 + \lambda^2\|z\|^2)$$

so

$$|2\lambda|\, |(Ay, z) + (z, Ay)| \le 2\gamma(\|y\|^2 + \lambda^2\|z\|^2) \quad \text{for all } y, z \in H\,.$$

Putting $z = Ay$, we obtain

$$|4\lambda|\, \|Ay\|^2 \le 2\gamma(\|y\|^2 + \lambda^2\|Ay\|^2)\,.$$

Take $\lambda = \|y\|/\|Ay\|$; then

$$4\|y\|\,\|Ay\| \le 2\gamma(\|y\|^2 + \|y\|^2)$$

or

$$\|Ay\| \le \gamma\|y\| \quad \text{for all } y \in H\,.$$

So $\|A\| \le \gamma$, and this completes the proof. □

3.3 Compact Operators

The study of linear operators in infinite dimensional spaces is complicated in comparison with the theory in finite dimensional spaces. However, some classes of these operators can be fully described; the first to note this was D. Hilbert. Among the most significant are the compact operators: they are close to finite dimensional operators in terms of their properties, and they play important roles in applications.

In this section we deal with a linear operator A acting from a normed space X to a Banach space Y.

Definition 3.3.1. A linear operator A is *compact* if it takes bounded sets of X into precompact subsets of Y.

Equivalently, A is compact if and only if every bounded sequence in X has a subsequence whose image under A is a Cauchy sequence in Y. Compact linear operators are sometimes called *completely continuous operators*.

Theorem 3.3.1. A compact linear operator A is bounded.

Proof. Suppose to the contrary that it is not bounded. Hence there is a bounded sequence $\{x_n\} \subset X$ such that $\|Ax_n\| \to \infty$ as $n \to \infty$. But the sequence $\{Ax_n\}$ does not contain a convergent subsequence, and this contradicts the definition of a compact operator. □

So a compact operator is continuous. The converse is, in general, false. For example, the identity operator $Ix = x$ is continuous but not compact if X is not finite dimensional, since a ball is precompact only in a finite dimensional space.

Problem 3.3.1. Show that, in order for a linear operator A to be compact, it is sufficient that A maps the closed unit ball of X into a precompact subset of Y. □

Problem 3.3.2. Let A and B be compact linear operators from X to Y and α a numerical constant. Show that the linear operator $C = \alpha A + B$ is compact. □

Let us consider an example of a compact operator. We show that the operator

$$(Bf)(t) = \int_0^1 K(t,s)f(s)\,ds$$

acting in $C(0, 1)$ is compact when $K(t, s) \in C([0, 1] \times [0, 1])$. It suffices to show that B takes the unit ball of $C(0, 1)$ into a precompact subset S of $C(0, 1)$. By boundedness of $|K(t, s)|$, the set S is uniformly bounded. By Arzelà's theorem on compactness in $C(0, 1)$, it remains to establish that the family of functions $(Bf)(t) \in S$ is equicontinuous, which we will demonstrate. As $K(t, s)$ is uniformly continuous on $[0, 1] \times [0, 1]$, for $\varepsilon > 0$ we can find $\tilde{\delta} > 0$ such that

$$|K(t + \delta, s) - K(t, s)| < \varepsilon \quad \text{for all } \delta < \tilde{\delta} \text{ and } t, t + \delta \in [0, 1] \,.$$

So for $t, t + \delta \in [0, 1]$, $\delta < \tilde{\delta}$, and $|f(s)| \leq 1$, the relations

$$\begin{aligned} |(Bf)(t + \delta) - (Bf)(t)| &= \left| \int_0^1 K(t + \delta, s) f(s)\, ds - \int_0^1 K(t, s) f(s)\, ds \right| \\ &\leq \max_{s \in [0,1]} |K(t + \delta, s) - K(t, s)| \leq \varepsilon \end{aligned}$$

demonstrate that S is equicontinuous. Hence B is compact in $C(0, 1)$.

Problem 3.3.3. Show that if $K(t, s) \in C([0, 1] \times [0, 1])$, then the integral operator B is also a compact operator in $L^2(0, 1)$. □

We shall see below that this restriction on $K(t, s)$ can be weakened.

Problem 3.3.4. Let A be an operator from $C^{(1)}(a, b)$ to $C(a, b)$ that maps a function from $C^{(1)}(a, b)$ to the same function but as an element of $C(a, b)$. Show that A is compact. □

The set of all compact linear operators is closed in $L(X, Y)$; this follows from

Theorem 3.3.2. If a sequence $\{A_n\} \subset L(X, Y)$ of compact linear operators converges to A in the norm of $L(X, Y)$, then $A \in L(X, Y)$ is compact.

Proof. Let $\|A_n - A\| \to 0$ as $n \to \infty$. Take a bounded sequence $\{x_n\} \subset X$. To prove that A is compact (its linearity is evident) we should show that there is a subsequence $\{x_{n_k}\}$ such that $\{Ax_{n_k}\}$ is a Cauchy sequence in Y. For this, from $\{x_n\}$, thanks to the compactness of A_k, we first select a subsequence $\{x_{n_1}\}$ such that $\{A_1 x_{n_1}\}$ is a Cauchy sequence in Y. From $\{x_{n_1}\}$ we select similarly a subsequence $\{x_{n_2}\}$ such that the sequence $\{A_2 x_{n_2}\}$ is also a Cauchy sequence. This can be repeated with $\{x_{n_2}\}$, $\{x_{n_3}\}$, and so on, producing a Cauchy sequence $\{A_k x_{n_k}\}$ each time. The diagonal subsequence $\{x_{n_n}\}$ is needed; indeed re-denoting x_{n_n} by z_n, we find that the $\{A_k z_n\}$ are Cauchy sequences for every fixed $k = 1, 2, 3, \ldots$, and

$$\begin{aligned} \|Az_{n+m} - Az_n\| &= \|(Az_{n+m} - A_k z_{n+m}) + (A_k z_{n+m} - A_k z_n) + (A_k z_n - Az_n)\| \\ &\leq \|A - A_k\| \, \|z_{n+m}\| + \|A_k(z_{n+m} - z_n)\| + \|A_k - A\| \, \|z_n\| \\ &\to 0 \quad \text{as } k, n \to \infty \,. \end{aligned}$$

This completes the proof. □

Now we can establish the promised result for the integral operator $Bf(t)$. We assume $K(t, s) \in L^2([0, 1] \times [0, 1])$ and show that $Bf(t)$ is a compact operator in $L^2(0, 1)$. By definition of $L^2(\Omega)$, there is a sequence of functions $\{K_n(t, s)\} \subset C([0, 1] \times [0, 1])$ such that

$$\int_0^1 \int_0^1 |K(t, s) - K_n(t, s)|^2 \, ds \, dt \to 0 \quad \text{as } n \to \infty . \tag{3.3.1}$$

Each of these kernels $K_n(t, s)$ defines an operator B_n,

$$(B_n f)(t) = \int_0^1 K_n(t, s) f(s) \, ds ,$$

which is a compact operator in $L^2(0, 1)$. From the inequality

$$\begin{aligned} \|Bf\|_{L^2(0,1)} &= \left(\int_0^1 \left| \int_0^1 K(t, s) f(s) \, ds \right|^2 dt \right)^{1/2} \\ &\le \left(\int_0^1 \int_0^1 |K(t, s)|^2 \, ds \, dt \right)^{1/2} \left(\int_0^1 |f(s)|^2 \, ds \right)^{1/2} \end{aligned}$$

it follows that

$$\|B\| \le \left(\int_0^1 \int_0^1 |K(t, s)|^2 \, ds \, dt \right)^{1/2} \quad \text{in } L^2(0, 1)$$

and by (3.3.1) we have $\|B - B_n\| \to 0$ as $n \to \infty$. By Theorem 3.3.2, this means that B is a compact linear operator in $L^2(0, 1)$.

Theorem 3.3.3. A compact operator $A \in L(X, Y)$ takes a weakly Cauchy sequence $\{x_n\} \subset X$ into a strongly Cauchy sequence $\{Ax_n\} \subset Y$. Moreover, if $x_n \rightharpoonup x_0$ then $Ax_n \to Ax_0$ in Y as $n \to \infty$.

Proof. A weakly Cauchy sequence $\{x_n\}$ is bounded in X so, by definition of compactness of A, we can take a strongly Cauchy subsequence $\{Ax_{n_1}\}$. As Y is a Banach space, $\{Ax_{n_1}\}$ converges to an element y of Y. By Lemma 3.2.6, $\{Ax_n\}$ is a weak Cauchy sequence in Y; because its subsequence converges strongly to y, the whole sequence $\{Ax_n\}$ converges weakly to $y \in Y$.

We now show that the whole sequence $\{Ax_n\}$ converges strongly to y. Suppose to the contrary that there is a subsequence $\{Ax_{n_2}\}$ which does not converge to y, i.e., there exists $\varepsilon > 0$ such that

$$\|Ax_{n_2} - y\| > \varepsilon . \tag{3.3.2}$$

From the sequence $\{Ax_{n_2}\}$ we can select a subsequence $\{Ax_{n_3}\}$ which is strongly Cauchy in Y and thus has a strong limit point $y_1 \in Y$. This subsequence converges weakly to the same element y_1. By the above, it converges weakly to y, too. By uniqueness of the weak limit, we get $y_1 = y$ and this contradicts (3.3.2). To complete the proof, we should add that by Lemma 3.2.6, if $x_n \rightharpoonup x_0$ then the above y is equal to Ax_0. □

In Sect. 2.1 we formulated the imbedding theorems in Sobolev spaces (Theorems 2.1.1–4). Now we can give a clear meaning to the term "compact operator": such an operator takes every sequence that converges weakly in $W^{k,p}(\Omega)$ into a sequence that converges strongly in a corresponding space shown by the condition of a theorem. In particular, in any $W^{k,2}(\Omega)$, $k \geq 1$, the imbedding operator into $L^2(\Omega)$ is compact.

Since all energy spaces introduced earlier can be considered as closed subspaces of $W^{k,2}(\Omega)$ for $k = 1$ or 2, we can say more about the operator K defined according to the Riesz representation theorem by the equality

$$(Ku, \varphi)_E = \int_\Omega \rho(\mathbf{x})\, u(\mathbf{x})\, \varphi(\mathbf{x})\, d\Omega$$

that was introduced in equation (3.2.5). We combine its properties into Lemma 3.3.1 below.

Note that in the case of a free boundary of a body, we should consider only the variants of energy spaces where the elements are "balanced" functions to avoid "rigid" motions. Subsequently, we may extend this property for the operator in a corresponding energy factor space where the zero element is the set of all rigid motions of the body.

Lemma 3.3.1. If Ω is a closed and bounded domain (in $\mathbb{R}^2$ or $\mathbb{R}^3$) and $\rho(\mathbf{x})$ is a bounded piecewise continuous function, then K is a compact self-adjoint operator in any of the spaces E_M, E_P, or E_E.

Proof. We begin with the inequalities

$$|(Ku, \varphi)_E| = \left| \int_\Omega \rho(\mathbf{x})\, u(\mathbf{x})\, \varphi(\mathbf{x})\, d\Omega \right|$$
$$\leq \sup_{\mathbf{x}\in\Omega} |\rho(\mathbf{x})|\, \|u\|_{L^2(\Omega)}\, \|\varphi\|_{L^2(\Omega)} \leq m\, \|u\|_{L^2(\Omega)}\, \|\varphi\|_E \qquad (3.3.3)$$

which follow from the Schwarz inequality and the imbedding theorems in an energy space.

Let $\{u_n\}$ be a bounded sequence in an energy space so that it contains a weakly Cauchy subsequence which we denote by $\{u_n\}$ again. As the imbedding operator from E to $L^2(\Omega)$ is compact, the latter is a strong Cauchy sequence in $L^2(\Omega)$. Setting $u = u_{n+m} - u_n$ and $\varphi = K(u_{n+m} - u_n)$ in (3.3.3), we get

$$(K(u_{n+m} - u_n), K(u_{n+m} - u_n))_E \leq m\, \|u_{n+m} - u_n\|_{L^2(\Omega)}\, \|K(u_{n+m} - u_n)\|_E\ .$$

Dividing both sides of this by $\|K(u_{n+m} - u_n)\|_E$, we get

$$\|Ku_{n+m} - Ku_n\|_E \leq m\, \|u_{n+m} - u_n\|_{L^2(\Omega)} \to 0 \text{ as } n \to \infty\ ,$$

which means that K is compact. Self-adjointness was shown in Problem 3.2.3. □

Definition 3.3.2. An operator A_n acting from X to Y is *finite dimensional* if it takes the form

$$A_n x = \sum_{k=1}^{n} \Phi_k(x) y_k$$

where the $\Phi_k(x)$ are functionals on X and $y_k \in Y$.

If the Φ_k are continuous linear functionals, then A_n is a continuous linear operator.

Problem 3.3.5. Confirm this. □

In this case the next result holds.

Theorem 3.3.4. A continuous finite dimensional linear operator A_n is compact.

Proof. Let S be a bounded set in X. By boundedness of the functionals Φ_k, the numerical set $\{\Phi_k(x) \colon x \in S\}$ is also bounded. Take a sequence $\{x_m\} \subset S$. By boundedness of the numerical sequence $\{\Phi_1(x_m)\}$, we can select a convergent numerical subsequence $\{\Phi_1(x_{m_1})\}$. Considering the numerical sequence $\{\Phi_2(x_{m_1})\}$, we can choose a convergent subsequence $\{\Phi_2(x_{m_2})\}$. Continuing this process, we get a subsequence $\{x_{m_n}\}$ for which each of the sequences $\{\Phi_k(x_{m_n})\}$ $(k = 1, \ldots, n)$ is convergent and thus $\{A_n x_{m_n}\}$ is a Cauchy sequence in Y. □

We apply this theorem to a matrix operator

$$\mathbf{y} = A\mathbf{x}\,, \qquad \mathbf{y} = (y_1, y_2, \ldots)\,, \qquad y_k = \sum_{l=1}^{\infty} a_{kl} x_l \qquad (k = 1, 2, 3, \ldots)$$

in ℓ^2 to show that A is compact if

$$\sum_{k=1}^{\infty} \sum_{l=1}^{\infty} |a_{kl}|^2 < \infty\,. \tag{3.3.4}$$

We have shown that

$$\|A\| \le \left(\sum_{k=1}^{\infty} \sum_{l=1}^{\infty} |a_{kl}|^2 \right)^{1/2}.$$

Consider an operator A_n defined as follows:

$$\mathbf{y} = A_n \mathbf{x}, \qquad y_k = \begin{cases} \sum_{l=1}^{\infty} a_{kl} x_l\,, & k \le n\,, \\ 0\,, & k > n\,. \end{cases}$$

A_n is a finite-dimensional operator and so is compact. Since

$$\|A - A_n\| \le \left(\sum_{k=n+1}^{\infty} \sum_{l=1}^{\infty} |a_{kl}|^2 \right)^{1/2} \to 0 \text{ as } n \to \infty$$

then, by Theorem 3.3.2, A is compact if (3.3.4) holds. Note the similarity to the above proof of compactness for the integral operator $Bf(t)$.

Problem 3.3.6. Assume $K(\mathbf{x}, \mathbf{y}) \in L^2(\Omega \times \Omega)$ where Ω is a closed bounded domain in $\mathbb{R}^n$. Show that an integral operator

$$(Af)(\mathbf{x}) = \int_\Omega K(\mathbf{x}, \mathbf{y})\, f(\mathbf{y})\, d\Omega_{\mathbf{y}}$$

is a compact operator in $L^2(\Omega)$. □

Problem 3.3.7. Using the Riesz representation theorem, introduce a nonlinear operator K_1 in E_{MC} by the equality

$$(K_1(u), \varphi)_{E_M} = \int_\Omega \rho(\mathbf{x})\, u^n(\mathbf{x})\, \varphi(\mathbf{x})\, d\Omega\,,$$

where $\rho(\mathbf{x})$ is a given bounded piecewise continuous function on a compact set Ω and n is a positive integer. Show that K_1 takes every weakly Cauchy sequence $\{u_m(\mathbf{x})\}$ into the strongly Cauchy sequence $\{K_1 u_m\}$ in E_{MC}. □

3.4 Compact Operators in Hilbert Space

In a Hilbert space, the statements of Theorems 3.3.3 and 3.3.4 can be sharpened.

Theorem 3.4.1. An operator A acting in a Hilbert space is compact if and only if it takes every weakly Cauchy sequence $\{x_n\}$ into the strong Cauchy sequence $\{Ax_n\}$ in H.

Proof. Necessity was proved in Theorem 3.3.3. Let us prove sufficiency. Let M be a bounded set in H and $A(M)$ its image under A. We need to show that $A(M)$ is precompact. Take a sequence $\{y_n\}$ belonging to $A(M)$ and consider the sequence $\{x_n\}$ lying in M such that $Ax_n = y_n$. Since M is bounded, so is $\{x_n\}$. Thus $\{x_n\}$ contains a subsequence $\{x_{n_k}\}$ that is a weak Cauchy sequence in H. By the condition of the theorem, its image, the sequence $\{Ax_{n_k}\}$, is a (strong) Cauchy sequence in H so $A(M)$ is precompact. □

Theorem 3.4.2. Assume A is a compact operator acting in a separable Hilbert space H. Then there is a sequence of finite dimensional continuous linear operators $\{A_n\}$ such that $\|A - A_n\| \to 0$ as $n \to \infty$.

Proof. The separable Hilbert space H has an orthonormal basis $\{g_n\}$ and any $f \in H$ can be represented as

$$f = \sum_{k=1}^{\infty} (f, g_k) g_k\,.$$

By continuity of A we have

$$Af = \sum_{k=1}^{\infty} (f, g_k) A g_k\,.$$

Denote by A_n a finite dimensional linear operator

$$A_n f = \sum_{k=1}^{n} (f, g_k) A g_k ,$$

and $R_n = A - A_n$. By Theorem 3.3.2, to prove that $A_n \to A$ as $n \to \infty$, it is sufficient to show that $\alpha_n = \|R_n\| \to 0$ as $n \to \infty$. We will demonstrate this.

By definition of the norm,

$$\alpha_n = \sup_{\|f\| \le 1} \|R_n f\| .$$

First we show that there is an element f_n^* such that

$$\|f_n^*\| \le 1 \quad \text{and} \quad \alpha_n = \|R_n f_n^*\| .$$

Indeed, let $\{f_k\}$ be a maximizing sequence such that $\|f_k\| \le 1$ and $\|R_n f_k\| \to \alpha_n$ as $k \to \infty$. Choosing a weakly convergent subsequence $\{f_{k_l}\}$ whose weak limit is f_n^*, thanks to Lemma 2.9.2, we get $\|f_n^*\| \le 1$. As R_n is a compact operator, the sequence $\{R_n f_{k_l}\}$ converges strongly to $R_n f_n^*$. So $\alpha_n = \|R_n f_n^*\|$. (Question for the reader: What is the value of $\|f_n^*\|$?)

But

$$R_n f_n^* = A\Big(\sum_{k=n+1}^{\infty} (f_n^*, g_k) g_k \Big)$$

so

$$\alpha_n = \|A\varphi_n\| , \qquad \varphi_n = \sum_{k=n+1}^{\infty} (f_n^*, g_k) g_k .$$

We show that the sequence $\{\varphi_k\} \subset H$ converges weakly to zero. Indeed, for an arbitrary continuous linear functional defined by an element

$$f = \sum_{m=1}^{\infty} (f, g_m) g_m ,$$

we get

$$\begin{aligned}
|(\varphi_n, f)| &= \left| \Big(\sum_{k=n+1}^{\infty} (f_n^*, g_k) g_k, \sum_{m=1}^{\infty} (f, g_m) g_m \Big) \right| \\
&= \left| \Big(\sum_{k=n+1}^{\infty} (f_n^*, g_k) g_k, \sum_{m=n+1}^{\infty} (f, g_m) g_m \Big) \right| \\
&\le \Big(\sum_{m=n+1}^{\infty} |(f, g_m)|^2 \Big)^{1/2} \|f_n^*\| \to 0 \quad \text{as } n \to \infty
\end{aligned}$$

since $\|f_n^*\| \le 1$ and the series

$$\sum_{m=1}^{\infty} |(f, g_m)|^2 = ||f||^2$$

is convergent. Since φ_n is weakly convergent to zero and A is compact we have $||A\varphi_n|| = \alpha_n \to 0$ as $n \to \infty$. This completes the proof. □

In the above proof we have constructed a sequence of finite dimensional operators $\{A_n\}$,

$$A_n f = \sum_{k=1}^{n} (f, g_k) A g_k \,,$$

that tends to a linear compact operator A. Note that frequently in an equation containing a compact operator A, we can replace A by some finite dimensional operator A_n to approximate the solution.

Theorems 3.4.1 and 3.4.2 state, in particular, that in energy spaces (which are separable Hilbert spaces) we can use other equivalent definitions of a compact linear operator: such an operator (1) takes every weak Cauchy sequence into a strong Cauchy sequence, or (2) can be approximated with any prescribed accuracy in the operator norm by a finite-dimensional continuous linear operator.

Theorem 3.4.3. If A is a compact linear operator acting in a Hilbert space, then A^* is compact.

Proof. Take a sequence $\{f_n\}$ converging weakly to f_0. It suffices to show that $\{A^* f_n\}$ converges strongly to $A^* f_0$. Indeed,

$$\begin{aligned}
||A^* f_n - A^* f_0||^2 &= (A^* f_n - A^* f_0, A^* f_n - A^* f_0) \\
&= (f_n - f_0, AA^*(f_n - f_0)) \\
&\le ||f_n - f_0|| \, ||AA^*(f_n - f_0)|| \to 0 \quad \text{as } n \to \infty
\end{aligned}$$

(here we used the fact that a product AB is compact if A is compact and B is bounded, and also $||f_n|| < M = \text{constant}$). □

Problem 3.4.1. Prove that if A is a compact linear operator and B is a bounded linear operator, then AB is a compact linear operator. □

3.5 Functions Taking Values in a Banach Space

We reserve the name "function" for a single-valued correspondence from a subset of $\mathbb{R}^n$ to a Banach space Y. This is a useful convention in many problems of mechanics.

We begin with definitions. A rule that assigns to each point of a domain in $\mathbb{R}^n$ a unique element of a Banach space Y, written $y = f(\mathbf{x})$, $y \in Y$, is called a *function* with values in Y. All notions relative to the functions of classical calculus that are based only on the properties of the metric are easily transferred to the present setting.

For example, $y = f(\mathbf{x})$ is continuous at $\mathbf{x}_0$ if for every $\varepsilon > 0$ there exists $\delta = \delta(\varepsilon) > 0$ such that $\|f(\mathbf{x}) - f(\mathbf{x}_0)\| < \varepsilon$ whenever $|\mathbf{x} - \mathbf{x}_0| < \delta$. If $f(\mathbf{x})$ is continuous at every point of an open domain Ω in $\mathbb{R}^n$, it is said to be continuous in Ω. Continuity on a closed domain is introduced in a manner similar to the parallel notion in calculus.

The set of all continuous functions given on a closed and bounded domain Ω whose values lie in a Banach space Y is also a Banach space. We denote this space $C(\Omega; Y)$; the norm on $C(\Omega; Y)$ is defined by

$$\|f(\mathbf{x})\|_{C(\Omega;Y)} = \max_{\mathbf{x}\in\Omega} \|f(\mathbf{x})\|_Y .$$

For a function $y = f(t)$, $y \in Y$, $t \in (a, b)$, the derivative df/dt at $t = t_0$ is

$$\frac{df(t_0)}{dt} = \lim_{t\to t_0} \frac{f(t) - f(t_0)}{t - t_0} .$$

Higher order derivatives are introduced similarly. We may also introduce spaces $C^{(k)}(a, b; Y)$ analogous to the corresponding spaces of calculus.

Finally, we can construct the Riemann integral

$$\int_a^b f(t)\, dt$$

for a function with values in a Banach space; the method parallels that of ordinary calculus. There is nothing analogous to the mean value theorem, but we do have, for example,

$$\left\| \int_a^b f(t)\, dt \right\| \le \int_a^b \|f(t)\|\, dt \le \max_{t\in[a,b]} \|f(t)\| \cdot (b - a) \qquad (a < b) .$$

These are consequences of passages to the limit in corresponding inequalities for Riemann sums.

The construction of the Lebesgue integral for functions whose values lie in a Banach space is introduced using the completion theorem in a manner similar to that used for scalar-valued functions in Sect. 1.10.

If functions $y = f(x)$ take their values in a Hilbert space H, we can introduce a Hilbert space $L^2(a, b; H)$ with inner product

$$(f(t), g(t))_{L^2(a,b;H)} = \int_a^b (f(t), g(t))_H\, dt$$

as the completion of $C(a, b; H)$ in the corresponding norm

$$\|f(t)\|_{L^2(a,b;H)} = \left(\int_a^b \|f(t)\|_H^2\, dt \right)^{1/2} .$$

The particular case $L^2(a, b; L^2(\Omega))$ coincides with the space $L^2([a, b] \times \Omega)$.

In some problems we meet a situation where, in addition to the main energy inner product, there is another inner product not depending on time — the product in $L^2(\Omega)$ for example. We denote such an additional inner product by $\langle \cdot, \cdot \rangle$. Assuming as above that

$$\langle f, f \rangle \le m(f, f)_H \equiv m(f, f) \tag{3.5.1}$$

where the constant m does not depend on $f \in H$, we can construct a space analogous to the Sobolev space $W^{1,2}(a, b)$; it is a Hilbert space $W^1(a, b)$ in which the inner product is

$$(f(t), g(t))_{W^1(a,b)} = \int_a^b \left[\left\langle \frac{df}{dt}, \frac{dg}{dt} \right\rangle + (f, g) \right] dt \,. \tag{3.5.2}$$

The space $W^1(a, b)$ is the completion of $C^{(1)}(a, b; H)$ in the norm corresponding to (3.5.2).

Thanks to (3.5.1), we get

$$\int_a^b \langle f(t), f(t) \rangle \, dt \le m \int_a^b (f(t), f(t)) \, dt \le m \|f\|^2_{W^1(a,b)} \,.$$

We can obtain some properties of the elements of $W^1(0, T)$ if we take into account the Newton–Leibnitz formula

$$f(s + \Delta) - f(s) = \int_s^{s+\Delta} \frac{df(t)}{dt} \, dt$$

which holds for f continuously differentiable. We have

$$\int_s^{s+\Delta} \left\langle \frac{df(t)}{dt}, f(t) \right\rangle dt = \frac{1}{2} \int_s^{s+\Delta} \frac{d}{dt} \langle f, f \rangle \, dt = \frac{1}{2} \langle f(s+\Delta), f(s+\Delta) \rangle - \frac{1}{2} \langle f(s), f(s) \rangle$$

and

$$\begin{aligned} \|f(s + \Delta) - f(s)\|_0^2 &= \left\| \int_s^{s+\Delta} \frac{df}{dt} \, dt \right\|_0^2 \le \left(\int_s^{s+\Delta} 1 \cdot \left\| \frac{df}{dt} \right\|_0 dt \right)^2 \\ &\le \int_s^{s+\Delta} 1^2 \, dt \int_s^{s+\Delta} \left\| \frac{df}{dt} \right\|_0^2 dt \le \Delta \int_0^T \left\| \frac{df}{dt} \right\|_0^2 dt \end{aligned}$$

where $\|f\|_0^2 = \langle f, f \rangle$. Passage to the limit for a representative sequence of the element of $W^1(0, T)$ shows that this inequality remains valid for elements of $W^1(0, T)$. Hence the elements of $W^1(0, T)$ are continuous in t with respect to the norm $\| \cdot \|_0$. This is an imbedding theorem for $W^1(0, T)$.

We now turn to the problem of holomorphic functions with values in a Banach space.

A function $f(\lambda)$ given on an open domain G of the complex plane $\mathbb{C}$ to a Banach space X is called *holomorphic* in G if for each point $\lambda_0 \in G$ there is a neighborhood $D(\lambda_0)$ of λ_0 in which there is a power series expansion

$$f(\lambda) = f(\lambda_0) + \sum_{k=1}^{\infty} c_k(\lambda - \lambda_0)^k , \qquad \lambda \in D(\lambda_0)$$

converging uniformly by the norm of X in $D(\lambda_0)$.

A holomorphic vector-valued function has properties similar to those of a scalar-valued function. For example, if $f(\lambda)$ is holomorphic in $|\lambda - \lambda_0| < R$ and $\|f(\lambda)\| \leq M$, then it is infinitely differentiable in this disk, the Taylor expansion

$$f(\lambda) = \sum_{n=0}^{\infty} \frac{f^{(n)}(\lambda_0)}{n!} (\lambda - \lambda_0)^n , \qquad |\lambda - \lambda_0| < R ,$$

exists, and

$$\|f^{(n)}(\lambda_0)\| \leq n!\, MR^{-n} .$$

Next, if $f(\lambda)$ is a holomorphic function, $f(\lambda) \in X$, on the domain G, then

$$\oint_C f(\lambda)\, d\lambda = 0$$

for every simple closed rectifiable contour C in G such that the interior of C belongs to G. The Cauchy representation

$$f(\lambda) = \frac{1}{2\pi i} \oint_C \frac{f(z)}{z - \lambda}\, dz$$

is also valid for λ lying in the interior of C.

These and similar results can be found in Yosida [44].

3.6 Spectrum of a Linear Operator

In problems of the mechanics of continuous media there arises an operator equation

$$x - A(\mu)x = f \tag{3.6.1}$$

in a Banach space X, where $A(\mu)$ is a linear operator depending on a real or complex parameter. A typical example is an equation describing steady forced vibrations of elastic bodies, which has the form

$$x - \mu Ax = f .$$

In particular, eigen-oscillations of a string with fixed ends are governed by the boundary value problem

$$\lambda x + x'' = 0 , \qquad x(0) = x(1) = 0 ,$$

with $\mu = 1/\lambda$. Another instance of (3.6.1) is the equation

$$\left(I + \sum_{k=1}^{n} \mu^k A_k\right) x = f$$

which appears, for example, in the theory of an elastic band.

Let us introduce some notation. A value μ_0 is called a *regular point* of the operator $A(\mu)$ if there is a bounded inverse $(I - A(\mu_0))^{-1}$ whose domain is dense in X; otherwise, μ_0 belongs to the *spectrum* of $A(\mu)$.

The same terms will be used for an operator A: λ is a regular point of A if there is a bounded inverse $R(\lambda, A) = (\lambda I - A)^{-1}$ with domain dense in X; otherwise, λ is a point of the spectrum of A.

The set of all regular points of A is called the *resolvent set* of A. It is denoted by $\rho(A)$.

A point λ may fail to be a regular point of A for several reasons, and the set of spectrum points of A can be classified into three types.

1. The *point spectrum* is the set of all complex λ such that $(\lambda I - A)$ does not have an inverse. The equation $(\lambda I - A)x = 0$ then has a nontrivial solution called an *eigenfunction* or *eigenvector*, and λ is an *eigenvalue* of A.
2. The *continuous spectrum* is the set of all $\lambda \in \mathbb{C}$ such that there exists $(\lambda I - A)^{-1}$ whose domain $D(R(\lambda, A))$ is dense in X, but such that the operator $(\lambda I - A)^{-1}$ is not bounded.
3. The *residual spectrum* is the set of $\lambda \in \mathbb{C}$ such that $R(\lambda, A) = (\lambda I - A)^{-1}$ exists but with domain not dense in X.

We consider some examples.

1. A matrix operator acting in the n-dimensional Euclidean space has only a point spectrum consisting of no more than n points called the eigenvalues of the matrix. Other points of the complex plane are regular. Here continuous and residual parts of the spectrum cannot exist.

2. Any point of the complex plane belongs to the point spectrum of the differentiation operator d/dt acting in $C(a, b)$, since for every λ the equation

$$\frac{df}{dt} - \lambda f = 0$$

has a solution $f(t) = ce^{\lambda t}$ where c is a constant. So the operator d/dt in $C(a, b)$ has no regular points. (Question for the reader: What happens to the spectrum if we consider the equation in the subspace of $C(a, b)$ consisting of functions $f(x)$ that satisfy $f(a) = 0$?)

3. On the square $[0, \pi] \times [0, \pi]$ we consider the boundary value problem

$$\frac{\partial^2 u}{\partial x^2} + \lambda \frac{\partial^2 u}{\partial y^2} = f(x, y), \qquad u\big|_{\partial\Omega} = 0. \tag{3.6.2}$$

The operator $B(\lambda)$ on the left-hand side of this equation is considered in $L^2(\Omega)$, $\Omega = [0,\pi]\times[0,\pi]$. Note that we keep the terminology for the spectrum although this operator does not have the form (3.6.1).

The Fourier expansion of $f(x,y)$ is

$$f(x,y) = \sum_{m,n=1}^{\infty} f_{mn} \sin mx \sin ny\,, \qquad \sum_{m,n=1}^{\infty} |f_{mn}|^2 < \infty\,.$$

Let λ be a point of $\mathbb{C}$ — not on the negative real axis, but otherwise arbitrary. Then the solution of (3.6.2) is

$$u(x,y) = \sum_{m,n=1}^{\infty} -\frac{f_{mn}}{m^2+\lambda n^2} \sin mx \sin ny\,.$$

If $\lambda = \lambda_1 + i\lambda_2$ is such that $\lambda_2 \neq 0$, or if $\lambda_2 = 0$ but $\lambda_1 \geq 0$, then $|m^2 + \lambda n^2| > \delta > 0$ for all integers $m, n \geq 1$. Therefore

$$\|u(x,y)\|^2_{L^2(\Omega)} = \frac{\pi^2}{4} \sum_{m,n=1}^{\infty} |f_{mn}|^2 \frac{1}{|m^2+\lambda n^2|^2} \leq \frac{\pi^2}{4\delta^2} \sum_{m,n=1}^{\infty} |f_{mn}|^2$$

and it follows that

$$\|u\|_{L^2(\Omega)} \leq \frac{1}{\delta} \|f\|_{L^2(\Omega)}\,. \tag{3.6.3}$$

There are no other solutions to (3.6.2) so the inequality (3.6.3) means that the inverse is bounded and thus these λ belong to the resolvent set of the operator $B(\lambda)$ acting in $L^2(\Omega)$. What can we say about $\lambda \in \mathbb{C}$ such that $\lambda = \operatorname{Re}\lambda < 0$?

First we consider λ of the form $\lambda = -p^2/q^2$ where p and q are integers. For these λ the corresponding boundary value problem is not solvable for some $f(x,y)$. To show this, take $f(x,y) = \sin px \sin qy$. As is easily seen, if there is a solution to (3.6.2) then it has the form $u(x,y) = c \sin px \sin qy$ and for $\lambda_0 = -p^2/q^2$ must satisfy the equation $c(p^2 + \lambda_0 q^2) = -1$, which is impossible. Moreover, $u = \sin px \sin qy$ is a solution to the homogeneous equation (3.6.2) at $\lambda = \lambda_0$. So all $\lambda = -p^2/q^2$, where p, q are integers, belong to the point spectrum of $B(\lambda)$.

We consider the remaining part M of the negative real axis, i.e., the set of λ such that $\lambda = \operatorname{Re}\lambda < 0$ and λ cannot be represented in the form $-p^2/q^2$ for some integers p, q. For $\lambda \in M$ we can seek a solution in the form of a Fourier series

$$u(x,y) = \sum_{m,n=1}^{\infty} \frac{f_{mn}}{m^2+\lambda n^2} \sin mx \sin ny\,.$$

The set S of functions $f(x,y)$ of the form

$$f(x,y) = \sum_{m=1}^{N_1} \sum_{n=1}^{N_2} f_{mn} \sin mx \sin ny$$

is dense in $L^2(\Omega)$; since solutions corresponding to these $f(x,y) \in S$ are also in $L^2(\Omega)$ and defined uniquely, the inverse of $B(\lambda)$ is determined on a dense subset of $L^2(\Omega)$. We show that it is unbounded for $\lambda \in M$. Indeed, the set of all points of the form $\lambda_{pq} = -p^2/q^2$ is dense in M. Take $\lambda \in M$; there is a sequence $\lambda_n = -p_n^2/q_n^2 \to \lambda$ as $n \to \infty$. Take the function $f_n(x,y) = \sin p_n x \sin q_n y$, which is the right-hand side of (3.6.2). To this, there corresponds the solution

$$u_n(x,y) = -\frac{1}{p_n^2 + \lambda q_n^2} \sin p_n x \sin q_n y \, .$$

Their norms are related by

$$\|u_n\|_{L^2(\Omega)} = \frac{1}{|p_n^2 + \lambda q_n^2|} \|f_n\|_{L^2(\Omega)}$$

where $|p_n^2 + \lambda q_n^2| \to 0$ as $n \to \infty$. So the inverse to the operator $B(\lambda)$ is unbounded when $\lambda \in M$, and thus M is a subset of the continuous spectrum.

4. Now we consider the *coordinate operator* in $C(a,b)$, defined as

$$(Qu)(t) = tu(t) \, .$$

This operator has no eigenvalues. If $\lambda \notin [a,b]$, then it belongs to the resolvent set of Q since the equation

$$\lambda u(t) - tu(t) = f(t)$$

has the unique solution

$$u(t) = \frac{f(t)}{\lambda - t}$$

in $C(a,b)$.

But if $\lambda \in [a,b]$, then there exists the inverse defined by

$$u(t) = f(t)/(\lambda - t)$$

whose domain consists of functions that can be represented in the form

$$f(t) = (\lambda - t)z(t) \, , \qquad z(t) \in C(a,b) \, .$$

This domain is not dense in $C(a,b)$, as all such functions vanish at $t = \lambda$. Hence the points of $[a,b]$ belong to the residual spectrum.

Problem 3.6.1. Show that for the coordinate operator acting in $L^2(a,b)$, the points of $[a,b]$ belong to the continuous spectrum. □

Problem 3.6.2. Show that if A is an invertible linear operator acting in a Banach space X, then A and A^{-1} have the same eigenvectors. □

Problem 3.6.3. Find the eigenvalues and eigenfunctions of the operator A given by $Af(x) = f(-x)$ and acting in the real Banach space $C(-\pi,\pi)$. □

3.7 Resolvent Set of a Closed Linear Operator

Theorem 3.7.1. Assume that A is a closed linear operator acting in a complex Banach space X. For any λ_0 belonging to the resolvent set of A, the resolvent $R(\lambda_0, A) = (\lambda_0 I - A)^{-1}$ is a continuous linear operator defined on the whole of X.

Proof. By definition of resolvent set, the domain of $R(\lambda_0, A)$ is dense in X and there is a constant $m > 0$ such that

$$||(\lambda_0 I - A)x|| \geq m\,||x|| \,. \tag{3.7.1}$$

Take $y \in X$; by definition, there is a sequence $\{x_n\}$ such that

$$\lim_{n\to\infty} (\lambda_0 I - A)x_n = y \quad \text{(strong limit)} \,.$$

By (3.7.1), we have $\lim_{n\to\infty} x_n = x$ and $(\lambda_0 I - A)x = y$ since A is closed. So the range of $(\lambda_0 I - A)^{-1}$ is X. □

Theorem 3.7.2. Assume that A is a closed linear operator acting in a complex Banach space X. Then the resolvent set $\rho(A)$ is an open domain of $\mathbb{C}$ and $R(\lambda, A)$ is holomorphic with respect to λ in $\rho(A)$.

Proof. By Theorem 3.7.1, $R(\lambda, A)$ is a continuous linear operator on X for any $\lambda \in \rho(A)$. So the series

$$R(\lambda_0, A)\left\{I + \sum_{n=1}^{\infty} (\lambda_0 - \lambda)^n R^n(\lambda_0, A)\right\}$$

is convergent in the disk $|\lambda - \lambda_0| < 1/||R(\lambda_0, A)||$ of $\mathbb{C}$ and thus is a holomorphic function in this disk. Multiplying this series by $(\lambda I - A) = (\lambda - \lambda_0)I + (\lambda_0 I - A)$, we get I, i.e., it is the inverse to $\lambda I - A$. □

Problem 3.7.1. Carry out the multiplication mentioned above. □

Theorem 3.7.3. Under the condition of Theorem 3.7.2, the *Hilbert identity*

$$R(\lambda, A) - R(\mu, A) = (\mu - \lambda)R(\lambda, A)R(\mu, A)$$

holds for any $\lambda, \mu \in \rho(A)$.

Proof. The identity follows from

$$\begin{aligned} R(\lambda, A) &= R(\lambda, A)(\mu I - A)R(\mu, A) \\ &= R(\lambda, A)\{(\mu - \lambda)I + (\lambda I - A)\}R(\mu, A) \\ &= (\mu - \lambda)R(\lambda, A)R(\mu, A) + R(\mu, A) \end{aligned}$$

since $R(\lambda, A)(\lambda I - A) = I$. □

Let B be a bounded linear operator in X. Then the series

$$\frac{1}{\lambda}\left(I + \sum_{n=1}^{\infty} \lambda^{-n} B^n\right)$$

is convergent if $|\lambda| > ||B||$. Multiplying it by $\lambda I - B$ we get I, i.e.,

$$R(\lambda, B) = \frac{1}{\lambda}\left(I + \sum_{n=1}^{\infty} \lambda^{-n} B^n\right) \text{ for } |\lambda| \geq ||B|| .$$

Problem 3.7.2. Confirm this. □

Lemma 3.7.1. The expansion

$$R(\lambda, B) = \frac{1}{\lambda}\left(I + \sum_{n=1}^{\infty} \lambda^{-n} B^n\right)$$

is valid in the domain $|\lambda| > r_\sigma(B)$, where $r_\sigma(B)$ is the spectral radius of B defined by

$$r_\sigma(B) = \lim_{n\to\infty} ||B^n||^{1/n} .$$

Proof. We show that $r_\sigma(B)$ exists. Denote

$$r_0 = \inf_n ||B^n||^{1/n} .$$

We establish that $r_\sigma(B) = r_0$. By definition of infimum, for any positive ε we can find an integer N such that

$$||B^N||^{1/N} \leq r_0 + \varepsilon .$$

For large n represented as

$$n = kN + l \qquad (0 \leq l < N)$$

we get

$$||B^n||^{1/n} \leq ||B^{kN}||^{1/n} \, ||B^l||^{1/n} \leq ||B^N||^{k/n} \, ||B^l||^{1/n} \leq (r_0 + \varepsilon)^{kN/n} \, ||B^l||^{1/n} \leq r_0 + \varepsilon + \varepsilon_1(n)$$

where $\varepsilon_1(n) \to 0$ as $n \to \infty$. Together with the inequality $||B^n||^{1/n} \geq r_0$, this proves $r_\sigma(B)$ exists and equals r_0. The rest of the proof is trivial. □

Problem 3.7.3. Let $A(\mu)$ be a continuous operator-function in X which is holomorphic with respect to μ in $\mathbb{C}$. Show that the resolvent set $\rho(A(\mu))$ of $A(\mu)$ is open and $(I - A(\mu))^{-1}$ is a holomorphic operator-function in $\rho(A(\mu))$. □

3.8 Spectrum of a Compact Operator in Hilbert Space

An important class of operators for which there is a full description of the spectrum is the class of compact linear operators. The first results in this direction were due to I. Fredholm; studying the integral operator, he established spectral properties similar to those of the matrix operator. The theory was then extended to the class of compact operators (F. Riesz, J. Schauder) which we now consider in a Hilbert space. The theory is of great interest as it describes eigen-oscillations of bounded elastic bodies.

So let A be a compact linear operator acting in a Hilbert space H. We are seeking eigenvectors and eigenvalues of A, i.e., nontrivial solutions to the equation

$$(I - \mu A)x = 0 . \tag{3.8.1}$$

In the previous section we used traditional spectrum terminology from functional analysis. From now on we use the term *eigenvalue* for μ from (3.8.1) but not for λ from the equation $(\lambda I - A)x = 0$. The reason is that in later applications of the theory we consider a nontraditional introduction of the operator equations for oscillations that are composed in energy spaces. For example, in the generalized setup of the eigenfrequency problem in our presentation, for a membrane which is governed by the equation

$$\Delta u = -\omega^2 u ,$$

the term Δu corresponds to the operator I, whereas the right-hand side term u corresponds to the operator A of (3.8.1). (Indeed, recall the development of Sect. 2.6, where the classical equation (2.6.1) was recast as (2.6.3) in the energy space E_{MC}.) Thus μ is equal to the squared eigenfrequency of the membrane, ω^2.

The Fredholm–Riesz–Schauder theory will be presented as a number of lemmas and theorems. We want to underline that the results on the properties of eigenvectors corresponding to a fixed eigenvalue are the same for compact linear operators $A(\mu)$ with general dependence on μ.

Let μ_0 be an eigenvalue of A. By continuity and linearity of A, the set of all eigenvectors corresponding to μ_0, when supplemented with the zero vector 0, is a closed subspace of H denoted by $H(\mu_0)$.

Problem 3.8.1. Prove that $H(\mu_0)$ is a closed subspace of H. □

Lemma 3.8.1. $H(\mu_0)$ is finite dimensional.

Proof. By definition of compact operator and the equality $x = \mu_0 Ax$, $x \in H(\mu_0)$, from any bounded sequence $\{x_k\} \subset H(\mu_0)$ we can choose a Cauchy subsequence. This means that every bounded subset of $H(\mu_0)$ is precompact and, by Theorem 1.17.3, $H(\mu_0)$ is finite dimensional. □

Suppose x belongs to both subspaces $H(\mu_k)$ and $H(\mu_n)$ where $\mu_k \neq \mu_n$. Then $x = \mu_k Ax$ and $x = \mu_n Ax$. It follows that $\mu_k x = \mu_n x$, and therefore $x = 0$. Consequently,

$$H(\mu_k) \cap H(\mu_n) = \{0\} \ \text{ if } \mu_k \neq \mu_n .$$

We have the following stronger statement.

Lemma 3.8.2. Assume $\{x_1^{(i)}, x_2^{(i)}, \ldots, x_{n_i}^{(i)}\}$ is a linearly independent system in the subspace $H(\mu_i)$ for each i. Then the union of the elements

$$\{x_1^{(1)}, x_2^{(1)}, \ldots, x_{n_1}^{(1)}\}, \ldots, \{x_1^{(k)}, x_2^{(k)}, \ldots, x_{n_k}^{(k)}\}$$

is a linearly independent system in the space $H(\mu_1)\dot{+}\cdots\dot{+}H(\mu_k)$ whose elements are linear combinations of elements of the spaces $H(\mu_i)$ $(i = 1, \ldots, k)$. If each of these systems $\{x_1^{(i)}, x_2^{(i)}, \ldots, x_{n_i}^{(i)}\}$ is a basis of the corresponding subspace $H(\mu_i)$ $(i = 1, \ldots, k)$, then their union is a basis of $H(\mu_1)\dot{+}\cdots\dot{+}H(\mu_k)$.

Proof. Since the dimension of the direct sum $H(\mu_1)\dot{+}\cdots\dot{+}H(\mu_k)$ is less than or equal to the sum of the dimensions of the $H(\mu_i)$, it will suffice to show that the union of the $x_{n_i}^{(j)}$ is a linearly independent system. Let us renumber successively the eigenvectors $x_{n_i}^{(j)}$ and eigenvalues μ_i in such a way that μ_k corresponds to x_k: $x_k = \mu_k A x_k$ (For instance, if the original μ_1 corresponds to three eigenvectors $x_1^{(1)}, x_2^{(1)}, x_3^{(1)}$, we simply redenote the resulting pairs as (μ_1, x_1), (μ_2, x_2), (μ_3, x_3), and so on.) The proof of independence is carried out by induction. Assume a system $x_1, \ldots, x_n$ is linearly independent. Let us add the next eigenvector x_{n+1} and consider the equation

$$\sum_{k=1}^{n+1} c_k x_k = 0$$

with respect to the coefficients c_k. Applying A to both sides we get

$$\sum_{k=1}^{n+1} c_k A x_k = 0$$

or, since x_k is an eigenvector of A (i.e., $Ax_k = x_k/\mu_k$), we have

$$\mu_{n+1} \sum_{k=1}^{n+1} \frac{c_k}{\mu_k} x_k = 0 .$$

Subtracting this from $\sum_{k=1}^{n+1} c_k x_k = 0$, we have

$$\sum_{k=1}^{n} c_k \left(1 - \frac{\mu_{n+1}}{\mu_k}\right) x_k = 0 .$$

By assumption, $c_k = 0$ for those $k = 1, \ldots, n-s$ for which $\mu_k \neq \mu_{n+1}$. For the rest of the eigenvectors, that is for $x_{n-s+1}, \ldots, x_{n+1}$ corresponding to $\mu_k = \mu_{n+1}$, we have

$$\sum_{k=n-s+1}^{n+1} c_k x_k = 0 .$$

In this all x_k correspond to μ_{n+1} and so are linearly independent by the choice. Thus $c_k = 0$ for $k = 1, \ldots, n+1$. □

Lemma 3.8.3. The set of eigenvalues of a compact linear operator A has no finite limit points in $\mathbb{C}$.

Proof. Suppose there is a sequence of distinct eigenvalues $\mu_n \to \mu_0$, $|\mu_0| < \infty$. For each μ_n take an eigenvector x_n. Denote by H_n the subspace spanned by $x_1, x_2, \ldots, x_n$. By Lemma 3.8.2, $H_n \subset H_{n+1}$ and $H_n \neq H_{n+1}$. Hence, by the orthogonal decomposition theorem, we may decompose H_{n+1} into H_n and another nonempty subspace orthogonal to H_n, from which we may choose a normalized element. So there exists $y_{n+1} \in H_{n+1}$ such that $\|y_{n+1}\| = 1$ and y_{n+1} is orthogonal to H_n.

The sequence $\{\mu_n y_n\}$ is bounded in H so the sequence $\{A(\mu_n y_n)\}$ must contain a Cauchy subsequence. But this is impossible as is shown below. Indeed

$$A(\mu_{n+m} y_{n+m}) - A(\mu_n y_n) = y_{n+m} - (y_{n+m} - \mu_{n+m} A y_{n+m} + \mu_n A y_n) . \tag{3.8.2}$$

Now $y_{n+m} \in H_{n+m}$. We show that the term in parentheses on the right-hand side belongs to H_{n+m-1} $(m \geq 1)$; indeed

$$y_{n+m} = \sum_{k=1}^{n+m} c_k x_k$$

and so

$$y_{n+m} - \mu_{n+m} A y_{n+m} = \sum_{k=1}^{n+m} c_k x_k - \mu_{n+m} A\left(\sum_{k=1}^{n+m} c_k x_k\right) = \sum_{k=1}^{n+m-1} c_k \left(1 - \frac{\mu_{n+m}}{\mu_k}\right) x_k \in H_{n+m-1} .$$

Since $\mu_n A y_n \in H_n \subset H_{n+m-1}$ as well, we have proved the needed property for the term in parentheses.

Thus y_{n+m} and $(y_{n+m} - \mu_{n+m} A y_{n+m} + \mu_n A y_n)$ are mutually orthogonal. From (3.8.2) it follows that

$$\|A(\mu_{n+m} y_{n+m}) - A(\mu_n y_n)\|^2 = \|y_{n+m}\|^2 + \|y_{n+m} - \mu_{n+m} A y_{n+m} + \mu_n A y_n\|^2 \geq 1 ,$$

which contradicts the fact that $\{A(\mu_n y_n)\}$ contains a Cauchy sequence. □

Combining these three lemmas, we formulate

Theorem 3.8.1. There are no more than a countable number of eigenvalues of a compact linear operator acting in a Hilbert space; the set of eigenvalues has no finite limit point in $\mathbb{C}$. A subspace $H(\mu_k)$ of all eigenvectors of A corresponding to a μ_k is finite dimensional and $H(\mu_k) \cap H(\mu_n) = \{0\}$ if $\mu_k \neq \mu_n$.

Note that the final assertion is implied by Lemma 3.8.2.

Let us denote by $M(\mu_0)$ the orthogonal complement to $H(\mu_0)$ in H. By the decomposition Theorem 1.20.1, $M(\mu_0)$ is closed and therefore complete.

Lemma 3.8.4. There are constants $m_1 > 0$ and $m_2 > 0$ such that

$$m_1 \|x\| \le \|x - \mu_0 Ax\| \le m_2 \|x\| \tag{3.8.3}$$

for all $x \in M(\mu_0)$.

Proof. To obtain the right-hand inequality we write

$$\|x - \mu_0 Ax\| \le \|x\| + \|\mu_0 Ax\| \le (1 + |\mu_0| \|A\|) \|x\| .$$

Let us prove the left-hand inequality. Suppose there is no $m_1 > 0$ such that the inequality holds for all $x \in M(\mu_0)$. This means there is a sequence $\{x_n\} \subset M(\mu_0)$ such that $\|x_n\| = 1$ and $\|x_n - \mu_0 Ax_n\| \to 0$ as $n \to \infty$. Because A is compact, the sequence $\{Ax_n\}$ contains a Cauchy subsequence. By the identity

$$x_n = \mu_0 Ax_n + (x_n - \mu_0 Ax_n)$$

the sequence $\{x_n\}$ also contains a Cauchy subsequence. Let us denote this Cauchy subsequence by $\{x_n\}$ again and let it converge to an element $x_0 \in M(\mu_0)$ such that $\|x_0\| = \lim_{n\to\infty} \|x_n\| = 1$. By continuity of A we have $Ax_n \to Ax_0$ as $n \to \infty$, which gives

$$\|x_0 - \mu_0 Ax_0\| = \lim_{n\to\infty} \|x_n - \mu_0 Ax_n\| = 0$$

and hence $x_0 = \mu_0 Ax_0$. Therefore $x_0 \in H(\mu_0)$, which contradicts the fact that $x_0 \in M(\mu_0)$ and $x_0 \neq 0$. □

The inequalities (3.8.3) state that on $M(\mu_0)$ we can introduce the inner product

$$(x, y)_1 = (x - \mu_0 Ax, y - \mu_0 Ay)$$

and its induced norm $\|x\|_1 = \|x - \mu_0 Ax\|$, which is equivalent to the norm of H.

Remark 3.8.1. Although the left hand side of (3.8.3) implies that $(I - \mu_0 A)^{-1}$ is bounded, it does not follow that its domain is all of $M(\mu_0)$. So when μ_0 is not an eigenvalue, we still cannot conclude from Lemma 3.8.4 that it is not a point of the spectrum: it may belong to the residual part of the spectrum. □

Now we begin to treat the problem of solvability of the equation

$$x - \mu Ax = f$$

in detail. To avoid complication and to cover the case of a general dependence on μ of $A(\mu)$ at once, we denote $\mu A = B$ (or $A(\mu) = B$) and study the equation

$$x - Bx = f \tag{3.8.4}$$

with a compact linear operator B acting in a Hilbert space.

We denote by N a subspace of eigenvectors of B corresponding to $\mu = 1$, i.e., all solutions to

$$x = Bx\ ,$$

and by M the orthogonal complement to N in H. For B^*, the adjoint of B, which is also a compact operator, we denote by N^* the space of all eigenvectors $x = B^*x$ and by M^* the orthogonal complement to N^* in H.

Lemma 3.8.5. The equation

$$x - B^*x = f \tag{3.8.5}$$

has a solution if and only if $f \in M$.

Proof. Necessity. Let (3.8.5) have a solution x_0. Then for an arbitrary element y of N we get

$$(f, y) = (x_0 - B^*x_0, y) = (x_0, y - By) = (x_0, 0) = 0\ ,$$

i.e., f does belong to M.

Sufficiency. Let $f \in M$. We mentioned that $\|x\|_1 = \|x - Bx\|$ is a norm in M which is equivalent to the norm of H in M. This norm is induced by the inner product $(x, y)_1 = (x - Bx, y - By)$ in M with which M is a Hilbert space as well as with the initial inner product $(\cdot, \cdot)$.

The functional (x, f) is linear and continuous on H, so it is linear and continuous on M. By the Riesz representation theorem, it can be represented on M using $(\cdot, \cdot)_1$ as

$$(x, f) = (x, f^*)_1 = (x - Bx, f^* - Bf^*)\ .$$

As $f \in M$, this equality, being valid for $x \in M$, holds for all $x \in H$ as well. Indeed taking $x = x_1 + x_2$ where $x_1 \in N$ and $x_2 \in M$, we have

$$x - Bx = x_1 - Bx_1 + x_2 - Bx_2 = x_2 - Bx_2$$

as $x_1 - Bx_1 = 0$. So, for any $x \in H$,

$$(x - Bx, f^* - Bf^*) = (x_2 - Bx_2, f^* - Bf^*) = (x_2, f^*)_1 = (x_2, f) = (x, f)$$

since $(x_1, f) = 0$. We denote $f^* - Bf^*$ by g; then the last equality takes the form

$$(x - Bx, g) = (x, f) \ \text{ for all } x \in H\ .$$

It follows that

$$(x, g - B^*g) = (x, f) \ \text{ for all } x \in H\ .$$

So $g - B^*g = f$ and g is the needed solution to (3.8.5). □

Since B and B^* are mutually adjoint, we get

Corollary 3.8.1. The equation

$$x - Bx = f \tag{3.8.6}$$

has a solution if and only if $f \in M^*$.

Inequality (3.8.3) yields

Corollary 3.8.2. For all $f \in M^*$ there is a positive constant m not depending on f such that

$$\|x\| \leq m \|f\|$$

where $x \in M$ is a solution to (3.8.6).

Clearly, in the Lemma, $m = 1/m_1$.

Lemma 3.8.5 and Corollary 3.8.1 can be reformulated in other terms as

$$R(I - B) = M^* , \qquad R(I - B^*) = M ,$$

where $R(\cdot)$ denotes the range of an operator.

Lemma 3.8.6. Let N_n be the space of all solutions of the equation

$$(I - B)^n x = 0$$

(i.e., the null space of $(I - B)^n$). Then

(i) N_n is a finite dimensional subspace of H;
(ii) the relation $N_n \subseteq N_{n+1}$ holds for $n = 1, 2, \ldots$;
(iii) there is an integer k such that $N_n = N_k$ if $n > k$.

Proof. Since $(I - B)^n = I - nB + \cdots$, then $(I - B)^n$ has a structure $I - B_1$ where B_1 is a compact linear operator. Hence by Lemma 3.8.1, (i) is fulfilled. Property (ii) is evident because $(I - B)^n x = 0$ implies $(I - B)^{n+1} x = 0$.

To check (iii) we first mention that if for some k we have $N_{k+1} = N_k$ then $N_{k+m} = N_k$ for all $m = 1, 2, 3, \ldots$; indeed, in this case take $x_0 \in N_{k+2}$ so that

$$0 = (I - B)^{k+2} x_0 = (I - B)^{k+1}((I - B)x_0) ,$$

i.e., $(I - B)x_0 \in N_{k+1}$. But this means $(I - B)x_0 \in N_k$ or $(I - B)^{k+1} x_0 = 0$, and so $x_0 \in N_{k+1} = N_k$. Thus $N_{k+2} = N_k$. This can be repeated for any $m > 2$: $N_{k+m} = N_k$.

Now suppose to the contrary that there is no k such that $N_k = N_{k+1}$. Then there is a sequence of elements $\{x_n\}$ such that $x_n \in N_n$, $\|x_n\| = 1$, and x_n is orthogonal to N_{n-1}. Let us consider the sequence $\{Bx_n\}$. By compactness of B it must contain a Cauchy subsequence but this leads to a contradiction. Indeed, we have

$$Bx_{n+m} - Bx_n = x_{n+m} - (x_{n+m} - Bx_{n+m} + Bx_n) .$$

Note that $x_{n+m} \in N_{n+m}$ while $(x_{n+m} - Bx_{n+m} + Bx_n) \in N_{n+m-1}$. Indeed, as $Bx_n \in N_n$,

$$(I - B)^n Bx_n = B(I - B)^n x_n = 0$$

and

$$(I - B)^{n+m-1}(x_{n+m} - Bx_{n+m}) = (I - B)^{n+m} x_{n+m} = 0 .$$

Therefore x_{n+m} is orthogonal to $(x_{n+m} - Bx_{n+m} + Bx_n)$ and so

$$\|Bx_{n+m} - Bx_n\|^2 = \|x_{n+m}\|^2 + \|x_{n+m} - Bx_{n+m} + Bx_n\|^2 \geq 1 .$$

This means that $\{Bx_n\}$ does not contain a Cauchy subsequence. □

Theorem 3.8.2. $R(I - B) = H$ if and only if $N = \{0\}$.

Proof. Necessity. Let $R(I - B) = H$ and suppose that $N \neq \{0\}$. Take $x_0 \in N$, $x_0 \neq 0$. Since $R(I - B) = H$ we can solve successively the following infinite system of equations:

$$(I - B)x_1 = x_0 \; ; \qquad (I - B)x_2 = x_1 \; ; \qquad \ldots \qquad (I - B)x_{n+1} = x_n \; ; \qquad \ldots$$

The sequence of solutions has the following property:

$$(I - B)^n x_n = x_0 \neq 0 \quad \text{but} \quad (I - B)^{n+1} x_n = (I - B)x_0 = 0 \, ,$$

i.e., there is no finite k such that $N_{k+1} = N_k$. This contradicts statement (iii) of Lemma 3.8.6.

Sufficiency. Let $N = \{0\}$. Then $M = H$ and so, by Lemma 3.8.5, $R(I - B^*) = M = H$. By the necessity part of the proof given above, $N^* = 0$ and thus $M^* = H$. By Corollary 3.8.1, we get $R(I - B) = M^* = H$. □

Corollary 3.8.3. If $R(I - B) = H$, then the inverse $(I - B)^{-1}$ is continuous.

This follows from (3.8.3) written in terms of B. A consequence of this and Theorem 3.8.2 is

Theorem 3.8.3. A compact linear operator A in a Hilbert space can have a point spectrum only.

To show this we should only reformulate the above results in terms of A: $B = \mu A$.

Another formulation of this theorem is

Theorem 3.8.4. If $N(\mu_0) = \{0\}$ then μ_0 is a regular point of operator A and so the spectrum of a linear compact operator A can consist only of the point spectrum.

Note that the zero operator $A = 0$ is an example of a linear compact operator whose spectrum is empty.

Theorem 3.8.5. The spaces N and N^* have the same dimension.

Proof. Let the dimensions of N and N^* be n and m, respectively, and suppose that $m > n$. Choose orthonormal bases $x_1, \ldots, x_n$ and $y_1, \ldots, y_m$ of N and N^*, respectively. Let us introduce an auxiliary operator C by

$$(I - C)x = (I - B)x + \sum_{k=1}^{n} (x, x_k) y_k \, ,$$

where C is a compact linear operator as the sum of the compact operator $-B$ and a finite dimensional operator. So C satisfies the above lemmas for B.

First we show that the kernel of $I - C$ does not contain nonzero elements. Indeed if $(I - C)x_0 = 0$ then

$$(I - B)x_0 + \sum_{k=1}^{n}(x_0, x_k)y_k = 0 .$$

Since $R(I - B) = M^* \perp N^*$, all terms of the sum on the left-hand side of the equality are mutually orthogonal and so each of them equals zero:

$$(I - B)x_0 = 0 , \qquad (x_0, x_k)y_k = 0 \quad (k = 1, \ldots, n) .$$

From $(I - B)x_0 = 0$ it follows that $x_0 \in N$, the remainder means x_0 is orthogonal to all basis elements of N, thus $x_0 = 0$.

By Theorem 3.8.2, the range of $I - C$ is H and thus the equation $(I - C)x = y_{n+1}$ has a solution x_0. But we get

$$1 = (y_{n+1}, y_{n+1}) = (y_{n+1}, (I - C)x_0) = (y_{n+1}, (I - B)x_0) + \left(y_{n+1}, \sum_{k=1}^{n}(x_0, x_k)y_k\right)$$
$$= ((I - B^*)y_{n+1}, x_0) = 0 .$$

This contradiction shows that $m \leq n$. On the other hand, B is adjoint to B^* and, by the above, $n \leq m$. Thus $n = m$. □

Remark 3.8.2. In the last proof we used the operator $I - C$, which was continuously invertible on H. The same property holds for an operator $I - C_\varepsilon$ defined by

$$(I - C_\varepsilon)x = (I - B)x + \varepsilon \sum_{k=1}^{n}(x, x_k)y_k$$

with any $\varepsilon \neq 0$. The operator $I - C_\varepsilon$ has a continuous inverse and

$$\|B - C_\varepsilon\| = O(\varepsilon) .$$

So for small $|\varepsilon| \neq 0$, the operator $I - C_\varepsilon$ is close to $I - B$. However we can solve the equation $(I - C_\varepsilon)x = f$ for any $f \in H$, whereas the original equation $(I - B)x = f$ has no solution for some f. The operator $I - C_\varepsilon$ allows us to regularize the equation $(I - B)x = f$. Such regularizations are widely used in applications. □

Remark 3.8.3. The results of this section are known as the *Fredholm alternative*. For instance, Theorem 3.8.2 (i.e., that $R(I - B) = H$ if and only if $N = \{0\}$) implies that $R(I - B) \neq H$ if and only if $N \neq \{0\}$. Since we have either $R(I - B) = H$ or $R(I - B) \neq H$, we have the *alternative* that *either*

(a) $R(I - B) = H$ and $N = \{0\}$

or

(b) $R(I - B) \subset H$ and $N \neq \{0\}$.

In case (a), the equation $x - Bx = f$ has a unique solution x for each $f \in H$. Furthermore, since N^* contains only the zero element by Theorem 3.8.5, the equation

$x - B^*x = 0$ has only the trivial solution $x = 0$. In case (b), there exists $f \in H$ for which the equation $x - Bx = f$ has no solution x. Furthermore, Corollary 3.8.1 implies that this equation has a solution if and only if f is orthogonal to all solutions of the equation $x - B^*x = 0$.

From the other theorems it follows that in case (a) we also have $R(I - B^*) = H$ and $N^* = \{0\}$. So, in that case, the equation $x - B^*x = f^*$ also has a unique solution x for each $f^* \in H$. In case (b) we have $R(I - B^*) \neq H$ and $N^* \neq \{0\}$. So, in that case, there exists f^* for which $x - B^*x = f^*$ has no solution. It has a solution if and only if f^* is orthogonal to all solutions of the equation $x - Bx = 0$. □

3.9 Analytic Nature of the Resolvent of a Compact Linear Operator

For a linear operator A we know that the resolvent $(I - \mu A)^{-1}$ is a holomorphic operator-function in μ in non-spectral points. But what is its behavior near the spectrum? We can answer this question for a compact operator.

We begin the study with the case of a continuous finite-dimensional operator acting in a Hilbert space. The general form of such an operator is

$$A_n x = \sum_{k=1}^{n} (x, a_k) x_k$$

where the system $x_1, \ldots, x_n$ is assumed to be linearly independent.

We consider the equation

$$x - \mu \sum_{k=1}^{n} (x, a_k) x_k = f. \tag{3.9.1}$$

Its solution has the form

$$x = f + \sum_{k=1}^{n} c_k x_k .$$

Placing this into (3.9.1), we get

$$f + \sum_{k=1}^{n} c_k x_k - \mu \sum_{k=1}^{n} \left(f + \sum_{j=1}^{n} c_j x_j, a_k \right) x_k = f$$

or

$$\sum_{k=1}^{n} x_k \left(c_k - \mu \sum_{j=1}^{n} c_j (x_j, a_k) \right) = \mu \sum_{k=1}^{n} (f, a_k) x_k .$$

Since $x_1, \ldots, x_n$ is a linearly independent system we get an algebraic system

$$c_k - \mu \sum_{j=1}^{n} (x_j, a_k) c_j = \mu(f, a_k) \qquad (k = 1, \ldots, n)\,, \tag{3.9.2}$$

which can be solved using Cramer's rule:

$$c_k = \frac{D_k(\mu, f)}{D(\mu)} \qquad (k = 1, \ldots, n)\,.$$

Thus a solution to (3.9.1) is

$$x = \frac{1}{D(\mu)} \left[D(\mu) f + \sum_{k=1}^{n} D_k(\mu, f) x_k \right].$$

In this case the solution to (3.9.1) is a ratio of two polynomials in μ of degree no more than n. All μ which are not eigenvalues of A_n are points where the resolvent is holomorphic, hence they cannot be zeros of $D(\mu)$. But if μ_0 is an eigenvalue of A_n then $D(\mu_0) = 0$. If it is not true then for any $f \in H$ there is a solution to (3.9.1) and this means μ_0 is not an eigenvalue. So the set of all zeros of $D(\mu)$ coincides with the set of all eigenvalues of A_n, and so each eigenvalue of A_n is a pole of finite multiplicity of the resolvent $(I - \mu A_n)^{-1}$.

Now we consider a general case:

Theorem 3.9.1. Every eigenvalue of a compact linear operator A acting in a Hilbert space is a pole of finite multiplicity of the resolvent $(I - \mu A)^{-1}$.

Proof. By Theorem 3.4.2, for any small $\varepsilon > 0$ the operator A can be represented as

$$A = A_n + A_\varepsilon\,, \qquad \|A_\varepsilon\| \le \varepsilon\,,$$

where A_n is a finite dimensional operator. Fix $\varepsilon > 0$. The equation under consideration takes the form

$$x - \mu(A_n + A_\varepsilon)x = f\,. \tag{3.9.3}$$

Consider the operator $(I - \mu A_\varepsilon)^{-1}$. In the disk $|\mu| < 1/\varepsilon$ it can be represented in the form

$$(I - \mu A_\varepsilon)^{-1} = I + \sum_{k=1}^{\infty} \mu^k A_\varepsilon^k\,.$$

(The series is majorized by the series $1 + \sum_{k=1}^{\infty} |\mu^k|\, \|A_\varepsilon\|^k$. The fact that it is inverse to $(I - \mu A_\varepsilon)$ is checked directly.) So in the disk $|\mu| < 1/\varepsilon$ the operator $(I - \mu A_\varepsilon)^{-1}$ is a holomorphic operator-function in μ.

We apply this operator to equation (3.9.3):

$$x - \mu(I - \mu A_\varepsilon)^{-1} A_n x = (I - \mu A_\varepsilon)^{-1} f$$

wherein, as above,

$$A_n x = \sum_{k=1}^{n} (x, a_k) x_k\,.$$

Let us denote

$$f^* = (I - \mu A_\varepsilon)^{-1} f, \qquad x_k^* = (I - \mu A_\varepsilon)^{-1} x_k ;$$

then the equation takes the form

$$x - \mu \sum_{k=1}^{n} (x, a_k) x_k^* = f^*$$

which looks like (3.9.1) — the difference is that f^* and x_k^* are holomorphic functions in μ in the disk $|\mu| < 1/\varepsilon$. When $|\mu| < 1/\varepsilon$, the system $x_1^*, \ldots, x_n^*$ is linearly independent since $x_1, \ldots, x_n$ is linearly independent and $(I - \mu A_\varepsilon)$ is continuously invertible. So, by analogy with (3.9.1), we can point out that for $|\mu| < 1/\varepsilon$ the solution to (3.9.3) is

$$x = \frac{1}{D(\mu)} \left[D(\mu) f^* + \sum_{k=1}^{n} D_k(\mu, f^*) x_k^* \right] \tag{3.9.4}$$

wherein all zeros of $D(\mu)$ have multiplicity no more than n.

If μ_0 is not an eigenvalue of A then the solution (3.9.4) is holomorphic in μ in a neighborhood of μ_0 and so $D(\mu_0) \neq 0$. But if μ_0 is an eigenvalue then $D(\mu_0) = 0$ since otherwise the equation would be solvable for all f^*, which is impossible.

So the set of eigenvalues of A belonging to the disk $|\mu| < 1/\varepsilon$ coincides with the set of zeros of $D(\mu)$ lying in this disk. □

3.10 Spectrum of a Holomorphic Compact Operator Function

Let $A(\mu)$ be an operator-function whose value, for any $\mu \in G$, an open domain in $\mathbb{C}$, is a compact linear operator in a Hilbert space and $A(\mu)$ be holomorphic in G. We know that the spectrum of such operator-functions is a point spectrum. Following I.Ts. Gokhberg and M.G. Krein [13] we study the distribution of eigenvalues.

Lemma 3.10.1. For $\mu_0 \in G$ there is a positive ε such that for all μ in a domain $0 < |\mu - \mu_0| < \varepsilon$ the equation

$$(I - A(\mu))x = 0 \tag{3.10.1}$$

has the same number of linearly independent solutions.

Proof. Let $x_1, \ldots, x_n$ be an orthonormal basis of the space of solutions of (3.10.1) when $\mu = \mu_0$. By Theorem 3.8.5 there is an orthonormal basis $y_1, \ldots, y_n$ of the space of solutions of the equation $(I - A^*(\mu_0))x = 0$ and, by the proof of Theorem 3.8.5, the operator

$$Q(\mu_0)x = (I - A(\mu_0))x + \sum_{k=1}^{n} (x, x_k) y_k$$

has a continuous inverse. As $A(\mu)$ depends on μ continuously, there is some neighborhood $|\mu - \mu_0| < \rho$ of μ_0 in which $Q(\mu)$ has a continuous inverse.

Equation (3.10.1) is equivalent to

$$Q(\mu)x = \sum_{k=1}^{n} (x, x_k) y_k$$

and, in the above-mentioned neighborhood, to a system

$$x = \sum_{k=1}^{n} \xi_k Q^{-1}(\mu) y_k\,, \quad |\mu - \mu_0| < \rho\,,$$
$$\xi_k = (x, x_k) \quad (k = 1, \ldots, n)\,,$$

which, in turn, can be reduced to an equivalent system of algebraic equations (substituting x from the first equation into the others)

$$\xi_k - \sum_{j=1}^{n} (Q^{-1}(\mu) y_j, x_k) \xi_j = 0 \qquad (k = 1, \ldots, n) \tag{3.10.2}$$

whose number of linearly independent solutions coincides with the number for (3.10.1) when $|\mu - \mu_0| < \rho$. In this domain, all terms in (3.10.2) are holomorphic as are all elements of its determinant.

If all elements of the main determinant of the system (3.10.2) equal zero identically, then the system (3.10.2) has n linearly independent solutions in the disk $|\mu - \mu_0| < \rho$, and the lemma is proven.

Otherwise, let $\Delta_p(\mu)$ be a minor of highest order p which is nonzero at some point of the disk $|\mu - \mu_0| < \rho$. Being holomorphic, $\Delta_p(\mu)$ is nonzero in this disk except, perhaps, at a finite number of points. This means that in this disk, except at those points, the number of linearly independent solutions of (3.10.2) is $n - p$. Therefore, we can exhibit a disk $|\mu - \mu_0| < \varepsilon$ such that for all its points μ, except perhaps $\mu = \mu_0$, the system (3.10.2) and thus (3.10.1) has the same number $n - p$ of linearly independent solutions. □

Theorem 3.10.1. Assume $A(\mu)$ is an operator-function, being holomorphic on a connected open domain $G \subset \mathbb{C}$, whose values are compact linear operators in a Hilbert space. Then $\alpha(\mu)$, the number of linearly independent solutions of (3.10.1), is the same, $\alpha(\mu) = n$, for all points of G, except some isolated points of G at which $\alpha(\mu) > n$. In particular, if there exists $\mu_0 \in G$ such that $\alpha(\mu_0) = 0$, then the spectrum of $A(\mu)$ consists of isolated points of G. (This happens if, for example, there exists $\mu_0 \in G$ such that $A(\mu_0) = 0$.)

Proof. Consider $\alpha(\mu)$. Assume its minimal value is n and that it is taken at $\mu = \mu_0$. Let μ_1 be a point at which $\alpha(\mu_1) > n$. We shall show that this point is isolated and, moreover, that there is an $\varepsilon > 0$ such that for all $\mu \neq \mu_1$, $|\mu - \mu_1| < \varepsilon$, the number $\alpha(\mu) = n$. Draw a curve lying in G from μ_0 to μ_1. By Lemma 3.10.1, for any $\mu^* \in G$ there is a positive number $\varepsilon(\mu^*)$ such that for any $\mu \in G$, $0 < |\mu - \mu^*| < \varepsilon(\mu^*)$, the number of linearly independent solutions of (3.10.1) is constant. These disks make up a covering of G from which we can choose a finite covering of G. As neighboring disks of the covering are mutually intersecting, then for all points of the disks of the finite covering, except perhaps for their centers, the number of linearly independent solutions of (3.10.1) is constant and equals n. □

3.11 Spectrum of Self-Adjoint Compact Linear Operator in Hilbert Space

We did not touch on the problem of existence of the spectrum. The example of the zero operator shows that there are compact operators having no finite spectral points. But there is a class of operators having eigenvalues; it is the class shown in the title of this section.

Lemma 3.11.1. All eigenvalues of a self-adjoint continuous linear operator A acting in a Hilbert space are real, as are all values of the form (Ax, x). Eigenvectors x_1, x_2 of A corresponding to μ_1, μ_2, respectively, $\mu_1 \neq \mu_2$, are mutually orthogonal; moreover, $(Ax_1, x_2) = 0$.

Proof. The functional (Ax, x) is real-valued because

$$(Ax, x) = (x, Ax) = \overline{(Ax, x)}\ .$$

So if $x_0 = \mu_0 A x_0$, then $(x_0, x_0) = \mu_0(Ax_0, x_0)$ and μ_0 is real as well. Now let

$$x_1 = \mu_1 A x_1 \quad \text{and} \quad x_2 = \mu_2 A x_2\ .$$

Multiply the first equation by x_2 from the right, and the second equation by x_1 from the left:

$$(x_1, x_2) = \mu_1(Ax_1, x_2)\ , \qquad (x_1, x_2) = (x_1, \mu_2 A x_2) = \mu_2(Ax_1, x_2)\ .$$

It follows that

$$(\mu_2 - \mu_1)(x_1, x_2) = 0$$

so $x_1 \perp x_2$ if $\mu_1 \neq \mu_2$. Returning to $(x_1, x_2) = \mu_1(Ax_1, x_2)$, we get $(Ax_1, x_2) = 0$. Note that in the theory of elasticity the last equality is called the relation of generalized orthogonality of eigenvectors. □

Definition 3.11.1. A functional $F(x)$ given on a Hilbert space is called *weakly continuous* if for every weakly convergent sequence $\{x_n\}$, $x_n \rightharpoonup x_0$, we have $F(x_n) \to F(x_0)$.

Note that a continuous linear functional is weakly continuous by the definition of weak convergence of a sequence of elements.

Lemma 3.11.2. A real-valued weakly continuous functional $F(x)$ given on a Hilbert space takes its maximal and minimal values on a ball $\|x\| \leq a$.

Proof. Let

$$\sup_{\|x\|\le a} F(x) = M\ .$$

Then there is a sequence $\{x_n\}$, with $\|x_n\| \le a$, such that $F(x_n) \to M$ as $n \to \infty$. From the bounded sequence $\{x_n\}$ we can choose a subsequence $\{x_{n_k}\}$ which is a weak Cauchy sequence and its weak limit x_0 (the existence of which is guaranteed because H is weakly complete) satisfies $\|x_0\| \le a$ (because the closed ball is weakly closed). By Definition 3.11.1, $F(x_{n_k}) \to F(x_0) = M$. The corresponding result for the minimum point is obtained by replacing F by $-F$. □

Lemma 3.11.3. Assume A is a self-adjoint compact linear operator in a Hilbert space. Then (Ax, x) is a real-valued weakly continuous functional on this space.

Proof. By Lemma 3.11.1, (Ax, x) is real-valued. Let $\{x_k\}$ be weakly convergent to x_0. Then

$$\begin{aligned}(Ax_k, x_k) - (Ax_0, x_0) &= (Ax_k, x_k) - (Ax_0, x_k) + (Ax_0, x_k) - (Ax_0, x_0)\\ &= (Ax_k - Ax_0, x_k) + (Ax_0, x_k - x_0) \to 0 \ \text{ as } k \to \infty\end{aligned}$$

since $\|Ax_k - Ax_0\| \to 0$ by compactness of A and $(Ax_0, x_k - x_0) \to 0$ as (x, Ax_0) is a continuous linear functional with respect to x in H. □

Denote

$$\lambda_+ = \sup_{\|x\|\le 1}(Ax, x)\ , \qquad \lambda_- = \inf_{\|x\|\le 1}(Ax, x)\ .$$

Theorem 3.11.1. Assume $A \neq 0$ is a self-adjoint compact linear operator acting in a Hilbert space. There is at least one eigenvalue μ of A. If both λ_+ and λ_- are nonzero, then there are at least two eigenvalues of A, namely, $\mu_1 = 1/\lambda_+$ and $\mu_2 = 1/\lambda_-$.

Proof. Since

$$\|A\| = \sup_{\|x\|\le 1} |(Ax, x)| \neq 0\ ,$$

at least one of λ_+, λ_- is nonzero. Without loss of generality, assume that $\lambda_+ \neq 0$. By Lemmas 3.11.2 and 3.11.3, (Ax, x) takes this maximal value in the unit ball at an element x_0: $(Ax_0, x_0) = \lambda_+$. By homogeneity of (Ax, x) in x, it is evident that $\|x_0\| = 1$.

Now consider a functional

$$\Phi(x) = \frac{(Ax, x)}{\|x\|^2} = (A\xi, \xi)\ , \qquad \xi = \frac{x}{\|x\|}\ ,$$

whose range of values coincides with the set of values of (Ax, x) when x runs over the sphere $\|x\| = 1$. Thus

$$\sup \Phi(x) = \sup_{\|x\|=1}(Ax, x) = (Ax_0, x_0) = \Phi(x_0)\ .$$

We shall show that x_0 is an eigenvector of A. Consider $\Phi(x_0 + \alpha y)$, where y is an arbitrary but fixed element of H, as a function of a real variable α; it is differentiable in some neighborhood of $\alpha = 0$ and takes its maximal value at $\alpha = 0$ so

$$\left.\frac{d\Phi(x_0 + \alpha y)}{d\alpha}\right|_{\alpha=0} = 0 . \tag{3.11.1}$$

This gives

$$\operatorname{Re}(Ax_0, y) - \frac{(Ax_0, x_0)}{\|x_0\|^2}\operatorname{Re}(x_0, y) = 0 \tag{3.11.2}$$

or

$$\operatorname{Re}[(Ax_0, y) - \lambda_+(x_0, y)] = 0 .$$

Replacing y by iy, we get

$$\operatorname{Im}[(Ax_0, y) - \lambda_+(x_0, y)] = 0$$

so

$$(Ax_0, y) - \lambda_+(x_0, y) = 0 . \tag{3.11.3}$$

Since y is an arbitrary element of H, we have

$$Ax_0 - \lambda_+ x_0 = 0 \quad \text{or} \quad x_0 - \frac{1}{\lambda_+}Ax_0 = 0 ,$$

i.e., x_0 is an eigenvector of A. If $\lambda_- \neq 0$, then we can show similarly that $\mu = 1/\lambda_-$ is also an eigenvalue of A. □

Problem 3.11.1. Show that (3.11.1) implies (3.11.2). □

Definition 3.11.2. A self-adjoint continuous linear operator A is called *strictly positive* if (1) $(Ax, x) \geq 0$ for all $x \in H$, and (2) $(Ax, x) = 0$ implies $x = 0$.

By Lemma 3.11.1, any two eigenvectors corresponding to different eigenvalues of a self-adjoint operator are mutually orthogonal. But we can orthonormalize a linearly independent system of eigenvectors corresponding to the same eigenvalue and so we can consider a basis of the set of all eigenvectors of a self-adjoint operator to be orthonormal. This and the method used in the proof of Theorem 3.11.1 allow us to prove

Theorem 3.11.2. Assume A is a strictly positive self-adjoint compact linear operator acting in a separable Hilbert space. Then

(i) A possesses infinitely many eigenvalues $\mu_1, \mu_2, \mu_3, \ldots$; the sequence of eigenvalues does not contain subsequences having finite limit points;

(ii) there is a system of eigenvectors $x_1, x_2, x_3, \ldots$ of A which is an orthonormal basis of H;

(iii) A has a representation

$$Ax = \sum_{k=1}^{\infty} \frac{(x, x_k)}{\mu_k} x_k \,, \qquad x_k - \mu_k A x_k = 0 \,.$$

Proof. (i) Assume $x_1, x_2, \ldots, x_n$ is an orthonormal system of eigenvectors of A corresponding to eigenvalues $\mu_1, \mu_2, \ldots, \mu_n$ respectively. Some of μ_i and μ_j can coincide. We show how to construct the next eigenpair (x_{n+1}, μ_{n+1}). Denote by H_n the orthogonal complement in H to a subspace spanned by $x_1, x_2, \ldots, x_n$. Considering A on H_n, we can repeat the proof of Theorem 3.11.1 and, denoting

$$\mu_{n+1} = 1/\lambda_{n+1} \,, \qquad \lambda_{n+1} = \sup_{\substack{\|x\| \le 1 \\ x \in H_n}} (Ax, x) \,,$$

obtain a vector x_{n+1} such that

$$x_{n+1} \in H_n \,, \quad \|x_{n+1}\| = 1 \,, \quad \text{and} \quad \lambda_{n+1} = (Ax_{n+1}, x_{n+1}) \,;$$

this vector x_{n+1} satisfies the equation (an analogy to (3.11.3))

$$(Ax_{n+1}, y) - \lambda_{n+1}(x_{n+1}, y) = 0 \ \text{ for all } y \in H_n \,.$$

In fact, this equality holds for all $y \in H$; if $y = x_k$ ($k = 1, \ldots, n$) then

$$\begin{aligned}
(Ax_{n+1}, x_k) - \lambda_{n+1}(x_{n+1}, x_k) &= (Ax_{n+1} - \lambda_{n+1} x_{n+1}, x_k) \\
&= (x_{n+1}, Ax_k - \lambda_{n+1} x_k) \\
&= (x_{n+1}, Ax_k) - \lambda_{n+1}(x_{n+1}, x_k) \\
&= (x_{n+1}, \lambda_k x_k) - \lambda_{n+1}(x_{n+1}, x_k) \\
&= (\lambda_k - \lambda_{n+1})(x_{n+1}, x_k) = 0
\end{aligned}$$

because $(x_{n+1}, x_k) = 0$, so

$$Ax_{n+1} - \lambda_{n+1} x_{n+1} = 0 \,.$$

Thus x_{n+1} is an eigenvector and $\mu_{n+1} = 1/\lambda_{n+1}$ an eigenvalue of A.

Now we can realize the process of successive construction of eigenvalues and eigenvectors of A, which could be disrupted only if for some n,

$$\sup_{\substack{\|x\| \le 1 \\ x \in H_n}} (Ax, x) = 0 \,.$$

But, since A is strictly positive, this would imply that $H_n = \{0\}$, i.e., that H is finite dimensional. The remainder of statement (i) is evident.

(ii) Let y be an arbitrary element of H. Consider

$$y_n = y - \sum_{k=1}^{n} (y, x_k) x_k$$

where $\{x_k\}$ is a sequence of eigenvectors defined in (i) above. We recall that y_n is the nth Fourier remainder of y. It is clear that $y_n \in H_n$. In the theory of Fourier expansions (Sect. 2.8) it was shown that $\{y_n\}$ is a Cauchy sequence. Assume its strong limit is $y_0 \neq 0$. For y_n, thanks to $y_n \in H_n$, we get

$$\frac{(Ay_n, y_n)}{\|y_n\|^2} \leq \lambda_{n+1} .$$

But $\lambda_n \to 0$ as $n \to \infty$ since the set $\{\mu_n\}$ has no finite limit points. Passage to the limit gives

$$\frac{(Ay_0, y_0)}{\|y_0\|^2} = 0$$

so $y_0 = 0$. Therefore

$$\sum_{k=1}^{\infty} (y, x_k)x_k = y ,$$

which completes the proof of (ii).

(iii) In the proof of Theorem 3.4.2, we showed that the operator A is a uniform limit of a sequence of finite-dimensional operators A_n,

$$A_n x = \sum_{k=1}^{n} (x, g_k)Ag_k ,$$

where $g_1, g_2, \ldots, g_n, \ldots$ is an orthonormal basis of H. We take as a basis the set $x_1, x_2, \ldots$ constructed in part (i). Then

$$A_n x = \sum_{k=1}^{n} (x, x_k)Ax_k = \sum_{k=1}^{n} \frac{(x, x_k)}{\mu_k} x_k$$

and so

$$Ax = \sum_{k=1}^{\infty} \frac{(x, x_k)}{\mu_k} x_k . \tag{3.11.4}$$

The proof is thereby completed. □

Remark 3.11.1. According to (3.11.4), the action of a strictly positive self-adjoint compact linear operator A on a vector x is determined solely by the *spectral characteristics* of A (i.e., its eigenpairs (μ_k, x_k)) and by the projections of x along the eigenvectors x_k. □

Let us note that under the conditions of Theorem 3.11.2, by Parseval's equality, we get

$$\|x\|^2 = \sum_{k=1}^{\infty} |(x, x_k)|^2 = \sum_{k=1}^{\infty} |(x, \mu_k Ax_k)|^2 = \sum_{k=1}^{\infty} \mu_k^2 \, |(x, Ax_k)|^2$$

for any $x \in H$.

Since A is strictly positive, we can introduce a new norm

$$\|x\|_A = (Ax, x)^{1/2}$$

and the corresponding scalar product $(x, y)_A = (Ax, y)$. When H with this norm is incomplete, we introduce its completion H_A with respect to this norm.

Let

$$y_k = \sqrt{\mu_k}\, x_k$$

where x_k is an eigenvector of part (i).

Lemma 3.11.4. Under the conditions of Theorem 3.11.2, the set of vectors

$$y_k = \sqrt{\mu_k}\, x_k \qquad (k = 1, 2, \ldots)$$

is an orthonormal basis of H_A.

Proof. The system $y_1, y_2, y_3, \ldots$ is orthonormal in H_A; indeed

$$(y_k, y_n)_A = (Ay_k, y_n) = \sqrt{\mu_k}\,\sqrt{\mu_n}\,(Ax_k, x_n) = \frac{\sqrt{\mu_k \mu_n}}{\mu_k}\,(x_k, x_n) = \delta_{kn}\ .$$

For any $x \in H$, Parseval's equality holds in H_A:

$$(x, x)_A = (Ax, x) = \left(A \sum_{k=1}^{\infty} (x, x_k) x_k, x\right) = \sum_{k=1}^{\infty} (x, x_k)(Ax_k, x) = \sum_{k=1}^{\infty} (x, \mu_k A x_k)(Ax_k, x)$$
$$= \sum_{k=1}^{\infty} (x, Ay_k)(Ay_k, x) = \sum_{k=1}^{\infty} |(Ax, y_k)|^2 = \sum_{k=1}^{\infty} |(x, y_k)_A|^2\ .$$

This means that the system $y_1, y_2, y_3, \ldots$ is an orthonormal basis of the set of elements $x \in H$ in H_A. But this set is dense in H_A, and that completes the proof. □

Problem 3.11.2. Confirm that if Parseval's equality (2.8.7) holds for all $f \in H$, then $\{g_k\}$ is a basis. □

As an example, consider the eigenvalue problem

$$y'' + \mu^2 y = 0\ , \qquad y(0) = y(\pi) = 0\ ,$$

with the well-known set of eigenfunctions

$$\left\{\tfrac{\sqrt{2}}{\pi} \sin kx\right\} \qquad (k = 1, 2, \ldots)\ . \tag{3.11.5}$$

We are interested in its properties. If E is a Hilbert space of functions $y \in \dot{W}^{1,2}(0, \pi)$ with scalar product

$$(y, z)_E = \int_0^{\pi} y'(x)\,\overline{z'(x)}\,dx$$

then the problem can be posed as the problem of finding nontrivial y satisfying the equation

$$(y, z)_E - \mu^2 (Ay, z)_E = 0$$

for any $z \in E$, where

$$(Ay, z)_E = \int_0^\pi y(x)\,\overline{z(x)}\,dx\,.$$

It was shown that A satisfies all conditions of Theorem 3.11.2, so the system (3.11.5) is an orthonormal basis in $L^2(0, \pi)$, which is the space H_A from Lemma 3.11.4. The same system is an orthogonal basis in $\dot{W}^{1,2}(0, \pi)$.

3.12 Some Applications of Spectral Theory

First we recall the spectrum for the different elastic bodies we considered. These are the problems of membranes, plates, and bodies in the framework of two- and three-dimensional linear elasticity. In generalized form, we have

$$(u, v)_E = \mu \int_\Omega \rho(\mathbf{x})\,u(\mathbf{x})\,\overline{v(\mathbf{x})}\,d\Omega\,. \tag{3.12.1}$$

Here E is an energy space for the corresponding elastic object occupying a bounded domain Ω, and $u(\mathbf{x})$ is a function for a membrane or a plate and a vector-function of displacements for an elastic body. The spaces E were introduced as real spaces — here we use their complex versions (i.e., the complex conjugate is applied to v in the integrand of $(u, v)_E$). Now we introduce an operator K using the Riesz representation theorem (Sects. 2.5 and 3.3)

$$(Ku, v)_E = \int_\Omega \rho(\mathbf{x})\,u(\mathbf{x})\,\overline{v(\mathbf{x})}\,d\Omega\,.$$

For free-boundary problems E consists of "balanced" elements; for a membrane, say, they satisfy $\int_\Omega u(\mathbf{x}, \mathbf{y})\,d\Omega = 0$. It was shown (Lemma 3.3.1) that K is a self-adjoint compact linear operator in an energy space. Moreover, if $\rho_0 \le \rho(\mathbf{x}) \le \rho_1$, with ρ_0, ρ_1 positive constants, then K is a strictly positive operator; indeed,

$$(Ku, u)_E = \int_\Omega \rho(\mathbf{x})\,|u(\mathbf{x})|^2\,d\Omega \ge \rho_0 \int_\Omega |u(\mathbf{x})|^2\,d\Omega$$

and if $(Ku, u)_E = 0$ then $u = 0$ in the sense of $L^2(\Omega)$ (almost everywhere).

So for any of the models of bounded elastic bodies that we considered, we get

Theorem 3.12.1. In the framework of all main (Dirichlet, Neumann, and mixed) spectral boundary value problems in the generalized statement for bounded membranes, plates, or linear elastic bodies, the spectrum of each problem contains only eigenvalue points μ_k, and:

(i) All μ_k are positive, $\mu_k \ge \mu_0 > 0$.

(ii) The set $\{\mu_k\}$ is infinite and does not contain a finite limit point.

(iii) To each μ_k there corresponds no more than a finite number of linearly independent eigenvectors which are assumed to be orthonormalized; the set of all these eigenvectors $\{u_k(\mathbf{x})\}$ is an orthonormal basis in the corresponding energy space and the set $\{\sqrt{\mu_k}u_k(\mathbf{x})\}$ is an orthonormal basis in $L^2(\Omega)$ with scalar product

$$(u, v)_{L^2(\Omega)} = \int_\Omega \rho(\mathbf{x})\, u(\mathbf{x})\, \overline{v(\mathbf{x})}\, d\Omega\, .$$

In Sect. 3.2 we considered a stability problem for a thin plate. In generalized terms this problem can be stated as

$$(w, \varphi)_{E_P} = \mu\, (Cw, \varphi)_{E_P}$$

where

$$(Cw, \varphi)_{E_P} = \iint_\Omega \left[T_x \frac{\partial w}{\partial x}\overline{\frac{\partial \varphi}{\partial x}} + T_{xy}\left(\frac{\partial w}{\partial y}\overline{\frac{\partial \varphi}{\partial x}} + \frac{\partial w}{\partial x}\overline{\frac{\partial \varphi}{\partial y}} \right) + T_y \frac{\partial w}{\partial y}\overline{\frac{\partial \varphi}{\partial y}} \right] d\Omega$$

with given functions $T_x, T_{xy}, T_y \in L^2(\Omega)$. Then we get a spectral problem

$$w = \mu C w\, .$$

For a clamped plate, it was shown that C is a self-adjoint continuous operator. Since the imbedding operator from $\dot{W}^{2,2}(\Omega)$ to $W^{1,4}(\Omega)$ is compact, the inequality

$$|(Cw, \varphi)_{E_P}| \le m \left[\iint_\Omega (T_x^2 + T_{xy}^2 + T_y^2)\, d\Omega \right]^{1/2} \cdot \left[\iint_\Omega \left(\left|\frac{\partial w}{\partial x}\right|^4 + \left|\frac{\partial w}{\partial y}\right|^4 \right) d\Omega \right]^{1/4} \cdot \left[\iint_\Omega \left(\left|\frac{\partial \varphi}{\partial x}\right|^4 + \left|\frac{\partial \varphi}{\partial y}\right|^4 \right) d\Omega \right]^{1/4}$$

shows that C is also compact.

Finally we assume the external tangential load to be compressible in total, which is expressed by the inequality

$$T_x w_1^2 + 2T_{xy} w_1 w_2 + T_y w_2^2 \ge c_0 (w_1^2 + w_2^2) \tag{3.12.2}$$

which is valid with a positive constant c_0 for all real w_1, w_2 and $\mathbf{x} \in \Omega$. Under the condition (3.12.2), C is a strictly positive operator on E_{PC}, and so we can say that this spectral problem is similar to one considered in Theorem 3.12.1. Thus its spectrum has properties as stated in Theorem 3.12.1.

We mentioned that the spectral theory of compact operators began with the study of integral equations originated by I. Fredholm. We found an integral operator to be compact in $L^2(\Omega)$ if its kernel belongs to $L^2(\Omega \times \Omega)$ and so we can reformulate all general results for these equations (try to do this yourself). Now we wish to consider another important class of integral operators, the so-called operators with kernels having weak singularities. These are kernels of the form

$$\mathcal{K}(\mathbf{x},\mathbf{y}) = \frac{R(\mathbf{x},\mathbf{y})}{r^{\alpha}}, \qquad r = |\mathbf{x}-\mathbf{y}|, \qquad \mathbf{x},\mathbf{y} \in \Omega \subset \mathbb{R}^n$$

where $\alpha < n$ and $R(\mathbf{x},\mathbf{y}) \in C(\Omega \times \Omega)$.

Lemma 3.12.1. A linear integral operator whose kernel has a weak singularity is compact in $L^2(\Omega)$.

Proof. First we show that this operator is bounded in $L^2(\Omega)$:

$$\begin{aligned}\|Au\|^2_{L^2(\Omega)} &= \int_\Omega \left| \int_\Omega \frac{R(\mathbf{x},\mathbf{y})}{r^\alpha} u(\mathbf{y})\, d\Omega_\mathbf{y} \right|^2 d\Omega_\mathbf{x} \\ &\le m \int_\Omega \left(\int_\Omega \frac{1}{r^\alpha} d\Omega_\mathbf{y} \int_\Omega \frac{|u(\mathbf{y})|^2}{r^\alpha}\, d\Omega_\mathbf{y} \right) d\Omega_\mathbf{x} ;\end{aligned}$$

since $\left| \int_\Omega \frac{d\Omega_\mathbf{y}}{r^\alpha} \right| \le M$ when $\mathbf{x} \in \Omega$, we further get

$$\|Au\|^2_{L^2(\Omega)} \le mM \int_\Omega \int_\Omega \frac{|u(\mathbf{y})|^2}{r^\alpha}\, d\Omega_\mathbf{x}\, d\Omega_\mathbf{y} \le mM^2 \int_\Omega |u(\mathbf{y})|^2\, d\Omega_\mathbf{y} ,$$

which demonstrates the continuity of A.

To show that A is compact, we introduce an auxiliary operator with kernel

$$\mathcal{K}_\varepsilon(\mathbf{x},\mathbf{y}) = \frac{R(\mathbf{x},\mathbf{y})}{r_\varepsilon^\alpha}, \qquad r_\varepsilon = \begin{cases} r, & r = |\mathbf{x}-\mathbf{y}| \ge \varepsilon , \\ \varepsilon, & r = |\mathbf{x}-\mathbf{y}| < \varepsilon . \end{cases}$$

The corresponding integral operator A_ε is compact because its kernel is continuous on $\Omega \times \Omega$. Now it suffices to prove that $\|A - A_\varepsilon\| \to 0$ as $\varepsilon \to 0$. Denote by $B(\mathbf{x})$ a ball about $\mathbf{x}$ of radius $\varepsilon > 0$. We have

$$\begin{aligned}\|Au - A_\varepsilon u\|^2_{L^2(\Omega)} &= \int_\Omega \left| \int_\Omega R(\mathbf{x},\mathbf{y}) \left(\frac{1}{r^\alpha} - \frac{1}{r_\varepsilon^\alpha} \right) u(\mathbf{y})\, d\Omega_\mathbf{y} \right|^2 d\Omega_\mathbf{x} \\ &= \int_\Omega \left| \int_{\Omega\cap B(\mathbf{x})} R(\mathbf{x},\mathbf{y}) \left(\frac{1}{r^\alpha} - \frac{1}{r_\varepsilon^\alpha} \right) u(\mathbf{y})\, d\Omega_\mathbf{y} \right|^2 d\Omega_\mathbf{x} \\ &\le 4 \int_\Omega \left(\int_{\Omega\cap B(\mathbf{x})} \frac{|R(\mathbf{x},\mathbf{y})|}{r^\alpha} |u(\mathbf{y})|\, d\Omega_\mathbf{y} \right)^2 d\Omega_\mathbf{x} \\ &\le 4m \int_\Omega \left(\int_{\Omega\cap B(\mathbf{x})} \frac{d\Omega_\mathbf{y}}{r^\alpha} \int_{\Omega\cap B(\mathbf{x})} \frac{|u(\mathbf{y})|^2}{r^\alpha}\, d\Omega_\mathbf{y} \right) d\Omega_\mathbf{x} .\end{aligned}$$

Since

$$\int_{\Omega\cap B(\mathbf{x})} \frac{d\Omega_\mathbf{y}}{r^\alpha} \le m_1\, \varepsilon^{n-\alpha} ,$$

we have

$$\|Au - A_\varepsilon u\|^2_{L^2(\Omega)} \le m_2\, \varepsilon^{n-\alpha}\, \|u\|^2_{L^2(\Omega)}$$

and thus $\|A - A_\varepsilon\| \to 0$ as $\varepsilon \to 0$. □

Note that integral operators with weakly singular kernels appear in the theory of Sobolev spaces: they participate in the integral representation of functions, and their properties led Sobolev to his famous imbedding theorems.

We leave it to the reader to formulate properties of the spectrum of a linear integral operator with kernel having weak singularity.

3.13 Courant's Minimax Principle

R. Courant proposed a way to determine the nth eigenvalue of a strictly positive self-adjoint compact linear operator A, by which this eigenvalue could be found independently of the other eigenvalues.

In Sect. 3.11 we have shown that μ_n given by

$$\mu_{n+1} = \frac{1}{\lambda_{n+1}}, \qquad \lambda_{n+1} = \sup_{\substack{\|u\|\le 1 \\ u\in H_n}} (Ax, x),$$

is the nth eigenvalue of A determined successively:

$$0 < \mu_1 \le \mu_2 \le \mu_3 \le \cdots .$$

As the corresponding orthonormal system of eigenvectors $x_1, x_2, x_3, \ldots$ is a basis of H, an element $x \in H$ can be represented as

$$x = \sum_{k=1}^{\infty} c_k x_k, \qquad \|x\|^2 = \sum_{k=1}^{\infty} |c_k|^2,$$

and thus

$$(Ax, x) = \left(A \sum_{k=1}^{\infty} c_k x_k, \sum_{k=1}^{\infty} c_k x_k\right) = \left(\sum_{k=1}^{\infty} c_k \lambda_k x_k, \sum_{k=1}^{\infty} c_k x_k\right) = \sum_{k=1}^{\infty} \lambda_k |c_k|^2 .$$

Let us take any n elements $y_1, \ldots, y_n$ of H and denote by Q_n the subspace spanned by these elements and by S_n its orthogonal complement in H. Let $Q_{n,\text{eig}}$ be the subspace of H spanned by the eigenvectors $x_1, x_2, \ldots, x_n$ of A that were determined in Sect. 3.11, and H_n the orthogonal complement of $Q_{n,\text{eig}}$ in H. We shall now prove the so-called minimax principle of Courant, which is

Theorem 3.13.1. Eigenvalue μ_{n+1} of A is

$$\mu_{n+1} = \frac{1}{\lambda_{n+1}}\,, \qquad \lambda_{n+1} = \inf_{Q_n} \sup_{\substack{\|x\|\le 1 \\ x\in S_n}} (Ax, x) \tag{3.13.1}$$

and $\lambda_{n+1} = \sup_{\|x\|\le 1}(Ax, x)$ when $x \in H_n$.

Proof. We first recall that we have shown (Sect. 3.11) that

$$\mu_{n+1} = \frac{1}{\lambda_{n+1}}\,, \qquad \lambda_{n+1} = \sup_{\substack{\|x\|\le 1 \\ x\in H_n}} (Ax, x)$$

is an eigenvalue of A. Then we recall that $\lambda_n \ge \lambda_{n+1}$ for all $n \ge 1$.

Suppose $Q_n \ne Q_{n,\mathrm{eig}}$. Then there exists

$$x_0 = \sum_{k=1}^{n} c_k^0 x_k \in Q_{n,\mathrm{eig}}$$

such that $\|x_0\| = 1$, $x_0 \in S_n$, and

$$(Ax_0, x_0) = \sum_{k=1}^{n} \lambda_k \, |c_k^0|^2 \ge \lambda_n \sum_{k=1}^{n} |c_k^0|^2 = \lambda_n \ge \lambda_{n+1}$$

and so for any S_n

$$\sup_{\substack{\|x\|\le 1 \\ x\in S_n}} (Ax, x) \ge \lambda_{n+1}\,.$$

This completes the proof. □

Some consequences of this principle are very important in mechanics.

Theorem 3.13.2. If an elastic body (a membrane, plate, or two- or three- dimensional body) is subjected to some additional geometrical constraints (fixed lines, surfaces, or their parts), then all corresponding eigenvalues μ_n can only grow; if some geometrical constraints are broken, then all μ_n can only be less than or equal to the original ones.

Proof. Additional constraints are assumed to be some linear restrictions of the type, say,

$$u\big|_{\gamma} = 0 \quad \text{or} \quad u\big|_{\Gamma} = 0$$

where γ, Γ are a line and a surface, respectively. The value (3.13.1) of the supremum of (Ax, x) in the definition of λ_n becomes less or the same and so this holds for their infimum, i.e., λ_n cannot be more than the old one, and so μ_n cannot be less under new restrictions. The second part of the statement of the theorem is now evident. □

Remark 3.13.1. In the energy spaces E_M and E_E for membranes and elastic bodies, a restriction

$$u\big|_P = 0$$

where P is a point in Ω, in accordance with Sobolev's imbedding theorems, is neglected, and so the fixing of several points of a membrane or an elastic body cannot increase the corresponding eigenfrequencies (this was shown by Vitt and Shubin [39]); but for eigenfrequencies of a plate, fixing of finite number of points of the plate increases eigenfrequencies. □

Chapter 4
Elements of Nonlinear Functional Analysis

From a functional analytic standpoint, nonlinear problems of mechanics are more complicated than linear problems; as in mechanics, they require new approaches. Many, like the problems of nonlinear elasticity in the general case, provide a wide field of investigation for mathematicians (see Antman [2]); the problem of existence of solutions in nonlinear elasticity in general is still open.

But some nonlinear mechanics problems can be treated on a known background; as in the linear case, we consider only those results from functional analysis that are needed in what follows.

4.1 Fréchet and Gâteaux Derivatives

We begin the nonlinear analysis of operators with definitions pertaining to differentiation.

In calculus, the differential df of an ordinary function f at a point x is defined as the principal linear part of the increment $f(x + dx) - f(x)$; if $f(x)$ is differentiable at x, then $df = f'(x)\,dx$. The differential df depends on x, and the increment of the argument is denoted by dx. For operators, this can be extended as follows.

Let $F(x)$ be a nonlinear operator acting from $D(F) \subset X$ to $R(F) \subset Y$, where X and Y are real Banach spaces. Assume $D(F)$ is open.

Definition 4.1.1. $F(x)$ is *differentiable in the Fréchet sense* at $x_0 \in D(F)$ if there is a bounded linear operator, denoted by $F'(x_0)$, such that

$$F(x_0 + h) - F(x_0) = F'(x_0)h + \omega(x_0, h) \text{ for all } \|h\| < \varepsilon$$

with some $\varepsilon > 0$, where

$$\|\omega(x_0, h)\|/\|h\| \to 0 \text{ as } \|h\| \to 0\,.$$

Then $F'(x_0)$ is called the *Fréchet derivative* of $F(x)$ at x_0, and

$$dF(x_0, h) = F'(x_0)h$$

is its *Fréchet differential.* $F(x)$ is Fréchet differentiable in an open domain $S \subset D(F)$ if it is Fréchet differentiable at every point of S.

For a linear continuous operator A acting from X to Y, we have $A(x+h)-Ax = Ah$. In the Fréchet sense, this means that $A' = A$.

Problem 4.1.1. (a) Assume $\mathbf{y} = \mathbf{f}(\mathbf{x})$ is a vector function from $\mathbb{R}^m$ to $\mathbb{R}^n$ and $\mathbf{f}(\mathbf{x}) \in (C^{(1)}(\Omega))^n$. Show that its Fréchet derivative at $\mathbf{x}_0 \in \Omega$ is the Jacobian matrix

$$J_{\mathbf{f}}(\mathbf{x}_0) = \left(\frac{\partial f_i(\mathbf{x}_0)}{\partial x_j} \right)_{\substack{i=1,\ldots,n \\ j=1,\ldots,m}} .$$

(b) Show that the Fréchet derivative of a continuous linear operator is the same operator. □

In the construction of the Fréchet derivative, the reader can recognize a method of the calculus of variations, used to obtain the Euler equations of a functional. The following derivative, in the sense of Gâteaux, is yet closer to this.

Definition 4.1.2. Assume that for all $h \in D(F)$ we have

$$\lim_{t \to 0} \frac{F(x_0 + th) - F(x_0)}{t} = DF(x_0, h) , \qquad x_0 \in D(F) ,$$

where $DF(x_0, h)$ is a linear operator with respect to h. Then $DF(x_0, h)$ is called the *Gâteaux differential* of $F(x)$ at x_0, and $F(x)$ is said to be *Gâteaux differentiable* at x_0. Denoting

$$DF(x_0, h) = F'(x_0)h ,$$

we obtain the *Gâteaux derivative* $F'(x_0)$. An operator is Gâteaux differentiable in an open domain $S \subset X$ if it has a Gâteaux derivative at every point of S.

The definitions of derivatives are clearly valid for functionals. Suppose $\Phi(x)$ is a functional which is Gâteaux differentiable in a Hilbert space and that $D\Phi(x, h)$ is bounded at $x = x_0$ as a linear functional in h. Then, by the Riesz representation theorem, it can be represented in the form of an inner product; denoting the representing element by grad $\Phi(x_0)$, we get

$$D\Phi(x_0, h) = (\operatorname{grad} \Phi(x_0), h) .$$

Hence we have an operator grad $\Phi(x_0)$ called the *gradient* of $\Phi(x)$ at x_0.

Theorem 4.1.1. If an operator $F(x)$ from X to Y is Fréchet differentiable at $x_0 \in D(F)$, then $F(x)$ is Gâteaux differentiable at x_0 and the Gâteaux derivative coincides with the Fréchet derivative.

Proof. Replace h by th in Definition 4.1.1:

$$F(x_0 + th) - F(x_0) = F'(x_0)th + \omega(x_0, th) .$$

Now rearrange, take the norm of both sides, and let $t \to 0$:

$$\left\| \lim_{t\to 0} \frac{F(x_0+h) - F(x_0)}{t} - F'(x_0)h \right\| = \lim_{t\to 0} \frac{1}{|t|} \cdot \|\omega(x_0, th)\|$$

$$= \|h\| \cdot \lim_{t\to 0} \frac{1}{|t|} \cdot \frac{\|\omega(x_0, th)\|}{\|h\|} = \|h\| \cdot \lim_{t\to 0} \frac{\|\omega(x_0, th)\|}{\|th\|} = 0 .$$

It follows that

$$\lim_{t\to 0} \frac{F(x_0+th) - F(x_0)}{t} = F'(x_0)h ,$$

hence $F'(x_0)$ is a Gâteaux derivative of $F(x)$ at x_0. □

Gâteaux differentiability does not imply Fréchet differentiability. We formulate a sufficient condition as

Problem 4.1.2. Assume the Gâteaux derivative of $F(x)$ exists in a neighborhood of x_0 and is continuous at x_0 in the uniform norm of $L(X, Y)$. Show that the Fréchet derivative exists and is equal to the Gâteaux derivative. □

We consider an operator equation with a parameter μ from a real Banach space M:

$$F(x, \mu) = 0$$

where $D(F(x, \mu)) \subseteq X$ and $R(F(x, \mu)) \subseteq Y$.

In mechanics, μ can represent loads or some parameters of a body or a process (say, disturbances of the thickness of a plate or its moduli).

There are different abstract analogs of the implicit function theorem; we present two of them.

Denote by $N(x_0, r; \mu_0, \rho)$ the following neighborhood of a pair:

$$N(x_0, r; \mu_0, \rho) = \{x \in X, \mu \in M : \|x - x_0\| < r, \ \|\mu - \mu_0\| < \rho\} .$$

Theorem 4.1.2. Assume:

(i) $F(x_0, \mu_0) = 0$;

(ii) $F(x_0, \mu)$ is continuous with respect to μ in a ball $\|\mu - \mu_0\| < \rho_1$;

(iii) there exist $r_1 > 0$ and a continuous linear operator A from X to Y, being continuously invertible and such that in any neighborhood $N(x_0, r; \mu_0, \rho)$ with $r \le r_1$ and $\rho \le \rho_1$ we have

$$\|F(x, \mu) - F(y, \mu) - A(x - y)\| \le \alpha(r, \rho) \|x - y\| ,$$

where

$$\limsup_{r,\rho\to 0} |\alpha(r, \rho)| \|A^{-1}\| = q < 1 .$$

Then there exist $r_0 > 0$ and $\rho_0 > 0$ such that in $N(x_0, r_0; \mu_0, \rho_0)$ the equation

$$F(x, \mu) = 0 \tag{4.1.1}$$

has the unique solution $x = x(\mu)$ which depends continuously on μ: $x(\mu) \to x(\mu_0)$ as $\mu \to \mu_0$.

Proof. We reduce (4.1.1) to a form needed to apply the contraction mapping principle:

$$x = K(x, \mu) , \qquad K(x, \mu) = x - A^{-1}F(x, \mu) .$$

This is equivalent to (4.1.1) because A is continuously invertible. $K(x, \mu)$ is a contraction operator with respect to x in some neighborhood of (μ_0, x_0). Indeed

$$\begin{aligned} \|K(x, \mu) - K(y, \mu)\| &= \|x - y - A^{-1}(F(x, \mu) - F(y, \mu))\| \\ &\le \|A^{-1}\| \, \|A(x - y) - (F(x, \mu) - F(y, \mu))\| \\ &\le \|A^{-1}\| \, |\alpha(r, \rho)| \, \|x - y\| \le (q + \varepsilon) \, \|x - y\| ; \end{aligned}$$

by (iii), $q + \varepsilon < 1$ if r and ρ are sufficiently small and $r < r_1$, $\rho < \rho_1$. Then there exist r_0, ρ_0 with $r_0 \le r_1$ and $\rho_0 \le \rho_1$ such that $K(x, \mu)$ takes a ball $\|x - x_0\| \le r_0$ into itself when $\|\mu - \mu_0\| \le \rho_0$. To see this, we first write

$$\begin{aligned} \|K(x, \mu) - x_0\| &\le \|K(x, \mu) - K(x_0, \mu)\| + \|K(x_0, \mu) - x_0\| \\ &\le (q + \varepsilon) \, \|x - x_0\| + \|A^{-1}F(x_0, \mu)\| \\ &\le (q + \varepsilon) \, \|x - x_0\| + \|A^{-1}\| \, \|F(x_0, \mu)\| . \end{aligned} \tag{4.1.2}$$

But $F(x_0, \mu) \to F(x_0, \mu_0) = 0$ as $\mu \to \mu_0$, so

$$\|A^{-1}\| \, \|F(x_0, \mu)\| \le (1 - q - \varepsilon) r_1 \text{ when } \|\mu - \mu_0\| \le \rho_2 \text{ for some } \rho_2 < \rho_1 .$$

Hence for any $r_0 < r_1$ and $\rho_0 < \rho_2$, the ball $\|x - x_0\| \le r_0$ is taken by $K(x, \mu)$ into itself when $\|\mu - \mu_0\| \le \rho_0$.

By the contraction mapping principle, there is a solution $x = x(\mu)$ in the neighborhood $N(x_0, r_0; \mu_0, \rho_0)$. The continuity of $x(\mu)$ at μ_0 follows from the bound

$$\|x(\mu) - x_0\| \le \frac{\|A^{-1}\|}{1 - q - \varepsilon} \, \|F(x_0, \mu)\| ,$$

a consequence of the contraction mapping principle. □

Problem 4.1.3. Verify the last statement of the proof. □

To prove the other variant of the implicit function theorem, we need the properties of Fréchet derivatives provided in the next two lemmas.

Lemma 4.1.1. Assume an operator $F(x)$ from X to Y has a Fréchet derivative at $x = x_0$, and an operator $x = S(z)$ from a real Banach space Z to X also has a Fréchet derivative $S'(z_0)$ and $x_0 = S(z_0)$. Then their composition $F(S(z))$ has a Fréchet derivative at $z = z_0$ and

$$(F(S(z_0)))' = F'(x_0)S'(z_0) .$$

Proof. Substituting

$$x - x_0 = S(z) - S(z_0) = S'(z_0)(z - z_0) + \omega_1(z_0, z - z_0)$$

into

$$F(x) - F(x_0) = F'(x_0)(x - x_0) + \omega(x_0, x - x_0) ,$$

we get

$$F(x) - F(x_0) = F'(x_0)S'(z_0)(z - z_0) + F'(x_0)\omega_1(z_0, z - z_0) + \omega(x_0, S(z) - S(z_0)) .$$

This completes the proof, since the last two terms on the right-hand side are of the order $o(\|z - z_0\|)$. □

Problem 4.1.4. Verify the final statement of the proof. □

The next lemma is the *Lagrange identity*.

Lemma 4.1.2. Assume that $F(x)$, acting from X to Y, is Fréchet differentiable in a neighborhood Ω of x_0, and suppose the segment connecting x_0 and x, which is the set of points $x_0 + t(x - x_0)$ for $0 \le t \le 1$, lies in Ω. Then for $x \in \Omega$ we have

$$F(x) - F(x_0) = \int_0^1 F'(x_0 + \theta(x - x_0))\, d\theta \cdot (x - x_0) .$$

Proof. Let

$$S(\theta) = x_0 + \theta(x - x_0)$$

and consider the function $F(S(\theta))$ of the real variable $\theta \in [0, 1]$. By Lemma 4.1.1, the composition $F(S(\theta))$ has a Fréchet derivative and

$$\frac{d}{d\theta}F(S(\theta)) = F'(S(\theta))\frac{dS(\theta)}{d\theta} = F'(x_0 + \theta(x - x_0))(x - x_0) .$$

Integration over θ gives

$$\int_0^1 \frac{d}{d\theta}F(S(\theta))\, d\theta = \int_0^1 F'(x_0 + \theta(x - x_0))(x - x_0)\, d\theta$$

where the left-hand side is

$$F(S(\theta))\Big|_{\theta=0}^{\theta=1} = F(x_0 + \theta(x - x_0))\Big|_{\theta=0}^{\theta=1} = F(x) - F(x_0)$$

and the factor $(x - x_0)$ can be removed from the integral on the right. □

We can now present the more traditional version of the implicit function theorem. First we introduce a partial Fréchet derivative $F_x(x,\mu)$ of $F(x,\mu)$ with respect to x as its Fréchet derivative with respect to x when μ is fixed.

Theorem 4.1.3. Assume:

(i) $F(x_0,\mu_0) = 0$;

(ii) for some $r > 0$ and $\rho > 0$, the operator $F(x,\mu)$ is continuous on the set $N(x_0,r;\mu_0,\rho)$;

(iii) $F_x(x,\mu)$ is continuous at (x_0,μ_0);

(iv) $F_x(x_0,\mu_0)$ has an inverse operator that is linear and continuous.

Then there exist $r_0 > 0$, $\rho_0 > 0$ such that the equation $F(x,\mu) = 0$ has the unique solution $x = x(\mu)$ in a ball $\|x - x_0\| \le r_0$ when $\|\mu - \mu_0\| \le \rho_0$. If there also exists $F_\mu(x,\mu)$ which is continuous at (x_0,μ_0), then $x(\mu)$ has a Fréchet derivative at $\mu = \mu_0$ and

$$x'(\mu_0) = -F_x^{-1}(x_0,\mu_0)F_\mu(x_0,\mu_0)\,.$$

Proof. We verify that $A = F_x(x_0,\mu_0)$ satisfies condition (iii) of Theorem 4.1.2. Consider

$$\Psi(x,y,\mu) = \|F(x,\mu) - F(y,\mu) - F_x(x_0,\mu_0)(x-y)\|\,.$$

By Lemma 4.1.2,

$$F(x,\mu) - F(y,\mu) = \int_0^1 F_x(y + \theta(x-y),\mu)\,d\theta\,(x-y)$$

and so

$$\begin{aligned}\Psi(x,y,\mu) &= \left\| \int_0^1 (F_x(y + \theta(x-y),\mu) - F_x(x_0,\mu_0))\,d\theta\,(x-y) \right\| \\ &\le \int_0^1 \|F_x(y + \theta(x-y),\mu) - F_x(x_0,\mu_0)\|\,d\theta\,\|x-y\| \le \alpha(r,\rho)\,\|x-y\|\end{aligned}$$

where

$$\alpha(r,\rho) = \sup_{x,\mu} \|F_x(x,\mu) - F_x(x_0,\mu_0)\| \text{ on } N(x_0,r;\mu_0,\rho)$$

is such that $\alpha(r,\rho) \to 0$ as $r,\rho \to 0$ since $F_x(x,\mu)$ is continuous at (x_0,μ_0). The other conditions of Theorem 4.1.2 are also satisfied and so a solution $x = x(\mu)$ exists. We leave the second part of the theorem on differentiability of $x(\mu)$ without proof. □

Using the implicit function theorem, we can determine whether a solution to a problem depends continuously and uniquely on certain parameters.

We studied several linear problems of mechanics with constant parameters. The reader can now verify that small disturbances of elastic moduli or, say, the thickness of a plate, bring small disturbances in displacements (small in a corresponding energy norm). We note that for linear problems this can be shown more easily by using the contraction mapping principle, but in nonlinear problems the implicit function theorem is more convenient.

4.2 Liapunov–Schmidt Method

We call (x_0, μ_0) a *regular point* of the equation $F(x, \mu) = 0$ if there is a neighborhood of (x_0, μ_0), say $N(x_0, r; \mu_0, \rho)$, in which there is a unique solution $x = x(\mu)$. The implicit function theorem gives sufficient conditions for regularity of $F(x, \mu)$ at a point (x_0, μ_0).

In mechanics, the breakdown of the regularity property of a solution is of great importance; it is usually connected with some qualitative change of the properties of a system under consideration: its behavior, stability, or type of motion. We now consider an important class of non-regular points of an operator equation.

Definition 4.2.1. The pair (x_0, μ_0) is a *bifurcation point* of the equation $F(x, \mu) = 0$ if for any $r > 0$, $\rho > 0$, in the ball $\|\mu - \mu_0\| \le \rho$ there exists μ such that the ball $\|x - x_0\| \le r$ contains at least two solutions of the equation corresponding to μ.

Many mechanics problems (in particular, those of shell theory) are such that in an energy space a partial Fréchet derivative $F_x(x_0, \mu_0)$ of a corresponding operator of a problem may be reduced to the form $I - B$, where $B = B(x_0, \mu_0)$ is a compact linear operator (typically self-adjoint) and so the results of the Fredholm–Riesz–Schauder theory apply. In particular, $I - B$ is not continuously invertible if and only if there is a nontrivial solution to $(I - B)x = 0$, and this is the case when the implicit function theorem is not applicable. This case is now considered.

Without loss of generality, we assume $x_0 = 0$, $\mu_0 = 0$ (we can always change $x \mapsto x_0 + x$, $\mu \mapsto \mu_0 + \mu$) so let

$$F(0, 0) = 0 .$$

Suppose F is an operator acting from $H \times M$ to H, where H is a real Hilbert space and M is a real Banach space. As we said, we suppose that $F_x(0, 0)$ takes the form

$$F_x(0, 0) = I - B_0$$

with B_0 a compact self-adjoint linear operator in H.

Equation $F(x, \mu) = 0$ can be rewritten in the form

$$(I - B_0)x = -F(x, \mu) + (I - B_0)x$$

or

$$(I - B_0)x = R(x, \mu) , \qquad R(x, \mu) = -F(x, \mu) + (I - B_0)x . \tag{4.2.1}$$

We now consider the Liapunov–Schmidt method of determining the dependence of solution to (4.2.1) on μ when $\|\mu\|$ is small and there are nontrivial solutions to the equation $(I - B_0)x = 0$. As in Sect. 3.8, denote by N the set of these nontrivial solutions (augmented by the zero element) and let $x_1, \dots, x_n$ be an orthonormal basis of N.

In the beginning of the proof of Theorem 3.8.5 we saw that the operator

$$Q_0 x = (I - B_0)x + \sum_{k=1}^{n} (x, x_k) x_k$$

is continuously invertible. Equation (4.2.1) can be written in the form

$$Q_0 x = R(x, \mu) + \sum_{k=1}^{n} \alpha_k x_k \,, \qquad \alpha_k = (x, x_k) \,. \tag{4.2.2}$$

We now consider (4.2.2) as an equation with respect to x that has parameters μ and $\alpha_1, \ldots, \alpha_n$. Let us split x into the form

$$x = u + \sum_{k=1}^{n} \beta_k x_k \,, \qquad (u, x_j) = 0 \quad (j = 1, \ldots, n) \,,$$

where $u \in N^{\perp}$ and $N^{\perp}$ is the orthogonal complement of N in H. As $(x, x_k) = \alpha_k$, we have

$$x = u + \sum_{k=1}^{n} \alpha_k x_k$$

and (4.2.2) becomes

$$Q_0 u = R\left(u + \sum_{k=1}^{n} \alpha_k x_k, \mu\right). \tag{4.2.3}$$

This equation defines u as a function of the variables $(\mu, \alpha_1, \ldots, \alpha_n)$. Since

$$R_x(0, 0) = -F_x(0, 0) + (I - B_0) = 0$$

we get

$$\left(Q_0 x - R(u + \sum_{k=1}^{n} \alpha_k x_k,\, \mu)\right)_u \Bigg|_{\substack{u=0 \\ \mu=0,\, \alpha_1=\ldots=\alpha_n=0}} = Q_0 \,,$$

where Q_0 is a continuously invertible operator, so all the conditions of the implicit function theorem are fulfilled. Therefore (4.2.3) has a unique solution for every $(\mu,$ $\alpha_1, \ldots, \alpha_n)$ when $\|\mu\|$ and $|\alpha_k|$ are small:

$$u = u(\mu,\, \alpha_1, \ldots, \alpha_n) \,.$$

This solution must be orthogonal to all x_k $(k = 1, \ldots, n)$, and to define values $\alpha_1, \ldots, \alpha_n$ we have the system

$$(u(\mu,\, \alpha_1, \ldots, \alpha_n), x_k) = 0 \qquad (k = 1, \ldots, n) \tag{4.2.4}$$

which is called the Liapunov–Schmidt equation of branching.

Using the Liapunov–Schmidt method one can investigate so-called post-critical behavior of a system, say, post-buckling of a von Kármán plate.

4.3 Critical Points of a Functional

From now on, we shall consider operators and real-valued functionals given in a real Hilbert space H. So let $\Phi(x)$ be a functional on H.

Definition 4.3.1. We call $x_0 \in H$ a *local minimal* (*maximal*) point of $\Phi(x)$ if there is a ball $B = \{x \colon \|x - x_0\| \le \varepsilon\}$, $\varepsilon > 0$, such that for all $x \in B$ we have $\Phi(x) \ge \Phi(x_0)$ ($\Phi(x) \le \Phi(x_0)$). Minimal and maximal points are called *extreme points* of $\Phi(x)$. If $\Phi(x) \ge \Phi(x_0)$ for all $x \in H$, then x_0 is a point of *absolute minimum* of $\Phi(x)$.

We prove the following

Theorem 4.3.1. Assume:

(i) $\Phi(x)$ is given on an open set $S \subset H$;

(ii) grad $\Phi(x)$ exists at $x = x_0 \in S$;

(iii) x_0 is an extreme point of $\Phi(x)$.

Then grad $\Phi(x_0) = 0$.

Proof. Choose $h \in H$. For fixed x_0 and h, the functional $\Phi(x_0 + th)$ is a function in a real variable t that attains its extremum at $t = 0$. Since

$$\left. \frac{d\Phi(x_0 + th)}{dt} \right|_{t=0} = 0 ,$$

we have

$$(\operatorname{grad} \Phi(x_0), h) = 0 . \tag{4.3.1}$$

Because h is arbitrary, the conclusion follows. □

Definition 4.3.2. A point x_0 at which grad $\Phi(x_0) = 0$ is called a *critical point* (or *stationary point*) of $\Phi(x)$.

We made implicit use of Theorem 4.3.1 for linear problems where $\Phi(x)$ was a quadratic functional representing the total potential energy of an elastic body and (4.3.1) defined a generalized solution to the corresponding problem. Similar results will hold for some nonlinear problems below. First we introduce some definitions.

Definition 4.3.3. A functional $\Phi(x)$ is *weakly continuous* at $x = x_0$ if for every sequence $\{x_k\}$ converging weakly to x_0 the numerical sequence $\Phi(x_k)$ tends to $\Phi(x_0)$ as $k \to \infty$. It is said to be weakly continuous on an open set $S \subset H$ if it is weakly continuous at every point of S.

Definition 4.3.4. A functional $\Phi(x)$ given on H is *growing* if

$$\inf_{\|x\|=R} \Phi(x) \to \infty \quad \text{as } R \to \infty .$$

We obtained a necessary condition for the existence of critical points of a functional. Now we present some sufficient conditions that have important applications in mechanics.

Lemma 4.3.1. Let Q be a weakly compact set in H. A weakly continuous functional $\Phi(x)$ is bounded and attains its minimal and maximal values on Q.

Proof. Since Q is weakly compact, it is also bounded and weakly closed. First we prove that the values of $\Phi(x)$ on Q are bounded from above. If not, there is a sequence $\{x_n\} \subset Q$ such that $\Phi(x_n) \to \infty$ as $n \to \infty$. By hypothesis $\{x_n\}$ contains a subsequence $\{x_{n_k}\}$ weakly convergent to $x_0 \in Q$ and so

$$\Phi(x_{n_k}) \to \Phi(x_0) \neq \infty \quad \text{as } n_k \to \infty\,,$$

which contradicts the assumption. Boundedness from below is seen similarly.

Let $d = \inf_{x \in Q} \Phi(x)$. By definition of infimum, there is a sequence $\{z_n\}$ for which $\Phi(z_n) \to d$ as $n \to \infty$. As above, it contains a subsequence $\{z_{n_k}\}$ converging weakly to $z_0 \in Q$. By weak continuity of $\Phi(x)$, we get $\Phi(z_0) = d$. The proof for the maximal value is similar. □

We recall that a ball

$$B(R) = \{x\colon \|x\| \le R\} \tag{4.3.2}$$

is weakly compact, and therefore meets the requirements of Q in Lemma 4.3.1.

In what follows, some problems of mechanics can be reduced to a problem of finding critical points of the functional

$$\Psi(x) = \|x\|^2 + \Phi(x) \tag{4.3.3}$$

with $\Phi(x)$ a weakly continuous functional. The functional $\Psi(x)$ is not weakly continuous because of the term $\|x\|^2$ and so Lemma 4.3.1 does not apply directly.

Theorem 4.3.2. If $\Phi(x)$ is a weakly continuous functional, then the functional (4.3.3) attains its minimal value on the ball (4.3.2).

Proof. By Lemma 4.3.1, $\Phi(x)$ and hence $\Psi(x)$ is bounded from below on $B(R)$. Let $d = \inf \Psi(x)$ on $B(R)$ and let $\{x_n\}$ be a sequence in $B(R)$ such that $\Psi(x_n) \to d$ as $n \to \infty$. By weak compactness of $B(R)$, we can produce a subsequence $\{x_{n_k}\}$ that converges weakly to $x_0 \in B(R)$. Moreover, from the bounded numerical sequence $\{\|x_{n_k}\|\}$ we can take a subsequence that tends to a number $a \le R$. Redenote the last subsequence as $\{x_n\}$ again.

We show that $\|x_0\| \le a$. Indeed, since $x_n \rightharpoonup x_0$ then $(x_n, x_0) \to \|x_0\|^2$ and we have

$$\|x_0\|^2 = \lim_{n\to\infty} |(x_n, x_0)| \le \lim_{n\to\infty} \|x_n\|\,\|x_0\| = a\,\|x_0\|$$

which gives $\|x_0\| \le a$.

By weak continuity of $\Phi(x)$, we get $\Phi(x_n) \to \Phi(x_0)$ as $n \to \infty$ and $\Psi(x_n) \to d = a^2 + \Phi(x_0)$ simultaneously. Since $x_0 \in B(R)$, we get

$$\Psi(x_0) = \|x_0\|^2 + \Phi(x_0) \geq \inf_{x \in B(R)} \Psi(x) = d = a^2 + \Phi(x_0) ,$$

and so $\|x_0\| \geq a$. With the above, this implies $\|x_0\| = a$ and thus x_0 is a point at which $\Psi(x)$ takes its minimal value on $B(R)$. □

Remark 4.3.1. Since $\{x_{n_k}\}$ from the proof converges weakly to x_0 and the sequence $\{\|x_{n_k}\|\}$ converges to $\|x_0\| = a$, this sequence converges to x_0 strongly in H. □

Definition 4.3.5. Assume $\inf \Phi(x) = d > -\infty$ on H. A sequence $\{x_n\}$ is a *minimizing sequence* of $\Phi(x)$ on H if $\Phi(x_n) \to d$ as $n \to \infty$.

The proof of Theorem 4.3.2 shows that, under the conditions of that theorem, any sequence minimizing $\Psi(x)$ on $B(R)$ contains a subsequence that converges strongly to an element at which $\Psi(x)$ is minimized. Now we can formulate

Theorem 4.3.3. Assume that a functional

$$\Psi(x) = \|x\|^2 + \Phi(x) ,$$

where $\Phi(x)$ is weakly continuous on H, is growing. Then:

(i) there exists $x_0 \in H$ at which $\Psi(x)$ takes its minimal value;

(ii) any minimizing sequence $\{x_n\}$ of $\Psi(x)$ contains a subsequence which converges strongly to a point at which $\Psi(x)$ takes its minimal value: moreover, every weakly convergent subsequence of $\{x_n\}$ converges strongly to a minimizer of $\Psi(x)$;

(iii) if a point x_0 at which $\Psi(x)$ takes its minimal value is unique, then a minimizing sequence converges to x_0 strongly;

(iv) if grad $\Phi(x_0)$ exists, then

$$2x_0 + \operatorname{grad} \Phi(x_0) = 0 .$$

Proof. By Theorem 4.3.2, on a ball $\|x\| \leq R$ the functional $\Psi(x)$ assumes its minimal value. Since $\Psi(x)$ is growing, R can be taken so large that the minimum is attained inside the open ball $\|x\| < R$. So statements (i) and (ii) follow from Theorem 4.3.2 and Remark 4.3.1. Statement (iv) follows from Theorem 4.3.1. The proof of (iii) is carried out in a way similar to that given in Sect. 2.9. □

Problem 4.3.1. Prove statement (iv) in detail. □

Now we consider the application of the Ritz method to solve the problem of minimizing $\Psi(x)$ under the restrictions of Theorem 4.3.3. First we state the equations of Ritz's method. Let $g_1, g_2, g_3, \ldots$ be a complete system in H such that every finite subsystem is linearly independent. Denote by H_n a subspace of H which is spanned by $g_1, \ldots, g_n$.

Let $\Psi(x)$ have grad $\Psi(x)$ at any $x \in H$. The equations to find the nth Ritz approximation are

$$(\operatorname{grad} \Psi(x_n), g_k) = 0 \qquad (k = 1, \ldots, n, \quad x_n \in H_n) .$$

Theorem 4.3.4. Under the restrictions of Theorem 4.3.3, the following hold:

(i) for each n there exists a solution $x_n \in H_n$, the nth Ritz approximation of the minimizer of $\Psi(x)$;

(ii) $\{x_n\}$, the sequence of Ritz approximations, is a minimizing sequence of $\Psi(x)$, and thus

(iii) $\{x_n\}$ contains at least one weakly convergent subsequence whose weak limit is a minimizer of $\Psi(x)$ — in fact, this subsequence converges strongly to the minimizer;

(iv) every weakly convergent subsequence of $\{x_n\}$ converges strongly to a minimizer of $\Psi(x)$; if a minimizer of $\Psi(x)$ is unique, then the whole sequence $\{x_n\}$ converges to it strongly.

Proof. (i) Solvability of the problem for the nth approximation of solution by the Ritz method follows from Theorem 4.3.3.

(ii) Let x_0 be a solution to the main problem

$$\Psi(x_0) = d = \inf_{x \in H} \Psi(x) .$$

As the system $g_1, g_2, g_3, \ldots$ is complete, there is $x^{(n)} \in H_n$ such that

$$\|x_0 - x^{(n)}\| = \delta_n \to 0 \ \text{ as } n \to \infty .$$

Since $\Psi(x)$ is continuous we get

$$|\Psi(x^{(n)}) - \Psi(x_0)| = \varepsilon_n \to 0 \ \text{ as } n \to \infty .$$

But x_n is a minimizer of $\Psi(x)$ on H_n, so

$$d = \Psi(x_0) \le \Psi(x_n) = \inf_{x \in H_n} \Psi(x) \le \Psi(x^{(n)}) .$$

Therefore

$$|\Psi(x_n) - \Psi(x_0)| \le \varepsilon_n \to 0 \ \text{ as } n \to \infty$$

and thus $\{x_n\}$ is a minimizing sequence of $\Psi(x)$.

The other statements follow from Theorem 4.3.3. □

Note that Theorem 4.3.4 can be applied to linear and nonlinear problems of mechanics.

4.4 Von Kármán Equations of a Plate

Theorem 4.3.4 will be applied to the boundary value problem of equilibrium of a plate described by the von Kármán equations, which are

$$\Delta^2 w = [f, w] + q \ \text{ in } \Omega \subset \mathbb{R}^2 , \tag{4.4.1}$$

$$\Delta^2 f = -[w, w] \ \text{ in } \Omega , \tag{4.4.2}$$

where $w(x, y)$ is the normal displacement of the midsurface Ω, $f(x, y)$ is the Airy function, $q = q(x, y)$ is the transverse external load, and

$$[u, v] = \frac{\partial^2 u}{\partial x^2}\frac{\partial^2 v}{\partial x^2} + \frac{\partial^2 u}{\partial y^2}\frac{\partial^2 v}{\partial y^2} - 2\frac{\partial^2 u}{\partial x \partial y}\frac{\partial^2 v}{\partial x \partial y} .$$

We consider the Dirichlet problem for these equations:

$$w\big|_{\partial\Omega} = \frac{\partial w}{\partial n}\Big|_{\partial\Omega} = 0 , \tag{4.4.3}$$

$$f\big|_{\partial\Omega} = \frac{\partial f}{\partial n}\Big|_{\partial\Omega} = 0 . \tag{4.4.4}$$

Let us consider the integro-differential equations

$$a(w, \varphi) = B(f, w, \varphi) + \iint_\Omega q\varphi \, d\Omega , \tag{4.4.5}$$

$$a(f, \eta) = -B(w, w, \eta) , \tag{4.4.6}$$

where

$$a(w, \varphi) = \iint_\Omega \left\{ \frac{\partial^2 w}{\partial x^2}\left(\frac{\partial^2 \varphi}{\partial x^2} + \nu \frac{\partial^2 \varphi}{\partial y^2}\right) + 2(1 - \nu)\frac{\partial^2 w}{\partial x \partial y}\frac{\partial^2 \varphi}{\partial x \partial y} + \frac{\partial^2 w}{\partial y^2}\left(\frac{\partial^2 \varphi}{\partial y^2} + \nu \frac{\partial^2 \varphi}{\partial x^2}\right)\right\} d\Omega$$

and

$$B(f, w, \varphi) = \iint_\Omega \left\{ \left(\frac{\partial^2 f}{\partial x \partial y}\frac{\partial w}{\partial y} - \frac{\partial^2 f}{\partial y^2}\frac{\partial w}{\partial x}\right)\frac{\partial \varphi}{\partial x} + \left(\frac{\partial^2 f}{\partial x \partial y}\frac{\partial w}{\partial x} - \frac{\partial^2 f}{\partial x^2}\frac{\partial w}{\partial y}\right)\frac{\partial \varphi}{\partial y}\right\} d\Omega .$$

Here $0 < \nu < 1/2$ is Poisson's ratio.

Note that $a(u, v)$ is the scalar product (2.3.4) (with an omitted multiplier — the bending rigidity) of the energy space E_{PC} for an isotropic plate, and we shall use this notation in this section.

Suppose that (4.4.5) and (4.4.6), with respect to the unknown function w, f, being smooth (of $C^{(4)}(\overline{\Omega})$) and satisfying the boundary conditions (4.4.3) and (4.4.4), are valid for every φ, η which also satisfy (4.4.3) for these functions and their normal derivatives on the boundary. The usual tools of the calculus of variations show that the pair (w, f) is a classical solution to the von Kármán equations (4.4.1) and (4.4.2). This means that we can use (4.4.5) and (4.4.6) to define a generalized solution to the problem under consideration. We note that (4.4.5) expresses the virtual work principle for the plate, and (4.4.6) is the equation of compatibility. So we introduce

Definition 4.4.1. A pair (w, f), $w, f \in E_{PC}$, is a generalized solution to the problem (4.4.1)–(4.4.4) if it satisfies the integro-differential equations (4.4.5)–(4.4.6) for any (φ, η), $\varphi, \eta \in E_{PC}$.

To make sense in the definition, the load $q = q(x, y)$ must be such that the term $\iint_\Omega q\varphi \, d\Omega$ is a continuous linear functional in E_{PC}; it suffices, for example, to have $q \in L^1(\Omega)$ (cf., Sect. 2.4).

Under the restrictions of Definition 4.4.1, all terms in (4.4.5)–(4.4.6) make sense. Indeed, all elements of E_{PC} have second derivatives belonging to $L^2(\Omega)$, so the left-hand sides of (4.4.5)–(4.4.6) make sense. For the linear terms on the right-hand sides, it is evident that they have meaning under the given conditions. Let us consider the nonlinear terms. Each of the first derivatives of elements of E_{PC} belong to $L^p(\Omega)$ for any $p < \infty$. So for a typical nonlinear term we have

$$\left| \iint_\Omega \frac{\partial^2 f}{\partial x^2} \frac{\partial w}{\partial y} \frac{\partial \varphi}{\partial y} \, d\Omega \right| \leq \left(\iint_\Omega \left| \frac{\partial^2 f}{\partial x^2} \right|^2 d\Omega \right)^{1/2} \cdot \left(\iint_\Omega \left| \frac{\partial w}{\partial y} \right|^4 d\Omega \right)^{1/4} \left(\iint_\Omega \left| \frac{\partial \varphi}{\partial y} \right|^4 d\Omega \right)^{1/4} . \quad (4.4.7)$$

Hence it is bounded and therefore makes sense in the setup as well.

We could present a functional whose gradient in the space $E_{PC} \times E_{PC}$ is defined by (4.4.5)–(4.4.6); unfortunately it is not of the form required by Theorem 4.3.4. That is why we shall reformulate the problem with respect to the unique unknown function w, defining f as an operator with respect to w and constructing a functional of w whose critical point is a generalized solution of the problem. We now embark on this program.

So let w be a fixed but arbitrary element of E_{PC}. Consider $B(w, w, \eta)$ as a functional with respect to η in E_{PC}. It is clearly linear. By (4.4.7) written for a typical term with $f = w$, thanks to the imbedding theorem in E_{PC}, we get

$$|B(w, w, \eta)| \leq m \, \|w\|^2_{E_P} \, \|\eta\|_{E_P} ,$$

i.e., the linear functional is continuous with respect to $\eta \in E_{PC}$. Therefore we can apply the Riesz representation theorem to get

$$-B(w, w, \eta) = (c, \eta)_{E_P} \equiv a(c, \eta) .$$

Being uniquely defined by $w \in E_{PC}$, the element $c \in E_{PC}$ can be considered as a value of a nonlinear operator

$$c = C(w) , \qquad a(C(w), \eta) = -B(w, w, \eta) . \quad (4.4.8)$$

Before studying the properties of C we introduce

Definition 4.4.2. An operator A mapping from a Banach space X to a Banach space Y is *compact* if it is continuous in X and takes every bounded set of X into a precompact set in Y. An operator is *completely continuous* if it takes every weakly convergent sequence of X, $x_n \rightharpoonup x_0$, into a sequence $A(x_n)$ converging strongly to $A(x_0)$.

Lemma 4.4.1. A completely continuous operator F mapping a Hilbert space X into a Banach space Y is compact.

Proof. F is continuous since when a sequence $\{x_n\}$ converges to x_0 strongly in X then it converges to x_0 weakly, too.

Next we take a bounded set S in X and let $\{x_n\} \subset S$. From $\{x_n\}$, thanks to its boundedness, we can choose a subsequence $\{x_{n_k}\}$ converging weakly to $x_0 \in X$. Then, by definition of complete continuity, we get the sequence $\{F(x_{n_k})\}$ converging to $F(x_0)$ strongly. This means $F(S)$ is precompact, hence F is compact. □

It is known that there are compact operators in a Hilbert space which are not completely continuous.

Corollary 4.4.1. If $F(x)$ is a completely continuous operator, then the functional $\|F(x)\|^2$ is a weakly continuous functional in X.

The proof is evident. Now we can prove

Lemma 4.4.2. The operator $C(w)$ defined by (4.4.8) is completely continuous.

Proof. When the functions $u, v, w \in E_{PC}$ are smooth, direct integration by parts gives

$$B(u, v, w) = B(v, u, w) = B(v, w, u) = B(w, u, v) \; ; \tag{4.4.9}$$

the limit passage shows that this is valid for $u, v, w \in E_{PC}$. So

$$-B(w, w, \eta) = \iint_\Omega \left\{ \left(\frac{\partial w}{\partial x}\right)^2 \frac{\partial^2 \eta}{\partial y^2} + \left(\frac{\partial w}{\partial y}\right)^2 \frac{\partial^2 \eta}{\partial x^2} - 2\frac{\partial w}{\partial x}\frac{\partial w}{\partial y}\frac{\partial^2 \eta}{\partial x \partial y} \right\} d\Omega \, .$$

Next we take an arbitrary sequence $\{w_n\}$ converging weakly to w_0 in E_{PC} and consider

$$|a(C(w_n) - C(w_0), \eta)| = |B(w_n, w_n, \eta) - B(w_0, w_0, \eta)| \, .$$

Using Hölder's inequality, we bound a typical term of the right-hand side of this equality as follows:

$$\begin{aligned} d_n &= \left| \iint_\Omega \left[\left(\frac{\partial w_n}{\partial x}\right)^2 - \left(\frac{\partial w_0}{\partial x}\right)^2 \right] \frac{\partial^2 \eta}{\partial y^2} \, d\Omega \right| \\ &= \left| \iint_\Omega \left(\frac{\partial w_n}{\partial x} - \frac{\partial w_0}{\partial x}\right)\left(\frac{\partial w_n}{\partial x} + \frac{\partial w_0}{\partial x}\right) \frac{\partial^2 \eta}{\partial y^2} \, d\Omega \right| \\ &\le \left\| \frac{\partial w_n}{\partial x} - \frac{\partial w_0}{\partial x} \right\|_{L^4(\Omega)} \left(\left\| \frac{\partial w_n}{\partial x} \right\|_{L^4(\Omega)} + \left\| \frac{\partial w_0}{\partial x} \right\|_{L^4(\Omega)} \right) \left\| \frac{\partial^2 \eta}{\partial y^2} \right\|_{L^2(\Omega)} . \end{aligned}$$

By the imbedding theorem in E_{PC}, which is a subspace of $W^{2,2}(\Omega)$, we get

$$d_n \le m_1 \left\| \frac{\partial w_n}{\partial x} - \frac{\partial w_0}{\partial x} \right\|_{L^4(\Omega)} (\|w_n\|_{E_P} + \|w_0\|_{E_P}) \|\eta\|_{E_P}$$

and, thanks to the boundedness of a weakly convergent sequence,

$$d_n \le m_2 \|w_n - w_0\|_{W^{1,4}(\Omega)} \|\eta\|_{E_P}$$

where m_1 and m_2 are constants.

Gathering all such bounds, we obtain

$$|a(C(w_n) - C(w_0), \eta)| \le m_3 \|w_n - w_0\|_{W^{1,4}(\Omega)} \|\eta\|_{E_P} .$$

Putting $\eta = C(w_n) - C(w_0)$, we finally obtain

$$\|C(w_n) - C(w_0)\|_{E_P} \le m_3 \|w_n - w_0\|_{W^{1,4}(\Omega)} \to 0 \quad \text{as } n \to \infty$$

since the imbedding operator of $W^{2,2}(\Omega)$ into $W^{1,4}(\Omega)$ is completely continuous (a particular case of Sobolev's imbedding theorems in $W^{2,2}(\Omega)$). The last limit passage shows that C is completely continuous. □

From this lemma we see that (4.4.6) with a given $w \in E_{PC}$ has the unique solution

$$f = C(w) . \tag{4.4.10}$$

If $\{w_n\}$ converges to w_0 weakly in E_{PC}, then $\{f_n\} = \{C(w_n)\}$ converges to $f_0 = C(w_0)$ strongly in E_{PC}.

From now on we consider f in (4.4.5) to be determined by (4.4.10).

For a fixed $w \in E_{PC}$, by bounds of the type (4.4.7), we see that the functional

$$B(f, w, \varphi) + \iint_\Omega q\varphi \, d\Omega$$

is linear and continuous with respect to $\varphi \in E_{PC}$. So applying the Riesz representation theorem, we have a representation

$$B(f, w, \varphi) + \iint_\Omega q\varphi \, d\Omega = a(U, \varphi)$$

where $U \in E_{PC}$ is uniquely determined by $w \in E_{PC}$; so we define an operator G, $U = G(w)$, acting in E_{PC}, by

$$B(f, w, \varphi) + \iint_\Omega q\varphi \, d\Omega = a(G(w), \varphi) . \tag{4.4.11}$$

In much the same way that Lemma 4.4.2 is proved we can establish

Lemma 4.4.3. *G is a completely continuous operator in E_{PC}.*

Now the following is evident:

Lemma 4.4.4. The system of equations (4.4.5)–(4.4.6) defining a generalized solution of the problem under consideration is equivalent to the operator equation

$$w = G(w) \tag{4.4.12}$$

with a completely continuous operator G acting in E_{PC}.

Now we introduce a functional

$$I(w) = \frac{1}{2}a(w, w) + \frac{1}{4}a(f, f) - \iint_{\Omega} qw \, d\Omega$$

where, again, f is defined by (4.4.8).

The decisive point of this section is

Lemma 4.4.5. For every $w \in E_{PC}$, we have

$$\operatorname{grad} I(w) = w - G(w) \,. \tag{4.4.13}$$

Proof. In accordance with the definition of the gradient of a functional, we consider

$$\left.\frac{dI(w + t\varphi)}{dt}\right|_{t=0} = \frac{1}{2}\frac{d}{dt}a(w + t\varphi, w + t\varphi)\Big|_{t=0} + \frac{1}{2}a\left(f, \frac{df}{dt}\right)\Big|_{t=0} - \iint_{\Omega} q\varphi \, d\Omega \,,$$

where $f = C(w + t\varphi)$. It is clear that

$$\frac{1}{2}\frac{d}{dt}a(w + t\varphi, w + t\varphi)\Big|_{t=0} = a(w, \varphi) \,.$$

Using the definition (4.4.8) of C, with regard for the equality $B(w, \varphi, \eta) = B(\varphi, w, \eta)$, a special case of (4.4.9), we calculate directly that

$$a\left(\left.\frac{df}{dt}\right|_{t=0}, \eta\right) = -2B(w, \varphi, \eta)$$

and so

$$a\left(f, \frac{df}{dt}\right)\Big|_{t=0} = -2B(w, \varphi, f) = -2B(f, w, \varphi) \,.$$

It follows that

$$\left.\frac{dI(w + t\varphi)}{dt}\right|_{t=0} = a(w, \varphi) - B(f, w, \varphi) - \int_{\Omega} q\varphi \, d\Omega$$

and, thanks to (4.4.11),

$$\left.\frac{dI(w + t\varphi)}{dt}\right|_{t=0} = a(w, \varphi) - a(G(w), \varphi) = a(w - G(w), \varphi) \,.$$

This, by definition of the gradient of a functional, means that (4.4.13) holds. □

Combining Lemmas 4.4.3 and 4.4.4, we have

Lemma 4.4.6. A critical point w of $I(w)$ defines the pair $(w, G(w))$ that is a generalized solution of the problem under consideration.

So we reduce the problem of finding a generalized solution of the problem to the problem of the minimum of a functional (it is not equivalent as there are in general solutions which are not points of minimum of the functional).

To apply Theorem 4.3.3, it remains to verify

Lemma 4.4.7. The functional $2I(w)$ is growing and has the form

$$\|w\|_{E_P}^2 + \Phi_1(w)$$

where

$$\Phi_1(w) = \frac{1}{2}a(f,f) - 2\iint_\Omega qw\,d\Omega$$

is a weakly continuous functional, f being defined by (4.4.10).

Proof. $2I(w)$ is growing since

$$2I(w) \geq a(w,w) - 2\left|\iint_\Omega qw\,d\Omega\right| = \|w\|_{E_P}^2 - 2\left|\iint_\Omega qw\,d\Omega\right|$$

and

$$2I(w) \geq \|w\|_{E_P}^2 - m\|w\|_{E_P} \to \infty \text{ as } \|w\|_{E_P} \to \infty$$

since q is assumed to be such that $\iint_\Omega qw\,d\Omega$ is a continuous functional with respect to $w \in E_{PC}$.

Weak continuity of $\Phi_1(w)$ is a consequence of Corollary 4.4.1, Lemma 4.4.2 for $a(f,f) = \|C(w)\|_{E_P}^2$, and the fact that the continuous linear functional $\iint_\Omega qw\,d\Omega$ is weakly continuous (by definition). □

So we can reformulate Theorem 4.3.3 in the case of the plate problem as follows

Theorem 4.4.1. Assume q is such that $\iint_\Omega qw\,d\Omega$ is a continuous linear functional with respect to w in E_{PC}. Then any critical point of the growing functional $I(w)$ which has at least one point of absolute minimum is a generalized solution of the plate problem in the sense of Definition 4.4.1; any minimizing sequence of $I(w)$ contains at least one subsequence which converges strongly to a generalized solution of the problem; each of the weak limit points of the minimizing sequence, which are also strong limit points, is a generalized solution to the problem.

The reader can also reformulate Theorem 4.3.4 in the present case to justify application of the Ritz method (and thus the method of finite elements) to von Kármán equations. Note that in this modification of the method we must find f exactly from (4.4.6). But it is not too difficult to show that f can be found approximately, also by the Ritz method, and the corresponding theorem on convergence remains valid.

4.5 Buckling of a Thin Elastic Shell

Following [42] (and [43]), we now consider a buckling problem for a shallow elastic shell described by equations of von Kármán's type. We want to study stability of the momentless state (here $w = 0$) of the shell. Assume the external load is proportional to a parameter λ. Let us assume that for every λ, there exists a momentless state of the shell whose stability we will study. We formulate the equations of equilibrium as follows:

$$\Delta^2 w = -\lambda\Big(T_1 \frac{\partial^2 w}{\partial x^2} + T_2 \frac{\partial^2 w}{\partial y^2} + 2T_{12}\frac{\partial^2 w}{\partial x \partial y} - F_1 \frac{\partial w}{\partial x} - F_2 \frac{\partial w}{\partial y}\Big) + [f, w + z]\,,$$
$$\Delta^2 f = -\{2[z, w] + [w, w]\}\,. \tag{4.5.1}$$

We consider a problem with Dirichlet conditions

$$w\big|_{\partial\Omega} = \frac{\partial w}{\partial n}\Big|_{\partial\Omega} = f\big|_{\partial\Omega} = \frac{\partial f}{\partial n}\Big|_{\partial\Omega} = 0\,. \tag{4.5.2}$$

Here $z = z(x, y) \in C^{(3)}(\overline{\Omega})$ is the equation of the shell middle surface. It is supposed that the tangential stresses T_1, T_2, T_{12} are given, belong to $L^2(\Omega)$ and, as assumed during derivation of the equations, satisfy equations of the two-dimensional theory of elasticity with forces (F_1, F_2). Other bits of notation are taken from the previous section.

The equations of a generalized statement of the problem under consideration are as follows:

$$a(w, \varphi) = \lambda \iint_\Omega \Big[T_1 \frac{\partial w}{\partial x}\frac{\partial \varphi}{\partial x} + T_2 \frac{\partial w}{\partial y}\frac{\partial \varphi}{\partial y} + T_{12}\Big(\frac{\partial w}{\partial x}\frac{\partial \varphi}{\partial y} + \frac{\partial w}{\partial y}\frac{\partial \varphi}{\partial x}\Big)\Big]\,dx\,dy$$
$$+ B(f, w + z, \varphi)\,, \tag{4.5.3}$$

$$a(f, \eta) = -2B(z, w, \eta) - B(w, w, \eta)\,. \tag{4.5.4}$$

Using standard variational tools, we can derive from these the equations (4.5.1) if a solution is assumed to be sufficiently smooth; conversely, from (4.5.1) we can derive (4.5.3)–(4.5.4). So we can take the latter equations to formulate

Definition 4.5.1. A pair w, f from E_{PC} is a generalized solution to the problem (4.5.1)–(4.5.2) if it satisfies the integro-differential equations (4.5.3)–(4.5.4) for any $\varphi, \eta \in E_{PC}$

The problem under consideration has a trivial solution $w = f = 0$. We are interested in when there exists a nontrivial solution, i.e., in solving a nonlinear eigenvalue problem.

First we mention that, as in Sect. 4.4, we solve the equation (4.5.4) with respect to $f \in E_{PC}$ when $w \in E_{PC}$ is given, and then exclude f from the equation (4.5.3), getting an equation in the unknown $w \in E_{PC}$. It is clear that

$$f = f_1 + f_2 ,$$

where the f_i are defined by the equations

$$a(f_1, \eta) = -2B(z, w, \eta) , \qquad a(f_2, \eta) = -B(w, w, \eta) .$$

Using the Riesz representation theorem we can find from these that

$$f_1 = Lw , \qquad f_2 = C(w) . \tag{4.5.5}$$

In Sect. 4.4 it was shown that $C(w)$ is a completely continuous operator. The linear operator L is also completely continuous (we leave it to the reader to show this).

In Sect. 3.2, we introduced the self-adjoint bounded linear operator C that is now redenoted as K. It is defined by

$$a(Kw, \varphi) = \iint_{\Omega} \left[T_1 \frac{\partial w}{\partial x} \frac{\partial \varphi}{\partial x} + T_{12} \left(\frac{\partial w}{\partial x} \frac{\partial \varphi}{\partial y} + \frac{\partial w}{\partial y} \frac{\partial \varphi}{\partial x} \right) + T_2 \frac{\partial w}{\partial y} \frac{\partial \varphi}{\partial y} \right] dx\, dy .$$

The operator K is compact in E_{PC} by Sobolev's imbedding theorem.

Applying the Riesz representation theorem to the relation (4.5.3) wherein f is defined by (4.5.5), we find an operator equation for a generalized solution of the problem under consideration

$$w - G(\lambda, w) = 0 . \tag{4.5.6}$$

The next point is to define a functional whose critical points are solutions to (4.5.6). It is

$$\mathcal{I}(\lambda, w) = \frac{1}{2} a(w, w) + \frac{1}{4} a(f, f) - \lambda J(w) ,$$

where

$$J(w) = \frac{1}{2} \iint_{\Omega} \left[T_1 \left(\frac{\partial w}{\partial x} \right)^2 + 2T_{12} \frac{\partial w}{\partial x} \frac{\partial w}{\partial y} + T_2 \left(\frac{\partial w}{\partial y} \right)^2 \right] dx\, dy .$$

$\mathcal{I}(\lambda, w)$ is the total energy of the shell-load system.

Lemma 4.5.1. For every $w \in E_{PC}$ we have

$$\operatorname{grad} \mathcal{I}(\lambda, w) = w - G(\lambda, w) . \tag{4.5.7}$$

The proof is similar to that for Lemma 4.4.4 and is omitted, as is the proof that $G(\lambda, w)$ is a completely continuous operator in $w \in E_{PC}$.

Next we consider the functional $a(f, f)$. It is seen that

$$\begin{aligned} a(f, f) &= a(f_1, f_1) + A_3(w) + A_4(w) , \\ A_3(w) &= 2a(f_1, f_2) = -4B(z, w, f_2) , \\ A_4(w) &= a(f_2, f_2) = \tfrac{1}{2} B(f_2, w, w) . \end{aligned}$$

Here $A_k(w)$ is a homogeneous function of degree k with respect to w, i.e.,

$$A_k(tw) = t^k A_k(w) ,$$

for any numerical parameter t. We leave it to the reader to show that $a(f, f)$, along with each of its parts, is a weakly continuous functional on E_{PC} (for $a(f, f)$, this is a consequence of Corollary 4.4.1).

It is evident that $J(w)$ is a weakly continuous functional in E_{PC}. So we have

Lemma 4.5.2. For every real number λ, the functional $\mathcal{I}(\lambda, w)$ takes the form

$$\mathcal{I}(\lambda, w) = \tfrac{1}{2}\|w\|^2_{E_{PC}} + \Psi(\lambda, w) , \qquad \Psi(\lambda, w) = \tfrac{1}{4}a(f, f) - \lambda J(w) ,$$

where $\Psi(\lambda, w)$ is a weakly continuous functional.

From now on, we assume that

$$J(w) > 0 \;\text{ for any }\; w \in E_{PC},\; w \neq 0 .$$

This assumption has the physical implication that almost everywhere in the shell the stress state of the shell is compressive.

To study stability of the non-buckled state of the shell (that is, when $w = 0$), beginning from L. Euler's work on stability of a bar, one solves the linearized (here around the zero state) eigenvalue problem that is now

$$\operatorname{grad}\left[\tfrac{1}{2}a(w, w) + \tfrac{1}{4}a(f_1, f_1)\right] = \lambda \operatorname{grad} J(w) . \tag{4.5.8}$$

The lowest eigenvalue of the latter, denoted λ_E and called the Euler lowest critical value, is usually considered as a value when the main, trivial form of equilibrium of the shell becomes unstable. We shall analyze this method for the shell.

We begin with the eigenvalue problem (4.5.8).

Lemma 4.5.3. There is a countable set λ_k of eigenvalues $\lambda_k > 0$ of the equation (4.5.8) considered in E_{PC}.

Proof. We first mention that the scalar product

$$\langle w, \varphi\rangle = a(w, \varphi) + \tfrac{1}{2}a(Lw, L\varphi) , \qquad f_1 = Lw ,$$

induces the norm in E_{PC} which is equivalent to the usual one since

$$a(w, w) \le \langle w, w\rangle \le m\, a(w, w) .$$

Using the new norm, we can rewrite (4.5.8) in the form

$$w = \lambda K_1 w$$

where operator K_1 is determined, thanks to the Riesz representation theorem, by the equality

$$\langle K_1 w, \varphi\rangle = \iint_\Omega \left[T_1 \frac{\partial w}{\partial x}\frac{\partial \varphi}{\partial x} + T_{12}\left(\frac{\partial w}{\partial x}\frac{\partial \varphi}{\partial y} + \frac{\partial w}{\partial y}\frac{\partial \varphi}{\partial x}\right) + T_2 \frac{\partial w}{\partial y}\frac{\partial \varphi}{\partial y}\right] dx\,dy\,.$$

It is easily seen that both K_1 and K are linear, strictly positive, self-adjoint, and compact, and hence Theorem 3.11.2 yields even more than the lemma states. □

For the trivial solution $w = f = 0$, the total energy $\mathcal{I}(\lambda, w) = 0$. A state of the shell at which $\mathcal{I}(\lambda, w)$ takes its minimal value is, in a certain sense, stable. So it is of interest what is the range of λ in which $\mathcal{I}(\lambda, w)$ can take negative values.

Theorem 4.5.1. Assume $T_1, T_{12}, T_2 \in L^2(\Omega)$ and w_E is an eigenfunction of the linearized boundary value problem (4.5.8) corresponding to its smallest eigenvalue λ_E, the Euler critical value. Then for every λ of the half-line

$$\lambda > \lambda^* \equiv \lambda_E - \frac{A_3^2(w_E)}{4A_4(w_E)J(w_E)} \tag{4.5.9}$$

there exists at least one nontrivial solution of the nonlinear boundary value problem (4.5.6) at which $\mathcal{I}(\lambda, w)$ is negative.

The proof is a consequence of the following three lemmas. The first of them is auxiliary.

Lemma 4.5.4. Assume that $w \in E_{PC}$ satisfies

$$\frac{\partial^2 w}{\partial x^2}\frac{\partial^2 w}{\partial y^2} - \left(\frac{\partial^2 w}{\partial x \partial y}\right)^2 = 0 \tag{4.5.10}$$

in the sense of $L^1(\Omega)$ *(almost everywhere in Ω)*. Then $w = 0$.

Proof. If $w \in C^{(2)}(\Omega)$, then (4.5.10) means the Gaussian curvature of the surface $z = w(x, y)$ vanishes so the surface is developable and, thanks to the boundary conditions (4.5.2), $w = 0$.

If $w \notin C^{(2)}(\Omega)$, we take another route. For arbitrary $w \in E_{PC}$, $F \in W^{2,2}(\Omega)$, the following formula holds:

$$\iint_\Omega \left[\left(\frac{\partial^2 F}{\partial x \partial y}\frac{\partial w}{\partial y} - \frac{\partial^2 F}{\partial y^2}\frac{\partial w}{\partial x}\right)\frac{\partial w}{\partial x} + \left(\frac{\partial^2 F}{\partial x \partial y}\frac{\partial w}{\partial x} - \frac{\partial^2 F}{\partial x^2}\frac{\partial w}{\partial y}\right)\frac{\partial w}{\partial y}\right] dx\,dy$$
$$= 2\iint_\Omega \left[\frac{\partial^2 w}{\partial x^2}\frac{\partial^2 w}{\partial y^2} - \left(\frac{\partial^2 w}{\partial x \partial y}\right)^2\right] F\,dx\,dy\,. \tag{4.5.11}$$

(This is easily seen after integrating by parts for smooth functions; the limit passage shows that it is valid for the needed elements.) In (4.5.11) we put

$$F = \tfrac{1}{2}(x^2 + y^2)$$

which gives for w satisfying (4.5.10)

$$\iint_\Omega \left[\left(\frac{\partial w}{\partial x} \right)^2 + \left(\frac{\partial w}{\partial y} \right)^2 \right] dx\, dy = 0\,.$$

This, together with the boundary conditions for w, completes the proof. □

Lemma 4.5.5. The functional $\mathcal{I}(\lambda, w)$ is growing for every $\lambda > 0$; that is, we have $\mathcal{I}(\lambda, w) \to \infty$ as $\|w\|_{E_P} \to \infty$.

Proof. On the unit sphere $S = \{w \in E_{PC} \colon a(w, w) = 1\}$ consider the set S_1 defined by

$$\tfrac{1}{2}a(w, w) - \lambda J(w) > \tfrac{1}{4}\,.$$

Then on the image of S_1 under the mapping $w \mapsto Rw$, we get

$$\mathcal{I}(\lambda, Rw) \geq \tfrac{1}{2}a(Rw, Rw) - \lambda J(Rw) = R^2 \left[\tfrac{1}{2}a(w, w) - \lambda J(w) \right] > \tfrac{1}{4}R^2 \quad (w \in S_1). \tag{4.5.12}$$

Next consider $\mathcal{I}(\lambda, Rw)$ when $w \in S_2 = S \setminus S_1$. Here

$$\tfrac{1}{2}a(w, w) - \lambda J(w) \leq \tfrac{1}{4}\,. \tag{4.5.13}$$

Let us introduce the weak closure of S_2 in E_{PC}, denoted by $\mathrm{Cl}\, S_2$. First we show that $\mathrm{Cl}\, S_2$ does not contain zero. If to the contrary it does contain zero then there is a sequence $\{w_n\} \in \mathrm{Cl}\, S_2$ such that $a(w_n, w_n) = 1$ and $w_n \rightharpoonup 0$ in E_{PC} (or, equivalently, in $W^{2,2}(\Omega)$). By the imbedding theorem in $W^{2,2}(\Omega)$, the sequences of first derivatives of $\{w_n\}$ tend to zero strongly in $L^p(\Omega)$ for any $p < \infty$ and thus $J(w_n) \to 0$, which contradicts (4.5.13) since

$$\tfrac{1}{2} \equiv \tfrac{1}{2}a(w_n, w_n) \leq \tfrac{1}{4} + \lambda J(w_n)\,.$$

Next we show that for all $w \in \mathrm{Cl}\, S_2$,

$$A_4(w) \geq c_* \tag{4.5.14}$$

wherein c_* is a positive constant. Indeed, if (4.5.14) is not valid there is a sequence $\{w_n\} \in \mathrm{Cl}\, S_2$ such that $A_4(w_n) \to 0$ as $n \to \infty$. This sequence contains a subsequence which converges weakly to w_0 belonging to $\mathrm{Cl}\, S_2$ too. Since A_4 is a weakly continuous functional,

$$A_4(w_0) = 0\,.$$

This means that

$$a(f_2, f_2) = 0\,, \qquad f_2 = C(w_0)\,.$$

Returning to (4.5.5), we get

$$B(w_0, w_0, \eta) = 0$$

or, equivalently,

$$\iint_\Omega \left[\frac{\partial^2 w_0}{\partial x^2} \frac{\partial^2 w_0}{\partial y^2} - \left(\frac{\partial^2 w_0}{\partial x \partial y} \right)^2 \right] \eta\, dx\, dy = 0$$

for any $\eta \in E_{PC}$. As E_{PC} is dense in $L^2(\Omega)$,

$$\frac{\partial^2 w_0}{\partial x^2}\frac{\partial^2 w_0}{\partial y^2} - \left(\frac{\partial^2 w_0}{\partial x \partial y}\right)^2 = 0$$

almost everywhere in Ω and, by Lemma 4.5.3, it follows that $w_0(x,y) = 0$. This contradicts the fact that w_0 belongs to $\mathrm{Cl}\, S_2$ which does not contain zero.

Since $|A_3(w)| \le c_1$ on S, we get, thanks to (4.5.14),

$$\mathcal{I}(\lambda, Rw) \ge c_* R^4 - \left(\tfrac{1}{4}R^2 + c_1 R^3\right)$$

when $w \in \mathrm{Cl}\, S_2$ and so for sufficiently large R, with regard for (4.5.12), we obtain

$$\mathcal{I}(\lambda, Rw) \ge \tfrac{1}{4}R^2$$

for all $w \in S$. This means that $\mathcal{I}(\lambda, w)$ is growing. □

By Theorem 4.3.3 it follows that, for any λ, the functional $\mathcal{I}(\lambda, w)$ takes its minimal value in E_{PC}. But $w = 0$ is also a critical point of the functional, so to conclude the proof of Theorem 4.5.1 we formulate

Lemma 4.5.6. Under the conditions stated in Theorem 4.5.1, the minimal value of $\mathcal{I}(\lambda, w)$ is negative if λ satisfies (4.5.9).

Proof. Consider $\mathcal{I}(\lambda, cw_E)$ where c is a constant. It is seen that

$$\begin{aligned}\mathcal{I}(\lambda, cw_E) = c^2 &\left[\tfrac{1}{2}a(w_E, w_E) + \tfrac{1}{4}a(Lw_E, Lw_E) - \lambda J(w_E)\right] \\ &+ c^3 A_3(w_E) + c^4 A_4(w_E)\,, \qquad (f_1 = Lw_E)\,.\end{aligned}$$

Further, from (4.5.8) it follows that

$$\tfrac{1}{2}a(w_E, w_E) + \tfrac{1}{4}a(Lw_E, Lw_E) = \lambda_E J(w_E)\,.$$

Hence

$$\mathcal{I}(\lambda, cw_E) = c^2\left[(\lambda_E - \lambda)J(w_E) + cA_3(w_E) + c^2 A_4(w_E)\right].$$

The minimum of $\mathcal{I}(\lambda, cw_E)/c^2$ considered as a function of the real variable c is taken at

$$c_0 = -\tfrac{1}{2}A_3(w_E)/A_4(w_E)\,;$$

this minimum is equal to

$$\min_c(c^{-2}\mathcal{I}(\lambda, cw_E)) = (\lambda_E - \lambda)J(w_E) - A_3^2(w_E)/A_4(w_E)\,.$$

So for λ satisfying (4.5.9), we get

$$\mathcal{I}(\lambda, c_0 w_E) < 0$$

and thus at w_0, a minimizer of $\mathcal{I}(\lambda, w)$ at the same λ,

$$\mathcal{I}(\lambda, w_0) < 0\,.$$

This completes the proof of the lemma, and therefore of Theorem 4.5.1. □

An important result follows from Theorem 4.5.1.

Corollary 4.5.1. Assume that there is an eigenfunction w_E corresponding to the Euler critical value λ_E such that

$$A_3(w_E) \neq 0 .$$

In this case we have a sharp inequality $\lambda^* < \lambda_E$.

This result is of fundamental importance in the theory of stability of shells, as it implies that if $A_3(w_E) \neq 0$, then the problem of stability cannot be solved by linearization in the neighborhood of a momentless state of stress, used since Euler in the theory of stability of beams. If $A_3(w_E) \neq 0$, we must investigate the problem of stability of a shell in its nonlinear formulation.

Theorem 4.5.2. Let $T_1, T_2, T_{12} \in L^2(\Omega)$. Then there is a value $\lambda_l \leq \lambda^*$ such that for any $\lambda < \lambda_l$ the nonlinear problem (4.5.6) has the unique solution $w = 0$.

Proof. Assume w is a solution of (4.5.6), i.e., the pair $w, f = Lw + C(w)$ from E_{PC} satisfies (4.5.3)–(4.5.4) for arbitrary $\varphi, \eta \in E_{PC}$. Setting $\varphi = w$ and $\eta = f$ in (4.5.3)–(4.5.4) we get

$$a(w,w) = 2\lambda J(w) + B(f,w,w) + B(f,z,w) ,$$
$$a(f,f) = -2B(z,w,f) - B(w,w,f) .$$

Summing these equalities term by term, we have the identity

$$a(w,w) + a(f,f) = 2\lambda J(w) - B(z,f,w) . \tag{4.5.15}$$

Using the elementary inequality $|ab| \leq a^2 + \frac{1}{4}b^2$, we get an estimate

$$\begin{aligned}|B(z,f,w)| &= \left| \iint_\Omega \left(\frac{\partial^2 f}{\partial x^2}\frac{\partial^2 z}{\partial x^2} + \frac{\partial^2 f}{\partial y^2}\frac{\partial^2 z}{\partial y^2} - 2\frac{\partial^2 f}{\partial x \partial y}\frac{\partial^2 z}{\partial x \partial y} \right) w \, dx\, dy \right| \\ &\leq \iint_\Omega \left[\left(\frac{\partial^2 f}{\partial x^2}\right)^2 + \left(\frac{\partial^2 f}{\partial y^2}\right)^2 + 2\left(\frac{\partial^2 f}{\partial x \partial y}\right)^2 \right] dx\, dy \\ &\quad + \frac{1}{4}\iint_\Omega \left[\left(\frac{\partial^2 z}{\partial x^2}\right)^2 + \left(\frac{\partial^2 z}{\partial y^2}\right)^2 + 2\left(\frac{\partial^2 z}{\partial x \partial y}\right)^2 \right] w^2 \, dx\, dy .\end{aligned}$$

Integrating by parts in the expression for $a(f,f)$ gives

$$a(f,f) = \iint_\Omega \left[\left(\frac{\partial^2 f}{\partial x^2}\right)^2 + \left(\frac{\partial^2 f}{\partial y^2}\right)^2 + 2\left(\frac{\partial^2 f}{\partial x \partial y}\right)^2 \right] dx\, dy$$

and thus, from (4.5.15), it follows that

$$a(w,w) \le 2\lambda J(w) + \frac{1}{4}\iint_\Omega \left[\left(\frac{\partial^2 z}{\partial x^2}\right)^2 + \left(\frac{\partial^2 z}{\partial y^2}\right)^2 + 2\left(\frac{\partial^2 z}{\partial x \partial y}\right)^2\right] w^2\, dx\, dy\,. \quad (4.5.16)$$

Now we need a lemma which will be proved later.

Lemma 4.5.7. On the surface $S = \{w\colon J(w) = 1\}$ in E_{PC}, the functional

$$I_1(w) = a(w,w) - \frac{1}{4}\iint_\Omega \left[\left(\frac{\partial^2 z}{\partial x^2}\right)^2 + \left(\frac{\partial^2 z}{\partial y^2}\right)^2 + 2\left(\frac{\partial^2 z}{\partial x \partial y}\right)^2\right] w^2\, dx\, dy$$

has finite minimum denoted by $2\lambda^{**}$.

We are continuing the proof of Theorem 4.5.2. From Lemma 4.5.7, it follows that

$$I_1(w) \ge 2\lambda^{**} J(w)$$

since all of the functionals are homogeneous with respect to w of order 2. Thus, from (4.5.16), we get

$$(2\lambda^{**} - 2\lambda) J(w) \le 0\,,$$

from which it follows that if $\lambda \le \lambda^{**}$ then

$$J(w) \le 0\,.$$

This is possible only at $w = 0$, and the proof is complete. □

Proof (of Lemma 4.5.7). Assume $\{w_n\}$ is a minimizing sequence of $I_1(w)$ on S and, by contradiction, that the minimum on S is not finite, i.e., $I_1(w_n) \to -\infty$ as $n \to \infty$. It is obvious that $\|w_n\|_{E_P} \to \infty$.

Define $w_n^* = w_n/\|w_n\|_{E_P}$. We can consider the sequence $\{w_n^*\}$ to be weakly convergent to an element $w_0^* \in E_{PC}$. In this case

$$J(w_n) = \|w_n\|_{E_P}^2\, J(w_n^*)$$

so

$$J(w_n^*) = J(w_n)/\|w_n\|_{E_P}^2 \to 0 \quad \text{as } n \to \infty\,.$$

Since J is weakly continuous then $J(w_0^*) = 0$ and thus $w_0^* = 0$. This means that $w_n^* \rightharpoonup 0$.

By the imbedding theorem we get

$$\sup_\Omega |w_n^*(x,y)| \to 0$$

and so

$$a_n = \iint_\Omega \left[\left(\frac{\partial^2 z}{\partial x^2}\right)^2 + \left(\frac{\partial^2 z}{\partial y^2}\right)^2 + 2\left(\frac{\partial^2 z}{\partial x \partial y}\right)^2\right] (w_n^*)^2\, dx\, dy \to 0 \quad \text{as } n \to \infty\,.$$

Thus

$$\lim_{n\to\infty} I_1(w_n) = \lim_{n\to\infty} \|w_n\|_{E_P}^2 \left(1 - \tfrac{1}{4}a_n\right) = +\infty ,$$

a contradiction. Similar considerations demonstrate that a minimizing sequence $\{w_n\}$ of I_1 is bounded. Then there is a subsequence that converges weakly to an element w_0. This element belongs to S since $J(w)$ is weakly continuous. The structure of I_1 provides that $I_1(w_0) = \lambda^{**}$. □

As a result of Theorem 4.5.2 we get the estimates

$$-\infty < \lambda^{**} \le \lambda_l \le \lambda^* \le \lambda_E < \infty . \tag{4.5.17}$$

From the statement of Lemma 4.5.7, it is seen that λ^{**} can be defined as the lowest eigenvalue of the boundary value problem

$$\operatorname{grad} I_1(w) = 2\lambda \operatorname{grad} J(w) . \tag{4.5.18}$$

Let us consider a particular case, a von Kármán plate. Here $z(x, y) = 0$ and thus the problem (4.5.18) takes the form

$$\operatorname{grad}(a(w, w)) = 2\lambda \operatorname{grad} J(w) .$$

But the equation (4.5.8) determining the λ_E for the plate coincides with this one as $f_1 = Lw = 0$ for a plate. Thus $\lambda_E = \lambda^{**}$ and (4.5.17) states that $\lambda_l = \lambda_E$ for the plate. This implies an important

Theorem 4.5.3. In the case of a plate (i.e., when $z(x, y) = 0$), under the conditions of Theorem 4.5.1, the equality $\lambda_l = \lambda_E$ is satisfied. In other words, for $\lambda \le \lambda_E$ there is a unique generalized solution, $w = 0$, of the problem under consideration; if $\lambda > \lambda_E$ then there is another solution of the problem, at which the functional of total energy of the plate is definite negative.

This theorem establishes the possibility of applying Euler's method of linearization to the problem of stability of a plate.

We note that many works (not mentioned here) are devoted to mathematical questions in the theory of von Kármán's plates and shells. The corresponding boundary value problems of the theory are a touchstone of abstract nonlinear mathematical theory because of their importance in applications, as well as their not too complicated form.

4.6 Nonlinear Equilibrium Problem for an Elastic Shallow Shell

We consider another simple modification of the nonlinear theory of elastic shallow shells when the geometry of the midsurface of the shell is identified with the geometry of a plane. This modification of the theory is widely applied in engineering calculations. Nonlinear theory of shallow shells in curvilinear coordinates is considered in [41] in detail.

We express the equations describing the behavior of the shell in a notation which is commonly used along with this version of the theory. Namely, we let x, y denote the coordinates on the plane that is identified with the midsurface of the shell, u, v denote the tangential components of the vector of displacements of the midsurface, w denote the transverse displacement of the midsurface, and subscripts x, y denote partial derivatives with respect to x and y. The equations of equilibrium of the shell are

$$D\nabla^4 w + N_1(k_1 - w_{xx}) + N_2(k_2 - w_{yy}) - 2N_{12}w_{xy} - F = 0\,, \tag{4.6.1}$$

$$\begin{aligned}
&\nabla^2 u + (1+\mu)/(1-\mu)(u_x + v_y)_x \\
&\qquad + 2/(1-\mu)[(k_1 w)_x + w_x w_{xx} + \mu(k_2 w)_x + \mu w_y w_{xy}] + w_y w_{xy} + w_x w_{yy} = 0\,, \\
&\nabla^2 v + (1+\mu)/(1-\mu)(u_x + v_y)_y \\
&\qquad + 2/(1-\mu)[(k_2 w)_y + w_y w_{yy} + \mu(k_1 w)_y + \mu w_x w_{xy}] + w_x w_{xy} + w_y w_{xx} = 0\,,
\end{aligned} \tag{4.6.2}$$

D, E, μ being the elastic constants, $0 < \mu < 1/2$. We consider the shell under the action of a transverse load F. The components of the tangential strain tensor are

$$\varepsilon_1 = u_x + k_1 w + \tfrac{1}{2}w_x^2\,, \quad \varepsilon_2 = v_y + k_2 w + \tfrac{1}{2}w_y^2\,, \quad \varepsilon_{12} = u_y + v_x + w_x w_y\,. \tag{4.6.3}$$

Let us formulate the conditions under which we justify application of Ritz's method to a boundary value problem for the shell, and so for the finite element method as well, and establish an existence theorem.

We suppose Ω, the domain occupied by the shell, satisfies the same conditions we imposed earlier for the von Kármán plate. Let the shell be clamped against the transverse translation at three points (x_i, y_i), $i = 1, 2, 3$, that do not lie on the same straight line:

$$w(x_i, y_i) = 0\,. \tag{4.6.4}$$

It is sufficient (but not necessary) to assume that

$$w\big|_{\Gamma_1} = 0 \tag{4.6.5}$$

holds on a portion Γ_1 of the boundary.

Let us call C_4 the set of functions w belonging to $C^{(4)}(\Omega)$ and satisfying the conditions (4.6.4)–(4.6.5).

For the tangential displacements u, v, the minimal restrictions for Ω in this consideration must be such that Korn's inequality of two-dimensional elasticity holds. That is (see Mikhlin [26]), we must have

$$\iint_\Omega (u^2 + v^2 + u_x^2 + u_y^2 + v_x^2 + v_y^2)\,dx\,dy \le m \iint_\Omega [u_x^2 + (u_y + v_x)^2 + v_y^2]\,dx\,dy\,. \tag{4.6.6}$$

One of the possible restrictions under which (4.6.6) holds for all u, v with the unique constant m is

$$u\big|_{\Gamma_2} = 0\,, \qquad v\big|_{\Gamma_2} = 0\,, \tag{4.6.7}$$

Γ_2 being some part of the boundary of Ω.

Let us introduce the set C_2 of vector functions $\omega = (u, v)$ with the components belonging to $C^{(2)}(\Omega)$ and satisfying (4.6.7).

We may suppose that some part of the boundary of the shell is elastically supported (the corresponding term of the energy of the system should be included into the expression of the energy norm) or that on some part of the boundary there is given a transverse load (here the term that is the work of the load on the boundary must be included into the energy functional). We will not place these conditions in the differential form; they are well known and can be derived from the variational statement of the problem. The presence of these conditions has no practical impact on the way in which we consider the problem.

Let us introduce energy spaces. Let E_1 be a subspace of $W^{1,2}(\Omega) \times W^{1,2}(\Omega)$ that is the completion of the set C_2 in the norm of $W^{1,2}(\Omega) \times W^{1,2}(\Omega)$. Korn's inequality (4.6.6) implies that on E_1 the following norm is equivalent:

$$\|\omega\|^2_{E_1} = \frac{Eh}{2(1-\mu^2)} \iint_\Omega [e_1^2 + e_2^2 + 2\mu e_1 e_2 + \tfrac{1}{2}(1-\mu)e_{12}^2]\,dx\,dy\,,$$

where

$$e_1 = u_x\,, \qquad e_2 = v_y\,, \qquad e_{12} = u_y + v_x\,,$$

and h is the shell thickness.

E_2, a subspace of $W^{2,2}(\Omega)$, is the completion of C_4 in the norm of $W^{2,2}(\Omega)$. On E_2 there is an equivalent norm (the energy norm we introduced for the problem of bending of the plate):

$$\|w\|^2_{E_2} = \tfrac{1}{2}D \iint_\Omega [(\nabla^2 w)^2 + 2(1-\mu)(w_{xy}^2 - w_{xx}w_{yy})]\,dx\,dy\,.$$

The norms on E_i induce the inner products that are denoted with use of the names of corresponding spaces. Denote $E_1 \times E_2$ by E.

Definition 4.6.1. $\mathbf{u} = (u, v, w) \in E$ is a generalized solution of the equilibrium problem for a shallow shell if for an arbitrary $\delta\mathbf{u} = (\delta u, \delta v, \delta w) \in E$ it satisfies

$$\iint_\Omega (M_1\delta\kappa_1 + M_2\delta\kappa_2 + 2M_{12}\delta\chi + N_1\delta\epsilon_1 + N_2\delta\epsilon_2 + N_{12}\delta\epsilon_{12})\,dx\,dy$$
$$= \iint_\Omega F\delta w\,dx\,dy + \int_{\partial\Omega} f\delta w\,ds\,, \tag{4.6.8}$$

where

$$M_1 = D(\kappa_1 + \mu\kappa_2)\,, \qquad M_2 = D(\kappa_2 + \mu\kappa_1)\,, \qquad M_{12} = D(1-\mu)\chi\,,$$

$$N_1 = \frac{Eh}{1-\mu^2}(\epsilon_1 + \mu\epsilon_2)\,, \quad N_2 = \frac{Eh}{1-\mu^2}(\epsilon_2 + \mu\epsilon_1)\,, \quad N_{12} = \frac{Eh}{2(1+\mu)}\epsilon_{12}\,,$$

$$\kappa_1 = -w_{xx}\,, \qquad \kappa_2 = -w_{yy}\,, \qquad \chi = -w_{xy}\,,$$

f being the external load on the edge of the shell.

We note that on the part of the boundary where $\delta w = 0$, it is not necessary to show f. However, we shall assume that on this part of the boundary the function $f = 0$. Also note that (4.6.8) expresses the virtual work principle for the shell.

It is seen that all the stationary points of the energy functional

$$I(\mathbf{u}) = \|w\|_{E_2}^2 + \frac{1}{2}\iint_\Omega (N_1\epsilon_1 + N_2\epsilon_2 + N_{12}\epsilon_{12})\,dx\,dy - \iint_\Omega Fw\,dx\,dy - \int_{\partial\Omega} fw\,ds$$

are solutions to (4.6.8) since moving all the terms of (4.6.8) to the left-hand side we get on the left in (4.6.8) the expression for the first variation of the functional $I(\mathbf{u})$.

Let us note that for the correctness of Definition 4.6.1 it is necessary to impose an additional requirement: the terms

$$\iint_\Omega F\delta w\,dx\,dy + \int_{\partial\Omega} f\delta w\,ds$$

must make sense for any $\delta w \in E_2$. The set of these loads is called E^*. By Sobolev's imbedding theorems, sufficient conditions for the loads to belong to E^* are:

$$F = F_0 + F_1$$

where $F_0 \in L(\Omega)$ and F_1 is a finite sum of δ-functions (point transverse forces);

$$f = f_0 + f_1$$

where $f_0 \in L(\partial\Omega)$ and f_1 is a finite sum of δ-functions (point transverse forces on $\partial\Omega$). Under these conditions, the functional

$$\iint_\Omega F\delta w\,dx\,dy + \int_{\partial\Omega} f\delta w\,ds$$

is linear and continuous in $\delta w \in E_2$.

By the Riesz representation theorem, there exists a unique element $g \in E_2$ such that

$$\iint_\Omega F\delta w\,dx\,dy + \int_{\partial\Omega} f\delta w\,ds = (g, \delta w)_{E_2}\,.$$

Now we can represent $I(\mathbf{u})$ in a more compact form:

$$I(\mathbf{u}) = \|w\|_{E_2}^2 + \frac{1}{2}\iint_\Omega (N_1\epsilon_1 + N_2\epsilon_2 + N_{12}\epsilon_{12})\,dx\,dy - (g, w)_{E_2}\,.$$

Let us find the tangential displacements u_1, u_2 through w. For this consider the equation

$$\iint_\Omega (N_1\delta\epsilon_1 + N_2\delta\epsilon_2 + N_{12}\delta\epsilon_{12})\,dx\,dy = 0$$

in E_1. Reasoning as was done earlier, we can easily establish that this equation is uniquely solvable in E_1 with respect to $\omega = (u_1, u_2)$; the solution can be written as

$$\omega = G(w) ,$$

where G is a completely continuous operator. Let us put this ω into the expression of $I(\mathbf{u})$. After this substitution, the functional $I(\mathbf{u})$ depends only on w; it is denoted by $\aleph(w)$. Standard reasoning leads us to the statement that any stationary point of $\aleph(w)$ is a generalized solution of the problem under consideration.

The functional $\aleph(w)$ has a structure that is suitable for application of Theorem 4.3.4. To justify the Ritz method it is enough to show that $\aleph(w)$ is growing. Let us demonstrate this.

Lemma 4.6.1. Let the external load belong to E^*. Then $\aleph(w)$ is growing; that is, $\aleph(w) \to \infty$ when $\|w\|_{E_2} \to \infty$.

Proof. The proof follows from considering the form of $\aleph(w)$. Indeed, under the above assumptions, we have

$$N_1\epsilon_1 + N_2\epsilon_2 + 2N_{12}\epsilon_{12} \geq 0 .$$

Then

$$|(g, \delta w)_{E_2}| \leq \|g\|_{E_2} \|w\|_{E_2} ,$$

so

$$\aleph(w) \geq \|w\|^2_{E_2} - \|g\|_{E_2} \|w\|_{E_2} .$$

From this the lemma follows. □

Thus we have

Theorem 4.6.1. Let the conditions of Lemma 4.6.1 hold. Then

(i) there is a generalized solution of the problem of equilibrium of the shell that belongs to E_2 and admits a minimum of the functional $\aleph(w)$;

(ii) any sequence $\{w_n\}$ minimizing the functional $\aleph(w)$ in E_2 contains a subsequence that converges strongly to a generalized solution of the problem;

(iii) the equations of the Ritz method (and thus of Galerkin's method and so of any conforming modification of the finite element method) have a solution in each approximation; the set of approximations contains a subsequence that converges strongly to a generalized solution of the problem in E_2; moreover, any weakly converging subsequence of the Ritz approximations converges strongly to a generalized solution of the problem.

4.7 Degree Theory

This is only a sketch of degree theory of a map, which will be used in what follows. We begin with an illuminating example.

Let $f(z)$ be a function holomorphic on a closed domain D of the complex plane, and let ∂D, the boundary of D, be smooth and let it not contain zeros of $f(z)$. Then, as is well known, the number defined by the integral

$$n = \oint_{\partial D} \frac{f'(z)}{f(z)}\, dz$$

is equal to the number of zeros of $f(z)$ inside D with regard for their multiplicity.

This is extended to more general classes of maps; this is the so-called degree theory, a full presentation of which can be found in Schwartz [32].

The degree of a finite-dimensional vector-field $\boldsymbol{\Phi}(\mathbf{x})\colon \mathbb{R}^n \to \mathbb{R}^n$, originally due to L.E.J. Brouwer, is defined as follows. Let

$$\boldsymbol{\Phi}(\mathbf{x}) = (\Phi_1(\mathbf{x}), \ldots, \Phi_n(\mathbf{x}))$$

be continuously differentiable on a bounded open domain D with the boundary ∂D in $\mathbb{R}^n$. Suppose $\mathbf{p} \in \mathbb{R}^n$ does not belong to ∂D, then the set $\Phi^{-1}(\mathbf{p})$, the preimage of $\mathbf{p}$ in D, is discrete and, finally, at each $\mathbf{x} \in \boldsymbol{\Phi}^{-1}(\mathbf{p})$, the Jacobian

$$J_{\boldsymbol{\Phi}}(\mathbf{x}) = \det\left(\frac{\partial \Phi_i}{\partial x_j}\right)$$

does not vanish. Then the *degree* of $\boldsymbol{\Phi}$ with respect to $\mathbf{p}$ and D is

$$\deg(\mathbf{p}, \boldsymbol{\Phi}, D) = \sum_{\substack{\boldsymbol{\Phi}(\mathbf{x})=\mathbf{p} \\ \mathbf{x}\in D}} \operatorname{sign} J_{\boldsymbol{\Phi}}(\mathbf{x})$$

where $\operatorname{sign} J_{\boldsymbol{\Phi}}(\mathbf{x})$ is the signum of $J_{\boldsymbol{\Phi}}(\mathbf{x})$.

If $\deg(\mathbf{p}, \boldsymbol{\Phi}, D) \neq 0$, then there are solutions of the equation $\boldsymbol{\Phi}(\mathbf{x}) = \mathbf{p}$ in D. If $\mathbf{p} \notin \boldsymbol{\Phi}(D)$ then $\deg(\mathbf{p}, \boldsymbol{\Phi}, D) = 0$ and so $\deg(\mathbf{p}, \boldsymbol{\Phi}, D)$ determines, in a certain way, the number of solutions of the equation $\boldsymbol{\Phi}(\mathbf{x}) = \mathbf{p}$.

If there are points $\mathbf{x}$ at which $\boldsymbol{\Phi}(\mathbf{x}) = \mathbf{p}$ and $J_{\boldsymbol{\Phi}}(\mathbf{x}) = 0$, then we can introduce the degree of the map using the limit passage. We can always take a sequence of points $\mathbf{p}_k \to \mathbf{p}$ such that $J_{\boldsymbol{\Phi}}(\mathbf{x}) \neq 0$ at any $\mathbf{x} \in \boldsymbol{\Phi}^{-1}(\mathbf{p}_k)$; the degree of $\boldsymbol{\Phi}$ is now defined by

$$\deg(\mathbf{p}, \boldsymbol{\Phi}, D) = \lim_{k\to\infty} \deg(\mathbf{p}_k, \boldsymbol{\Phi}, D)\,.$$

It is shown that this number does not depend on the choice of the sequence $\{\mathbf{p}_k\}$ and also characterizes the number of solutions of the equation $\boldsymbol{\Phi}(\mathbf{x}) = \mathbf{p}$ in D.

The next step of the theory is to state it for $\boldsymbol{\Phi}(\mathbf{x})$ being of $C(\overline{D})$ (each of the components of $\boldsymbol{\Phi}(\mathbf{x})$ being of $C(\overline{D})$). This is done by using a limit passage. Namely, for $\boldsymbol{\Phi}(\mathbf{x})$, there is a sequence $\{\boldsymbol{\Phi}_k(\mathbf{x})\}$ such that $\boldsymbol{\Phi}_k(\mathbf{x}) \in C^{(1)}(\overline{D})$ and each component of $\boldsymbol{\Phi}_k(\mathbf{x})$ converges uniformly on $\overline{D}$ to a corresponding component of $\boldsymbol{\Phi}(\mathbf{x})$. Then as is shown, there exists

$$\lim_{k\to\infty} \deg(\mathbf{p}, \boldsymbol{\Phi}_k, D)$$

which does not depend on the choice of $\{\boldsymbol{\Phi}_k(\mathbf{x})\}$; it is, by definition, the degree of $\boldsymbol{\Phi}(x)$ with respect to $\mathbf{p}$ and D.

As there is a one-to-one correspondence between $\mathbb{R}^n$ and n-dimensional real Banach space, the notion of degree of a map is transferred to continuous maps in the latter space. Moreover, it is seen how it can be determined for a continuous map whose range is a finite-dimensional subspace of a Banach space.

In the case of general operator in a Banach space, the notion was extended to operators of the form $I + F$ with a compact operator F on a real Banach space X by J. Leray and J. Schauder [20]. To do this, they introduce an approximate operator as follows.

Let D be a bounded open domain in X with the boundary ∂D. As F is a compact operator, $F(\overline{D})$, the image of $\overline{D}$, is compact. So, by the Hausdorff criterion on compactness, there is a finite ε-net $N_\varepsilon = \{x_k : x_k \in F(\overline{D}); k = 1, \ldots, n\}$, such that for every $x \in \overline{D}$ there is an integer k such that $\|F(x) - x_k\| < \varepsilon$. Finally, the approximate operator F_ε is defined by

$$F_\varepsilon(x) = \frac{\sum_{k=1}^{n} \mu_k(x) x_k}{\sum_{k=1}^{n} \mu_k(x)} \qquad (x \in \overline{D})$$

where $\mu_k(x) = 0$ if $\|F(x) - x_k\| > \varepsilon$ and $\mu_k = \varepsilon - \|F(x) - x_k\|$ if $\|F(x) - x_k\| \le \varepsilon$. This operator is called the *Schauder projection operator*.

It is easily seen that the range of $F_\varepsilon(x)$ is a domain in a finite dimensional subspace X_n of X, the operator F_ε is continuous, and, moreover,

$$\|F(x) - F_\varepsilon(x)\| \le \varepsilon$$

when $x \in \overline{D}$.

By the above, we can introduce the degree of $I + F_\varepsilon$ with respect to p and $D_n = D \cap X_n$ if $p \notin (I + F_\varepsilon)(\partial D_n)$. As is shown in Schwartz [32], for sufficiently small $\varepsilon > 0$ the degree $\deg(p, I + F_\varepsilon, D_n)$ is the same and thus it is defined as the degree of the operator $I + F$ with respect to p and D.

The following properties of the degree of an operator $I+F$ with compact operator F hold:

1. If $x + F(x) \neq p$ in $\overline{D}$, then $\deg(p, I + F, D) = 0$;
2. if $\deg(p, I + F, D) \neq 0$, then in D there is at least one solution to the equation $x + F(x) = p$;
3. $\deg(p, I, D) = +1$ if $p \in D$;
4. if $D = \cup_i D_i$ where the family $\{D_i\}$ is disjoint and $\partial D_i \subset \partial D$, then

$$\deg(p, I + F, D) = \sum_i \deg(p, I + F, D_i) ;$$

5. $\deg(p, I + F, D)$ is continuous with respect to p and F;
6. (*invariance under homotopy*) Let $\Phi(x,t) = x + \Psi(x,t)$. Assume that for every $t \in [a,b]$ the operator $\Phi(x,t)$ is compact with respect to $x \in X$ and continuous in $t \in [a,b]$ uniformly with respect to $x \in \overline{D}$. Then the operators $\Psi_a = \Psi(\cdot, a)$ and $\Psi_b = \Psi(\cdot, b)$ are said to be *compact homotopic*. Let Ψ_a and Ψ_b be compact homotopic and $p \neq x + \Psi(x,t)$ for every $x \in \partial D$ and $t \in [a,b]$; then

$$\deg(p, I + \Psi_a, D) = \deg(p, I + \Psi_b, D) .$$

The sixth and third properties give a result that is frequently used to establish existence of solution of the equation

$$x + F(x) = 0 . \tag{4.7.1}$$

We formulate it as

Lemma 4.7.1. Assume $F(x)$ is a compact operator in a Banach space X and the equation $x + tF(x) = 0$ has no solutions on a sphere $\|x\| = R$ for any $t \in [a,b]$. Then in the ball $B = \{x \colon \|x\| < R\}$ there exists at least one solution to (4.7.1) and

$$\deg(0, I + F, B) = +1 .$$

In the next section we demonstrate an application of the lemma.

4.8 Steady-State Flow of a Viscous Liquid

Following I.I. Vorovich and V.I. Yudovich [42], we consider the steady-state flow of a viscous incompressible liquid described by the Navier–Stokes equations

$$\nu \Delta \mathbf{v} = (\mathbf{v} \cdot \nabla)\mathbf{v} + \nabla p + \mathbf{f} , \tag{4.8.1}$$

$$\nabla \cdot \mathbf{v} = 0 . \tag{4.8.2}$$

Let $\nu > 0$. We are treating a problem with boundary condition

$$\mathbf{v}\big|_{\partial\Omega} = \alpha . \tag{4.8.3}$$

From now on, we assume:

(i) Ω is a bounded domain in $\mathbb{R}^2$ or $\mathbb{R}^3$ whose boundary $\partial\Omega$ consists of r closed curves or surfaces S_k $(k = 1, \ldots, r)$ with continuous curvature.

(ii) There is a continuously differentiable vector-function

$$\mathbf{a}(\mathbf{x}) = (a_1(\mathbf{x}), a_2(\mathbf{x}), a_3(\mathbf{x}))$$

such that

$$a_k(\mathbf{x}) \in C^{(1)}(\overline{\Omega}), \qquad \nabla \cdot \mathbf{a} = 0 \text{ in } \Omega, \qquad \mathbf{a}\big|_{\partial\Omega} = \alpha .$$

(iii) On each S_k ($k = 1, \dots, r$) we have

$$\int_{S_k} \alpha \cdot \mathbf{n}\, dS = 0$$

where $\mathbf{n}$ is the unit outward normal at a point of S_k.

We note that the condition

$$\sum_{k=1}^{r} \int_{S_k} \alpha \cdot \mathbf{n}\, dS = 0$$

is necessary for solvability of the problem.

Let $H(\Omega)$ be the completion of the set $S^0(\Omega)$ of all smooth solenoidal vector-functions $\mathbf{u}(\mathbf{x})$ satisfying the boundary condition, in the norm induced by the scalar product

$$(\mathbf{u}, \mathbf{v})_{H(\Omega)} = \int_\Omega \nabla\mathbf{u} \cdot \nabla\mathbf{v}\, d\Omega \equiv \int_\Omega \operatorname{rot}\mathbf{u} \cdot \operatorname{rot}\mathbf{v}\, d\Omega$$

and so each of the components of $\mathbf{u}(\mathbf{x}) \in H(\Omega)$ is of $W^{1,2}(\Omega)$. Thus in the three dimensional case, the imbedding operator of $H(\Omega)$ into $(L^p(\Omega))^3$ is continuous when $1 \le p \le 6$ and compact when $1 \le p < 6$; in the two dimensional case, the imbedding operator is compact into $(L^p(\Omega))^2$ for any $1 \le p < \infty$.

We assume

(iv) $f_k(\mathbf{x}) \in L^p(\Omega)$, $p \ge 6/5$ in the three dimensional case ($k = 1, 2, 3$), $p > 1$ in the two dimensional case ($k = 1, 2$).

Definition 4.8.1. The quantity $\mathbf{v}(\mathbf{x}) = \mathbf{a}(\mathbf{x}) + \mathbf{u}(\mathbf{x})$ is a generalized solution to the problem (4.8.1)–(4.8.3) if $\mathbf{u}(\mathbf{x}) \in H(\Omega)$ satisfies the integro-differential equation

$$\nu(\mathbf{u}, \boldsymbol{\Phi})_{H(\Omega)} = -\int_\Omega [(\mathbf{u} \cdot \nabla)\mathbf{u} \cdot \boldsymbol{\Phi} + (\mathbf{u} \cdot \nabla)\mathbf{a} \cdot \boldsymbol{\Phi} + (\mathbf{a} \cdot \nabla)\mathbf{u} \cdot \boldsymbol{\Phi}$$
$$+ (\mathbf{a} \cdot \nabla)\mathbf{a} \cdot \boldsymbol{\Phi} + \nu \operatorname{rot}\mathbf{a} \cdot \operatorname{rot}\boldsymbol{\Phi} + \mathbf{f} \cdot \boldsymbol{\Phi}]\, d\Omega \qquad (4.8.4)$$

for any $\boldsymbol{\Phi} \in H(\Omega)$.

It is easily seen that if $\mathbf{a}(\mathbf{x})$ and $\mathbf{u}(\mathbf{x})$ belong to $C^{(2)}(\overline{\Omega})$ then $\mathbf{v}(\mathbf{x})$ is a classical solution to the problem (4.8.1)–(4.8.3).

Note that there are infinitely many vectors $\mathbf{a}(\mathbf{x})$ satisfying the assumption (ii) if there is one, but the set of generalized solutions does not depend on the choice of $\mathbf{a}(\mathbf{x})$.

To use Lemma 4.7.1, we reduce equation (4.8.4) to the operator form $\mathbf{u} + F(\mathbf{u}) = 0$, defining F with use of the Riesz representation theorem from the equality

$$\nu(F(\mathbf{u}), \boldsymbol{\Phi})_{H(\Omega)} = \int_{\Omega} [(\mathbf{u} \cdot \nabla)\mathbf{u} \cdot \boldsymbol{\Phi} + (\mathbf{u} \cdot \nabla)\mathbf{a} \cdot \boldsymbol{\Phi} + (\mathbf{a} \cdot \nabla)\mathbf{u} \cdot \boldsymbol{\Phi}$$
$$+ (\mathbf{a} \cdot \nabla)\mathbf{a} \cdot \boldsymbol{\Phi} + \nu \operatorname{rot} \mathbf{a} \cdot \operatorname{rot} \boldsymbol{\Phi} + \mathbf{f} \cdot \boldsymbol{\Phi}] \, d\Omega \,. \tag{4.8.5}$$

The estimates needed to prove that the right-hand side of (4.8.5) is a continuous linear functional in $H(\Omega)$ with respect to $\boldsymbol{\Phi}$ follow from traditional estimates of the terms using Hölder's inequality. But we now show a sharper result; namely,

Lemma 4.8.1. F is a completely continuous operator in $H(\Omega)$.

Proof. Let $\{\mathbf{u}_n(x)\}$ be a weakly convergent sequence in $H(\Omega)$. Then it converges strongly in $(L^4(\Omega))^k$ ($k = 2$ or 3). From (4.8.5), we get

$$\nu \, |(F(\mathbf{u}_m) - F(\mathbf{u}_n), \boldsymbol{\Phi})_{H(\Omega)}| =$$
$$= \left| \int_{\Omega} \{[(\mathbf{u}_m - \mathbf{u}_n) \cdot \nabla]\mathbf{u}_m \cdot \boldsymbol{\Phi} - (\mathbf{u}_n \cdot \nabla)(\mathbf{u}_m - \mathbf{u}_n) \cdot \boldsymbol{\Phi} \right.$$
$$\left. + [(\mathbf{u}_m - \mathbf{u}_n) \cdot \nabla]\mathbf{a} \cdot \boldsymbol{\Phi} + (\mathbf{a} \cdot \nabla)(\mathbf{u}_m - \mathbf{u}_n) \cdot \boldsymbol{\Phi}\} \, d\Omega \right|$$
$$\leq M \, \|\mathbf{u}_m - \mathbf{u}_n\|_{L^4(\Omega)} \, \|\boldsymbol{\Phi}\|_{H(\Omega)}$$

with a constant M which does not depend on m, n, or $\boldsymbol{\Phi}$. Setting

$$\boldsymbol{\Phi} = F(\mathbf{u}_m) - F(\mathbf{u}_n)$$

in the inequality, we obtain

$$\nu \, \|F(\mathbf{u}_m) - F(\mathbf{u}_n)\|_{H(\Omega)} \leq M \, \|\mathbf{u}_m - \mathbf{u}_n\|_{(L^4(\Omega))^k} \to 0 \ \text{ as } m, n \to \infty$$

so F is completely continuous. □

From Definition 4.8.1 it follows that

Lemma 4.8.2. A generalized solution of the problem under consideration in the sense of Definition 4.8.1 satisfies the operator equation

$$\mathbf{u} + F(\mathbf{u}) = 0 \, ; \tag{4.8.6}$$

conversely, a solution to (4.8.6) is a generalized solution of the problem.

By Lemma 4.7.1, it now suffices to show that all solutions of the equation $\mathbf{u} + tF(\mathbf{u}) = 0$, for all $t \in [0, 1]$, lie in a sphere $\|\mathbf{u}\|_{H(\Omega)} \leq R$ for some $R < \infty$. First we show this in the simpler case of homogeneous boundary condition (4.8.3). Here $\alpha = 0$ and thus $\mathbf{a}(\mathbf{x}) = 0$. □

Theorem 4.8.1. The problem (4.8.1)–(4.8.3) with $\alpha = 0$ has at least one generalized solution in the sense of Definition 4.8.1. Each generalized solution $\mathbf{u}(\mathbf{x})$ is bounded, $\|\mathbf{u}\|_{H(\Omega)} < R$ for some $R < \infty$ and the degree of $I + F$ with respect to 0 and $D = \{\mathbf{u} \in H(\Omega) \colon \|\mathbf{u}\| < R\}$ is +1.

Proof. As was said, it suffices to show an a priori estimate for solutions to the equation $\mathbf{u} + tF(\mathbf{u}) = 0$ for $t \in [0, 1]$. For a solution, there holds the identity

$$(\mathbf{u} + tF(\mathbf{u}), \mathbf{u})_{H(\Omega)} = 0$$

or, the same,

$$\nu(\mathbf{u}, \mathbf{u})_{H(\Omega)} + t \int_\Omega (\mathbf{u} \cdot \nabla)\mathbf{u} \cdot \mathbf{u}\, d\Omega = -t \int_\Omega \mathbf{f} \cdot \mathbf{u}\, d\Omega\,.$$

Integration by parts gives

$$\int_\Omega (\mathbf{u}\cdot\nabla)\mathbf{u}\cdot\mathbf{u}\, d\Omega = \frac{1}{2}\int_\Omega \sum_k u_k \frac{\partial}{\partial x_k}(\mathbf{u}\cdot\mathbf{u})\, d\Omega = -\frac{1}{2}\int_\Omega (\mathbf{u}\cdot\mathbf{u})(\nabla\cdot\mathbf{u})\, d\Omega = 0 \qquad (4.8.7)$$

since $\nabla \cdot \mathbf{u} = 0$ and thus, for a solution $\mathbf{u}$, we get

$$|\nu(\mathbf{u}, \mathbf{u})_{H(\Omega)}| = \left| t \int_\Omega \mathbf{f} \cdot \mathbf{u}\, d\Omega \right| \le \frac{\nu R}{2} \|\mathbf{f}\|_{L^p(\Omega)} \|\mathbf{u}\|_{H(\Omega)}$$

with some constant R, or

$$\|\mathbf{u}\|_{H(\Omega)} < R\,.$$

This completes the proof. □

Now we consider the more complicated case of nonhomogeneous boundary conditions (4.8.3). We need some auxiliary results.

Let ω_ε be a domain in $\overline{\Omega}$ consisting of points covered by all inward normals to $\partial\Omega$ having length ε. For sufficiently small $\varepsilon > 0$, these normals do not intersect and thus in ω_ε we can use a coordinate system pointing out for a $\mathbf{x} \in \omega_\varepsilon$ a point Q on $\partial\Omega$ and a number s, the distance from Q to $\mathbf{x}$ along the corresponding normal. So for a function $g(\mathbf{x})$ given on ω_ε, we write down $g(s, Q)$.

Lemma 4.8.3. There is a solenoidal vector function $\mathbf{a}_\varepsilon(\mathbf{x}) \in (C^{(1)}(\overline{\Omega}))^k$ such that $\mathbf{a}_\varepsilon(\mathbf{x}) = 0$ in $\Omega \setminus \omega_\varepsilon$,

$$\mathbf{a}_\varepsilon(\mathbf{x})\big|_{\partial\Omega} = \alpha\,, \quad \text{and} \quad |\mathbf{a}_\varepsilon(\mathbf{x})| \le M_1/\varepsilon \ \text{ in } \overline{\Omega} \qquad (4.8.8)$$

with a constant M_1 not depending on ε.

Proof. Let us introduce a function $q(\mathbf{x})$ by

$$q(s, Q) = \begin{cases} (\varepsilon^2 - s^2)^2/\varepsilon^4\,, & 0 \le s \le \varepsilon\,, \\ 0\,, & s > \varepsilon\,. \end{cases}$$

Let $\mathbf{a}(\mathbf{x})$ be a solenoidal vector-function satisfying assumption (ii) on p. 258. Under the taken assumptions, there is a vector-function $\mathbf{p}(\mathbf{x})$ such that

$$\mathbf{a}(\mathbf{x}) = \operatorname{rot} \mathbf{p}(\mathbf{x})\,.$$

It is seen that the vector function $\mathbf{a}_\varepsilon(\mathbf{x}) = \operatorname{rot}(q\mathbf{p})$ is needed.

Note that in the planar case, this is a vector $(0, 0, q\psi)$ where $\psi(x_1, x_2)$ is the flow function of $\mathbf{a}(\mathbf{x})$. □

Lemma 4.8.4. For $\mathbf{u} \in H(\Omega)$, we have

$$\int_{\omega_\varepsilon} |\mathbf{u}|^2 \, d\Omega \le M_2^2 \varepsilon^2 \int_{\omega_\varepsilon} \sum_{i,j} \left| \frac{\partial u_i}{\partial x_j} \right|^2 d\Omega \tag{4.8.9}$$

with a constant M_2 not depending on $\mathbf{u}$ or ε.

Proof. We show (4.8.9) for a smooth function. The limit passage will prove the general case. So for points of ω_ε we have

$$\mathbf{u}(s, Q) = \int_0^s \frac{\partial \mathbf{u}(t, Q)}{\partial t} \, dt \, .$$

By the Cauchy inequality,

$$\begin{aligned} \int_0^\varepsilon |\mathbf{u}(t, Q)|^2 \, dt &= \int_0^\varepsilon \left| \int_0^s \frac{\partial \mathbf{u}(t, Q)}{\partial t} \, dt \right|^2 ds \\ &\le \int_0^\varepsilon s \int_0^s \left| \frac{\partial \mathbf{u}(t, Q)}{\partial t} \right|^2 dt \, ds \le \frac{\varepsilon^2}{2} \int_0^\varepsilon \left| \frac{\partial \mathbf{u}(t, Q)}{\partial t} \right|^2 dt \, . \end{aligned}$$

It is easily seen that for any $g(\mathbf{x})$

$$m_1 \int_0^\varepsilon \int_{\partial\Omega} g^2(s, Q) \, ds \, dS \le \int_{\omega_\varepsilon} g^2 \, d\Omega \le m_2 \int_0^\varepsilon \int_{\partial\Omega} g^2(s, Q) \, ds \, dS$$

and so

$$\begin{aligned} \int_{\omega_\varepsilon} |\mathbf{u}|^2 \, d\Omega &\le m_2 \int_{\partial\Omega} \int_0^\varepsilon |\mathbf{u}(s, Q)|^2 \, ds \, dS \\ &\le m_2 \int_{\partial\Omega} \frac{\varepsilon^2}{2} \int_0^\varepsilon \left| \frac{\partial \mathbf{u}}{\partial t} \right|^2 dt \, dS \le \frac{m_2}{2m_1} \varepsilon^2 \int_{\omega_\varepsilon} \sum_{i,j} \left| \frac{\partial u_i}{\partial x_j} \right|^2 d\Omega \, . \end{aligned}$$

□

To apply degree theory to the problem under consideration, it remains to establish

Lemma 4.8.5. All solutions of the equation

$$\mathbf{u} + tF(\mathbf{u}) = 0 \tag{4.8.10}$$

for all $t \in [0, 1]$, are in a ball $\|\mathbf{u}\|_{H(\Omega)} < R$ whose radius R depends only on $\mathbf{f}$, $\partial\Omega$, $\mathbf{a}$, and ν.

Proof. Suppose that the set of solutions to (4.8.10) is unbounded. This means there is a sequence $\{t_k\} \subset [0, 1]$ and a corresponding sequence $\{\mathbf{u}_k\}$ such that $\mathbf{u}_k + t_k F(\mathbf{u}_k) = 0$ and

$$\|\mathbf{u}_k\|_{H(\Omega)} \to \infty \quad \text{as } k \to \infty \, .$$

Without loss of generality, we can consider $\{t_k\}$ to be convergent to $t_0 \in [0, 1]$ and, moreover, the sequence $\{\mathbf{u}_k^*\}$, $\mathbf{u}_k^* = \mathbf{u}_k / \|\mathbf{u}_k\|_{H(\Omega)}$, to be weakly convergent to an element $\mathbf{u}_0 \in H(\Omega)$ since $\{\mathbf{u}_k^*\}$ is bounded.

Let us consider the identity $(\mathbf{u}_k + t_k F(\mathbf{u}_k), \mathbf{u}_k) = 0$, namely,

$$-\nu \|\mathbf{u}_k\|^2_{H(\Omega)} = t_k \int_{\omega_\varepsilon} (\mathbf{a}_\varepsilon \cdot \nabla)\mathbf{u}_k \cdot \mathbf{u}_k \, d\Omega + t_k \int_\Omega [(\mathbf{a}_\varepsilon \cdot \nabla)\mathbf{a}_\varepsilon \cdot \mathbf{u}_k + \nu \operatorname{rot} \mathbf{a}_\varepsilon \cdot \operatorname{rot} \mathbf{u}_k + \mathbf{f} \cdot \mathbf{u}_k] \, d\Omega \qquad (4.8.11)$$

which is valid because of (4.8.7) and a similar equality

$$\int_{\omega_\varepsilon} (\mathbf{u}_k \cdot \nabla)\mathbf{a}_\varepsilon \cdot \mathbf{u}_k \, d\Omega = 0 \, .$$

The first integral on the right-hand side of (4.8.11) is a weakly continuous functional with respect to $\mathbf{u}_k$, and for the second integral we have

$$\left| \int_\Omega [(\mathbf{a}_\varepsilon \cdot \nabla)\mathbf{a}_\varepsilon \cdot \mathbf{u}_k + \nu \operatorname{rot} \mathbf{a}_\varepsilon \cdot \operatorname{rot} \mathbf{u}_k + \mathbf{f} \cdot \mathbf{u}_k] \, d\Omega \right| \le M_3 \, \|\mathbf{u}_k\|_{H(\Omega)}$$

where M_3 does not depend on $\mathbf{u}_k$. Dividing both sides of (4.8.11) by $\|\mathbf{u}_k\|^2_{H(\Omega)}$, it follows that

$$-\nu = t_0 \int_{\omega_\varepsilon} (\mathbf{a}_\varepsilon \cdot \nabla)\mathbf{u}_0 \cdot \mathbf{u}_0 \, d\Omega \, . \qquad (4.8.12)$$

We note that this holds for any small positive $\varepsilon < \varepsilon_0$ with a fixed ε_0 for which the above construction of the frame for ω_{ε_0} is valid. To prove it, take $\varepsilon = \eta < \varepsilon_0$

$$\mathbf{w}_k = \mathbf{u}_k + \mathbf{a}_{\varepsilon_0} - \mathbf{a}_\eta$$

and consider the identity

$$(\mathbf{u}_k + t_k F(\mathbf{u}_k), \mathbf{w}_k)_{H(\Omega)} = 0$$

which takes the form

$$-\nu \|\mathbf{w}_k\|^2_{H(\Omega)} = t_k \int_{\omega_\eta} (\mathbf{a}_\eta \cdot \nabla)\mathbf{w}_k \cdot \mathbf{w}_k \, d\Omega + t_k \int_\Omega [(\mathbf{a}_\eta \cdot \nabla)\mathbf{a}_\eta \cdot \mathbf{w}_k + \nu \operatorname{rot} \mathbf{a}_\eta \cdot \operatorname{rot} \mathbf{w}_k + \mathbf{f} \cdot \mathbf{w}_k] \, d\Omega \, .$$

Divide this equality by $\|\mathbf{u}_k\|^2_{H(\Omega)}$ term by term. Consider the sequence

$$\mathbf{w}_k^* = \mathbf{u}_k^* + (\mathbf{a}_\varepsilon - \mathbf{a}_\eta)/\|\mathbf{u}_k\|_{H(\Omega)} .$$

Since $\|\mathbf{u}_k\|_{H(\Omega)} \to \infty$, we have $(\mathbf{a}_\varepsilon - \mathbf{a}_\eta)/\|\mathbf{u}_k\|_{H(\Omega)} \to 0$ strongly. Since $\|\mathbf{u}_k^*\|_{H(\Omega)} = 1$, we have that $\|\mathbf{w}_k^*\|_{H(\Omega)} \to 1$. Besides, it is clear that $\mathbf{w}_k^* \to \mathbf{u}_0$ weakly and thus we get the needed equality (4.8.12) again.

Now we show that the limit of the integral on the right-hand side of (4.8.12) is zero. Thanks to (4.8.8) and (4.8.9), we obtain

$$\left| \int_{\omega_\varepsilon} (\mathbf{a}_\varepsilon \cdot \nabla)\mathbf{u}_0 \cdot \mathbf{u}_0 \, d\Omega \right| \le M_1 M_2 \int_{\omega_\varepsilon} |\operatorname{rot} \mathbf{u}_0|^2 \, d\Omega \to 0$$

as $\varepsilon \to 0$. Since $\nu > 0$, we have a contradiction. □

Now we can formulate

Theorem 4.8.2. Under assumptions (i)–(iv), there exists at least one generalized solution of the problem (4.8.1)–(4.8.3) in the sense of Definition 4.8.1. All generalized solutions of the problem are bounded in the energy space and the degree of the operator $I + F$ of the problem with respect to zero and a ball about zero with sufficiently large radius is $+1$.

Problem 4.8.1. Which of the assumptions (i)–(iv) are not necessary in proving Theorem 4.8.1? □

The reader can find a much more complete presentation of the mathematical theory of liquids in gases in [22, 35].

Appendix A
Summary of Inequalities and Imbeddings

A.1 Inequalities

Unless otherwise stated, p and q satisfy $p^{-1} + q^{-1} = 1$ with $p > 1$.

Algebraic inequalities

$$|ab| \le \tfrac{1}{2}(a^2 + b^2) \qquad |ab| \le a^2 + \tfrac{1}{4}b^2$$

$$(a + b)^2 \le 2(a^2 + b^2) \qquad (a, b \text{ real})$$

Young's inequality

$$ab \le \frac{a^p}{p} + \frac{b^q}{q} \qquad (a, b > 0)$$

Triangle inequality

Sums

$$\left|\sum_{k=1}^{n} x_k\right| \le \sum_{k=1}^{n} |x_k|$$

Series

$$\left|\sum_{k=1}^{\infty} x_k\right| \le \sum_{k=1}^{\infty} |x_k|$$

Integrals

$$\left|\int_\Omega f(\mathbf{x})\, d\Omega\right| \le \int_\Omega |f(\mathbf{x})|\, d\Omega$$

Abstract forms

$$d(x,y) \le d(x,z) + d(z,y)$$

$$\|x+y\| \le \|x\| + \|y\| \qquad \|x+y\| \ge \big|\, \|x\| - \|y\| \,\big|$$

Hölder's inequality

Sums

$$\left|\sum_{k=1}^{n} x_k y_k\right| \le \left(\sum_{k=1}^{n} |x_k|^p\right)^{1/p} \left(\sum_{k=1}^{n} |y_k|^q\right)^{1/q}$$

Series

$$\left|\sum_{k=1}^{\infty} x_k y_k\right| \le \left(\sum_{k=1}^{\infty} |x_k|^p\right)^{1/p} \left(\sum_{k=1}^{\infty} |y_k|^q\right)^{1/q}$$

Integrals

$$\left|\int_\Omega f(\mathbf{x}) g(\mathbf{x})\, d\Omega\right| \le \left(\int_\Omega |f(\mathbf{x})|^p\, d\Omega\right)^{1/p} \left(\int_\Omega |g(\mathbf{x})|^q\, d\Omega\right)^{1/q}$$

Minkowski's inequality $(p \ge 1)$

Sums

$$\left(\sum_{k=1}^{n} |x_k + y_k|^p\right)^{1/p} \le \left(\sum_{k=1}^{n} |x_k|^p\right)^{1/p} + \left(\sum_{k=1}^{n} |y_k|^p\right)^{1/p}$$

Series

$$\left(\sum_{k=1}^{\infty} |x_k + y_k|^p\right)^{1/p} \le \left(\sum_{k=1}^{\infty} |x_k|^p\right)^{1/p} + \left(\sum_{k=1}^{\infty} |y_k|^p\right)^{1/p}$$

Integrals

$$\left(\int_\Omega |f(\mathbf{x}) + g(\mathbf{x})|^p\, d\Omega\right)^{1/p} \le \left(\int_\Omega |f(\mathbf{x})|^p\, d\Omega\right)^{1/p} + \left(\int_\Omega |g(\mathbf{x})|^p\, d\Omega\right)^{1/p}$$

Schwarz inequality

Sums

$$\left|\sum_{k=1}^{n} x_k \overline{y_k}\right| \le \left(\sum_{k=1}^{n} |x_k|^2\right)^{1/2} \left(\sum_{k=1}^{n} |y_k|^2\right)^{1/2}$$

Series

$$\left|\sum_{k=1}^{\infty} x_k \overline{y_k}\right| \le \left(\sum_{k=1}^{\infty} |x_k|^2\right)^{1/2} \left(\sum_{k=1}^{\infty} |y_k|^2\right)^{1/2}$$

Integrals

$$\left|\int_{\Omega} f(\mathbf{x})\overline{g(\mathbf{x})}\, d\Omega\right| \leq \left(\int_{\Omega} |f(\mathbf{x})|^2\, d\Omega\right)^{1/2} \left(\int_{\Omega} |g(\mathbf{x})|^2\, d\Omega\right)^{1/2}$$

Abstract form

$$|(x, y)| \leq \|x\|\, \|y\|$$

Result for Lebesgue integral

$$\left|\int_{\Omega} F(\mathbf{x})\, d\Omega\right| \leq (\text{mes}\, \Omega)^{1/q} \left(\int_{\Omega} |F(\mathbf{x})|^p\, d\Omega\right)^{1/p}$$

Friedrichs inequality

$$\iint_{\Omega} u^2\, dx\, dy \leq m \iint_{\Omega} \left[\left(\frac{\partial u}{\partial x}\right)^2 + \left(\frac{\partial u}{\partial y}\right)^2\right] dx\, dy \qquad (u|_{\partial\Omega} = 0)$$

Poincaré's inequality

$$\iint_{\Omega} u^2\, dx\, dy \leq m \left\{\left(\iint_{\Omega} u\, dx\, dy\right)^2 + \iint_{\Omega} \left[\left(\frac{\partial u}{\partial x}\right)^2 + \left(\frac{\partial u}{\partial y}\right)^2\right] dx\, dy\right\}$$

Korn's inequality

$$\iiint_{\Omega} \left[|\mathbf{u}|^2 + \sum_{i,j=1}^{3} \left(\frac{\partial u_i}{\partial x_j}\right)^2\right] d\Omega \leq m \iiint_{\Omega} c^{ijkl} \epsilon_{kl}(\mathbf{u})\, \epsilon_{ij}(\mathbf{u})\, d\Omega \qquad (\mathbf{u}|_{\partial\Omega} = \mathbf{0})$$

A.2 Imbeddings

Imbedding concept; continuity and compactness of an imbedding

An *imbedding operator* from a normed space X to a normed space Y is a linear, injective operator i acting from X to Y. On this page, the notation $X \hookrightarrow Y$ is used to indicate the existence of such an operator. We say that the space X is imbedded in the space Y. It is common practice to *identify* the image $i(x)$ of an element $x \in X$ with x itself (whether or not the spaces X and Y contain elements of the same nature). Hence *continuity* of the imbedding $X \hookrightarrow Y$ means that there exists $c > 0$, independent of x, such that

$$\|x\|_Y \le c\,\|x\|_X \quad \text{for all } x \in X$$

where, on the left-hand side, x is regarded as an element of the space Y. *Compactness* of the imbedding means that every bounded sequence $\{x_n\} \subset X$ has a subsequence that can be regarded as a Cauchy sequence in Y. Recall that a compact linear operator is continuous (Theorem 3.3.1).

Simple examples

The symbol Ω denotes a compact region in $\mathbb{R}^n$ and $\partial\Omega$ its boundary.

$L^p(\Omega) \hookrightarrow L^q(\Omega) \quad (1 \le q \le p)$ $\qquad$ $C(\Omega) \hookrightarrow L^p(\Omega) \quad (1 \le p < \infty)$

$W^{l,p}(\Omega) \hookrightarrow W^{l,q}(\Omega) \quad (q < p)$ $\qquad$ $C^{(1)}(\Omega) \hookrightarrow C(\Omega)$

$\ell^p \hookrightarrow \ell^q \quad (1 \le p \le q)$ $\qquad$ $W^{1,1}(0,l) \hookrightarrow C(0,l)$

Special cases of Sobolev–Kondrashov imbedding theorem

domain	imbedding	conditions
$\Omega \subset \mathbb{R}^2$	$W^{1,2}(\Omega) \hookrightarrow L^q(\Omega)$	compact for $q \ge 1$
	$W^{1,2}(\Omega) \hookrightarrow L^q(\partial\Omega)$	compact for $q \ge 1$
$\Omega \subset \mathbb{R}^3$	$W^{1,2}(\Omega) \hookrightarrow L^q(\Omega)$	continuous for $1 \le q \le 6$; compact for $1 \le q < 6$
	$W^{1,2}(\Omega) \hookrightarrow L^q(\partial\Omega)$	continuous for $1 \le q \le 4$; compact for $1 \le q < 4$
$\Omega \subset \mathbb{R}^2$	$W^{2,2}(\Omega) \hookrightarrow C(\Omega)$	compact

Appendix B
Hints for Selected Problems

Note. In this appendix, equations are occasionally tagged with asterisks for reference (e.g., as (*) or (**)). All such equation labels are purely local and have meaning only within the hint for a given problem.

1.1.1 Setting $\mathbf{y} = \mathbf{x}$ in the triangle inequality we obtain $d(\mathbf{x}, \mathbf{x}) \leq d(\mathbf{x}, \mathbf{z}) + d(\mathbf{z}, \mathbf{x})$. By axioms D2 and D3 then, we have $0 \leq 2\,d(\mathbf{x}, \mathbf{z})$.

1.1.2 The metrics d_S and d_p are equivalent with

$$1 \leq \frac{d_p(\mathbf{x}, \mathbf{y})}{d_S(\mathbf{x}, \mathbf{y})} \leq N^{1/p},$$

because

$$\begin{aligned}\max_{1\leq i\leq n} |x_i - y_i| &= \left(\max_{1\leq i\leq n} |x_i - y_i|^p\right)^{1/p} \leq \left(\sum_{i=1}^{n} |x_i - y_i|^p\right)^{1/p} \\ &\leq \left[\sum_{i=1}^{n}\left(\max_{1\leq j\leq n} |x_j - y_j|\right)^p\right]^{1/p} = N^{1/p} \max_{1\leq j\leq n} |x_j - y_j| \,.\end{aligned}$$

The metrics d_E and d_k are equivalent with

$$\left(\min_{1\leq i\leq n} k_i\right)^{1/2} \leq \frac{d_k(\mathbf{x}, \mathbf{y})}{d_E(\mathbf{x}, \mathbf{y})} \leq \left(\max_{1\leq i\leq n} k_i\right)^{1/2}.$$

1.1.3 Axioms D1–D3 are obviously fulfilled in each case. For the expression (1.1.4), axiom D4 holds because $|x_i - y_i| \leq |x_i - z_i| + |z_i - y_i|$ for each i, and therefore

$$\begin{aligned}d(\mathbf{x}, \mathbf{y}) = \sup_i |x_i - y_i| &\leq \sup_i (|x_i - z_i| + |z_i - y_i|) \\ &\leq \sup_i |x_i - z_i| + \sup_i |z_i - y_i| = d(\mathbf{x}, \mathbf{z}) + d(\mathbf{z}, \mathbf{y}) \,.\end{aligned}$$

To verify D4 for the expressions (1.1.5) and (1.1.6), we need Minkowski's inequality

$$\left(\sum_{i=1}^{\infty}|a_i+b_i|^p\right)^{1/p}\le\left(\sum_{i=1}^{\infty}|a_i|^p\right)^{1/p}+\left(\sum_{i=1}^{\infty}|b_i|^p\right)^{1/p}.$$

For (1.1.5) we have, starting with the triangle inequality,

$$\begin{aligned}d(\mathbf{x},\mathbf{y})&=\left(\sum_{i=1}^{\infty}|x_i-z_i+z_i-y_i|^p\right)^{1/p}\le\left(\sum_{i=1}^{\infty}[|x_i-z_i|+|z_i-y_i|]^p\right)^{1/p}\\&\le\left(\sum_{i=1}^{\infty}|x_i-z_i|^p\right)^{1/p}+\left(\sum_{i=1}^{\infty}|z_i-y_i|^p\right)^{1/p}=d(\mathbf{x},\mathbf{z})+d(\mathbf{z},\mathbf{y}).\end{aligned}$$

For (1.1.6) we have

$$\begin{aligned}d(\mathbf{x},\mathbf{y})&=\left(\sum_{k=1}^{\infty}k^2|x_k-z_k+z_k-y_k|^2\right)^{1/2}\le\left(\sum_{k=1}^{\infty}k^2\,(|x_k-z_k|+|z_k-y_k|)^2\right)^{1/2}\\&=\left(\sum_{k=1}^{\infty}(k\,|x_k-z_k|+k\,|z_k-y_k|)^2\right)^{1/2}\\&\le\left(\sum_{k=1}^{\infty}(k\,|x_k-z_k|)^2\right)^{1/2}+\left(\sum_{k=1}^{\infty}(k\,|z_k-y_k|)^2\right)^{1/2}=d(\mathbf{x},\mathbf{z})+d(\mathbf{z},\mathbf{y}).\end{aligned}$$

1.2.1 Use a procedure similar to that used for Problem 1.1.3.

1.3.1 The only aspect of D1–D3 worthy of close examination is the implication $d(f,g)=0 \implies f=g$ of D2. Note that $d(f,g)=0$ implies

$$\max_{\mathbf{x}\in\Omega}|D^\alpha f(\mathbf{x})-D^\alpha g(\mathbf{x})|=0\quad\text{for all }\alpha\text{ such that }|\alpha|\le k.$$

In particular this holds for $\alpha=(0,\dots,0)$, giving

$$\max_{\mathbf{x}\in\Omega}|f(\mathbf{x})-g(\mathbf{x})|=0.$$

This implies that $f(\mathbf{x})=g(\mathbf{x})$ for all $\mathbf{x}\in\Omega$. Fulfillment of D4 follows from the triangle inequality

$$|D^\alpha f(\mathbf{x})-D^\alpha g(\mathbf{x})|\le|D^\alpha f(\mathbf{x})-D^\alpha h(\mathbf{x})|+|D^\alpha h(\mathbf{x})-D^\alpha g(\mathbf{x})|$$

for real numbers.

1.3.3 The constant functions $f(x)\equiv 0$ and $g(x)\equiv 1$ on $[0,1]$ are not equal, but the proposed distance function would give $d(f,g)=0$. Hence d is not a metric (cf., axiom D2). To generate a metric space under this distance function, we could narrow our consideration to the functions satisfying a condition such as $f(0)=0$.

1.5.1 Requested here is a general setup of the problem; we cannot, in the narrow sense, "find" a solution. Rather, we can investigate the setup of the variational problem in the form of a boundary value problem for a partial differential equation. Of course, from a logical viewpoint the initial setup as a problem of minimum is neither better nor worse than the latter one, since in both cases we cannot find explicit solutions in general. Historically, however, the theory of partial differential equations was well developed and variational problems were always reduced to the solution of corresponding boundary value problems.

Let us suppose that u has more smoothness than is required by the problem, namely that u belongs to $C^{(2)}$. We employ the usual methods of the calculus of variations. In the functional $J(u)$ we consider variations $u = u(x, y) + t\varphi(x, y)$ where t is a real variable and $\varphi|_{\partial\Omega} = 0$. For fixed φ the functional $J(u + t\varphi)$ can be regarded as an ordinary function of t, taking its minimum at $t = 0$. It is necessary that

$$0 = \frac{d}{dt} J(u + t\varphi)\Big|_{t=0} = \iint_\Omega \left[\left(\frac{\partial u}{\partial x} + t\frac{\partial \varphi}{\partial x} \right) \frac{\partial \varphi}{\partial x} + \left(\frac{\partial u}{\partial y} + t\frac{\partial \varphi}{\partial y} \right) \frac{\partial \varphi}{\partial y} - f\varphi \right]\Bigg|_{t=0} dx\, dy$$

$$= \iint_\Omega \left[\frac{\partial u}{\partial x}\frac{\partial \varphi}{\partial x} + \frac{\partial u}{\partial y}\frac{\partial \varphi}{\partial y} - f\varphi \right] dx\, dy\,.$$

We now integrate by parts using the formula

$$\iint_\Omega u \frac{\partial v}{\partial x_i}\, dx\, dy = - \iint_\Omega \frac{\partial u}{\partial x_i} v\, dx\, dy + \oint_{\partial\Omega} uv\, n_i\, ds\,,$$

where s parameterizes $\partial\Omega$ and n_i is the cosine of the angle between the outward normal $\mathbf{n}$ to $\partial\Omega$ and the x_i axis ($x_i = x, y$ for $i = 1, 2$, respectively). This gives us

$$- \iint_\Omega \left[\frac{\partial^2 u}{\partial x^2} + \frac{\partial^2 u}{\partial y^2} + f \right] \varphi\, dx\, dy + \oint_{\partial\Omega} \left[\frac{\partial u}{\partial x} n_x + \frac{\partial u}{\partial y} n_y \right] \varphi\, ds = 0\,.$$

Because $\varphi|_{\partial\Omega} = 0$, we have

$$\iint_\Omega \left[\frac{\partial^2 u}{\partial x^2} + \frac{\partial^2 u}{\partial y^2} + f \right] \varphi\, dx\, dy = 0$$

for any "admissible" φ. It follows (by the "fundamental lemma") that

$$\frac{\partial^2 u}{\partial x^2} + \frac{\partial^2 u}{\partial y^2} + f = 0\,.$$

This is Poisson's equation for the unknown u. It can be argued that this equation continues to hold even when the condition $\varphi|_{\partial S} = 0$ is removed, allowing us to write

$$\oint_{\partial\Omega} \left[\frac{\partial u}{\partial x} n_x + \frac{\partial u}{\partial y} n_y \right] \varphi\, ds = 0$$

and thereby deduce the natural boundary condition

$$\left[\frac{\partial u}{\partial x} n_x + \frac{\partial u}{\partial y} n_y \right]\Bigg|_{\partial\Omega} = \frac{\partial u}{\partial n}\Bigg|_{\partial\Omega} = 0\,.$$

For an introduction to the calculus of variations with a treatment of the membrane problem, see [19].

1.5.2 The energy integral for the plate is

$$\mathcal{E}_3(w) = \iint_\Omega \frac{D}{2} \left\{ (\Delta w)^2 + 2(1-\nu)(w_{xy}^2 - w_{xx}w_{yy}) \right\} dx\,dy$$

where

$$(\Delta w)^2 = \left(w_{xx} + w_{yy} \right)^2 = w_{xx}^2 + 2w_{xx}w_{yy} + w_{yy}^2 ,$$

so the quantity within braces in the integrand is

$$I \equiv w_{xx}^2 + w_{yy}^2 + 2(1-\nu)w_{xy}^2 + 2\nu w_{xx}w_{yy} .$$

Using the elementary inequality $2ab \geq -(a^2 + b^2)$ to write

$$\begin{aligned} I &\geq w_{xx}^2 + w_{yy}^2 + 2(1-\nu)w_{xy}^2 - \nu(w_{xx}^2 + w_{yy}^2) = (1-\nu)(w_{xx}^2 + w_{yy}^2) + 2(1-\nu)w_{xy}^2 \\ &= (1-\nu)(w_{xx}^2 + w_{yy}^2 + 2w_{xy}^2) \geq 0 \qquad (\text{since } \nu < 1/2) \end{aligned}$$

we see that

$$\mathcal{E}_3(w) \geq \iint_\Omega \frac{D(1-\nu)}{2} (w_{xx}^2 + w_{yy}^2 + 2w_{xy}^2)\, dx\,dy \geq 0 .$$

Therefore, $\mathcal{E}_3(w) = 0$ implies the three equations $w_{xx} = 0$, $w_{yy} = 0$, and $w_{xy} = 0$. As all second-order derivatives of w must vanish, w can be at most linear in x and y: $w = \alpha + \beta x + \gamma y$.

1.6.1 Consider the open ball $B(x_0, r)$. Any point $y_0 \in B(x_0, r)$ is the center of an open ball contained entirely in $B(x_0, r)$. Indeed, we have $B(y_0, a) \subseteq B(x_0, r)$ provided that $a < r - d(y_0, x_0)$. For if $z \in B(y_0, a)$, then $d(z, x_0) \leq d(z, y_0) + d(y_0, x_0) < a + d(y_0, x_0) < r$ so that $z \in B(x_0, r)$ as well.

1.6.2 (a) Let S_1 and S_2 be bounded sets in a metric space (X, d). There exists a ball $B(x_1, R_1)$ such that $S_1 \subseteq B(x_1, R_1)$, and a ball $B(x_2, R_2)$ such that $S_2 \subseteq B(x_2, R_2)$. The union $S_1 \cup S_2$ is contained in the ball $B(x_1, R)$, where $R = \max\{R_1, R_2 + d(x_2, x_1)\}$. (b) If $x, y \in B(0, r)$, then for any point $z = tx + (1-t)y$ on the line segment connecting them $(0 \leq t \leq 1)$, we have $\|z\| \leq t\|x\| + (1-t)\|y\| < tr + (1-t)r = r$. Hence $z \in B(0, r)$. The proof for the closed ball is similar, with the strict $<$ sign replaced by $\leq$. (c) Let $x, y \in A \cap B$. Then $x, y \in A$ and $x, y \in B$. Choose t such that $0 \leq t \leq 1$, and consider the point $z = tx + (1-t)y$. We have $z \in A$ and $z \in B$ by convexity of the individual sets. Therefore $z \in A \cap B$.

1.7.1 Suppose we have $x_n \to y$ as $n \to \infty$. Then, taking $\varepsilon = 1$ in the definition of convergence, we see that there exists N such that $n > N$ implies $d(x_n, y) < 1$. Let $R = \max\{d(x_1, y), d(x_2, y), \ldots, d(x_N, y), 1\}$. Then the entire sequence $\{x_n\}$ lies within distance R of the limit y.

1.7.2 The functions

$$f_n(x) = \begin{cases} 0\,, & 0 \le x \le \frac{1}{2}\,, \\ nx - \frac{n}{2}\,, & \frac{1}{2} \le x \le \frac{1}{2} + \frac{1}{n}\,, \\ 1\,, & \frac{1}{2} + \frac{1}{n} \le x \le 1\,, \end{cases} \qquad (n = 2, 3, 4, \ldots)$$

are each continuous on $[0, 1]$. To see that $\{f_n\}$ is a Cauchy sequence, we assume $m > n$ and calculate

$$\begin{aligned} d(f_n, f_m) &= \int_{\frac{1}{2}}^{\frac{1}{2}+\frac{1}{m}} \left| \left(mx - \frac{m}{2}\right) - \left(nx - \frac{n}{2}\right) \right| dx + \int_{\frac{1}{2}+\frac{1}{m}}^{\frac{1}{2}+\frac{1}{n}} \left| 1 - \left(nx - \frac{n}{2}\right) \right| dx \\ &= \frac{1}{2}\left(\frac{1}{n} - \frac{1}{m}\right) \to 0 \text{ as } m, n \to \infty\,. \end{aligned}$$

However, we have $f_n \to f$ where

$$f = \begin{cases} 0\,, & 0 \le x \le \frac{1}{2}\,, \\ 1\,, & \frac{1}{2} < x \le 1\,, \end{cases}$$

because

$$d(f_n, f) = \int_{\frac{1}{2}}^{\frac{1}{2}+\frac{1}{n}} \left| 1 - \left(nx - \frac{n}{2}\right) \right| dx = \frac{1}{2n} \to 0 \text{ as } n \to \infty\,.$$

1.7.3 (a) If $d(x_n, x) \to 0$, then $d(x_{n_k}, x) \to 0$ for any subsequence $\{x_{n_k}\}$ of $\{x_n\}$.

1.8.1 (a) Let $\overline{B}(x_0, r)$ denote the closed ball of radius r centered at x_0. Suppose y is an accumulation point of $\overline{B}(x_0, r)$. Choose a sequence of points $x_n \in \overline{B}(x_0, r)$ such that $d(x_n, y) < 1/n$ for each $n = 1, 2, 3, \ldots$. Then

$$d(x_0, y) \le d(x_0, x_n) + d(x_n, y) < r + 1/n \qquad (n = 1, 2, 3, \ldots)\,.$$

In the limit as $n \to \infty$, this inequality yields $d(x_0, y) \le r$ and hence $y \in \overline{B}(x_0, r)$. Since $\overline{B}(x_0, r)$ contains all its accumulation points, it is a closed set. (g) Let S_i $(i = 1, \ldots, n)$ be closed sets and suppose x is an accumulation point of the union $S = \cup_{i=1}^{n} S_i$. Then for every $k > 0$ there is a point $x_k \in S$ such that $d(x_k, x) < 1/k$. Since the S_i are finite in number, an infinite number of the x_k must belong to one of them, say S_j. Hence x is an accumulation point of S_j and, because S_j is closed, $x \in S_j$. Therefore $x \in S$.

1.8.2 The sequence of centers $\{x_n\}$ is a Cauchy sequence. Indeed, whenever $m > n$ we have $x_m \in \overline{B}(x_m, r_m) \subseteq \overline{B}(x_n, r_n)$ and hence $d(x_m, x_n) \le r_n$. By completeness then, $x_n \to x$ for some $x \in X$. For each n, the sequence $\{x_{n+p}\}_{p=1}^{\infty}$ lies in $\overline{B}(x_n, r_n)$ and converges to x; since the ball is closed we have $x \in \overline{B}(x_n, r_n)$. This proves existence of a point in the intersection of all the balls. If y is any other such point, then

$$d(y,x) \le d(y,x_n) + d(x_n,x) \le 2r_n \to 0 \text{ as } n \to \infty .$$

Hence $y = x$ and we have proved uniqueness.

1.9.1 Take any two other representatives $\{x'_n\}$ and $\{y'_n\}$ from X and Y, respectively, and use the inequality

$$|d(x_n,y_n) - d(x'_n,y'_n)| \le d(x_n,x'_n) + d(y_n,y'_n)$$

to get $|d(x_n,y_n) - d(x'_n,y'_n)| \to 0$ as $n \to \infty$. This shows that

$$\lim_{n\to\infty} d(x'_n,y'_n) = \lim_{n\to\infty} d(x_n,y_n) .$$

1.10.1 If $\{\tilde{f}_n(\mathbf{x})\}$ is another representative of $F(\mathbf{x})$, i.e., if $\|f_n(\mathbf{x}) - \tilde{f}_n(\mathbf{x})\|_p \to 0$, then we can set

$$\tilde{K} = \lim_{n\to\infty} \tilde{K}_n = \lim_{n\to\infty} \|\tilde{f}_n(\mathbf{x})\|_p$$

but subsequently find that

$$\begin{aligned} |K - \tilde{K}| &= \left|\lim_{n\to\infty} \|f_n(\mathbf{x})\|_p - \lim_{n\to\infty} \|\tilde{f}_n(\mathbf{x})\|_p\right| = \lim_{n\to\infty} \left|\|f_n(\mathbf{x})\|_p - \|\tilde{f}_n(\mathbf{x})\|_p\right| \\ &\le \lim_{n\to\infty} \|f_n(\mathbf{x}) - \tilde{f}_n(\mathbf{x})\|_p = 0 . \end{aligned}$$

1.10.2

$$\begin{aligned} \left|\int_\Omega f_n(\mathbf{x})\,d\Omega - \int_\Omega f_m(\mathbf{x})\,d\Omega\right| &= \left|\int_\Omega [f_n(\mathbf{x}) - f_m(\mathbf{x})]\;d\Omega\right| \\ &\le \int_\Omega |f_n(\mathbf{x}) - f_m(\mathbf{x})|\,d\Omega \le (\operatorname{mes}\Omega)^{1-1/p}\,\|f_n(\mathbf{x}) - f_m(\mathbf{x})\|_p \to 0 \text{ as } m,n \to \infty . \end{aligned}$$

1.10.3 If $1/p + 1/q = 1$, then Hölder's inequality gives

$$\left|\int_\Omega f_n(\mathbf{x})\,d\Omega\right| \le \int_\Omega 1 \cdot |f_n(\mathbf{x})|\,d\Omega \le (\operatorname{mes}\Omega)^{1/q}\left(\int_\Omega |f_n(\mathbf{x})|^p\,d\Omega\right)^{1/p} .$$

Now let $n \to \infty$. Note that if Ω is unbounded so that $\int_\Omega 1\,d\Omega$ diverges, we cannot use the trick of representing f as $1 \cdot f$ and using Hölder's inequality.

1.10.5 Use the Schwarz inequality for integrals.

1.11.1 By N3 with x replaced by $x - y$, we have

$$\|x\| - \|y\| \le \|x - y\| .$$

Swapping x and y we have, by N2,

$$\|y\| - \|x\| \le \|y - x\| = \|(-1)(x - y)\| = \|x - y\| .$$

Therefore $\|x - y\| \geq |\|x\| - \|y\||$.

1.11.2 Norm axioms N2 and N3 hold for all three expressions. We verify N1 for expressions (a) and (b). If $f(x) \equiv 0$ on $[a, b]$, then clearly $\|f\| = 0$ in each case. Now consider the converse statements. (a) If $\|f\| = 0$, then $f(a) = 0$ and $f'(x) \equiv 0$ on $[a, b]$. This implies that $f(x) =$ constant on $[a, b]$ with $f(a) = 0$, which means that $f(x) \equiv 0$ on $[a, b]$. (b) If $\|f\| = 0$, then $f(x) \equiv 0$ on $[a, b]$. For if $f(c) \neq 0$ for some $c \in [a, b]$, then by continuity the first term in the proposed norm expression (the integral term) would be nonzero. However, N1 fails for expression (c) because $\|f\| = 0$ implies only that $f =$ constant on $[a, b]$.

1.11.3 (c) Use the inequality

$$\|\alpha_n x_n - \alpha x\| \leq \|\alpha_n x_n - \alpha_n x\| + \|\alpha_n x - \alpha x\| \leq |\alpha_n| \, \|x_n - x\| + |\alpha_n - \alpha| \, \|x\|$$

and the fact that the convergent numerical sequence $\{\alpha_n\}$ is bounded. (d) Use the inequality

$$| \, \|x_n\| - \|x\| \, | \leq \|x_n - x\| \ .$$

(e) Use the inequality

$$\|y_n - x\| \leq \|y_n - x_n\| + \|x_n - x\| \ .$$

(f) Use the inequality

$$| \, \|x_n - y\| - \|x - y\| \, | \leq \|x_n - x\| \ .$$

(g) Use the inequality chain

$$| \, \|x_n - y_n\| - \|x - y\| \, | \leq \|(x_n - y_n) - (x - y)\| \leq \|x_n - x\| + \|y_n - y\| \ .$$

(h) Use the inequality $| \, \|x\| - \|y\| \, | \leq \|x - y\|$ and the triangle inequality to write

$$\begin{aligned} | \, \|x_n - y_n\| - \|x_m - y_m\| \, | &\leq \|(x_n - y_n) - (x_m - y_m)\| \\ &\leq \|x_n - x_m\| + \|y_n - y_m\| \to 0 \ \text{ as } n, m \to \infty \ . \end{aligned}$$

Therefore $\{\|x_n - y_n\|\}$ is a Cauchy sequence of real numbers. It is convergent because $\mathbb{R}$ is complete. (i) If A is bounded, then for any $\{x_n\} \subset A$ and any numerical sequence $\alpha_n \to 0$ we have $\|\alpha_n x_n\| = |\alpha_n| \, \|x_n\| \to 0$. Conversely, suppose that $\alpha_n x_n \to 0$ for any sequence $\{x_n\} \subset A$ and any numerical sequence $\{\alpha_n\}$ with $\alpha_n \to 0$. To show that A is bounded, we assume it is not bounded and obtain a contradiction. If A is not bounded, we can take a sequence $\{x_n\} \subset A$ such that $\|x_n\| > n$ for each n. Taking $\alpha_n = 1/n$, we have

$$\|\alpha_n x_n\| = |\alpha_n| \, \|x_n\| = (1/n) \, \|x_n\| > (1/n)(n) = 1 \ \text{ for all } n \ ,$$

which contradicts the assumption that $\alpha_n x_n \to 0$. (j) We have

$$||2x|| = ||(x + y) + (x - y)|| \leq ||x + y|| + ||x - y|| \leq 2\max\{||x + y||, ||x - y||\} .$$

1.12.2 For p in the interval $(0, 1)$ Minkowski's inequality is not valid.

1.13.1 To show that c is a Banach space, we use the completeness of $(\mathbb{R}, |x|)$. Let $\{\mathbf{x}^{(k)}\}$ be a Cauchy sequence in c. Its kth term is a numerical sequence:

$$\mathbf{x}^{(k)} = (x_1^{(k)}, x_2^{(k)}, x_3^{(k)}, \ldots) .$$

To each $\varepsilon > 0$ there corresponds $N = N(\varepsilon)$ such that

$$||\mathbf{x}^{(n+m)} - \mathbf{x}^{(n)}|| = \sup_i |x_i^{(n+m)} - x_i^{(n)}| \leq \varepsilon$$

whenever $n > N$ and $m > 0$. This implies that

$$|x_i^{(n+m)} - x_i^{(n)}| \leq \varepsilon \text{ for all } i \qquad (*)$$

whenever $n > N$ and $m > 0$. Hence, for each i, $\{x_i^{(j)}\}_{j=1}^{\infty}$ is a Cauchy sequence in $\mathbb{R}$ and has a limit, say x_i^*. Now let

$$\mathbf{x}^* = (x_1^*, x_2^*, x_3^*, \ldots) .$$

By definition of x_k^* we have $||\mathbf{x}^*|| \leq \lim_{k\to\infty} ||\mathbf{x}^{(k)}||$ and so $\mathbf{x}^* \in m$ (i.e., the norm is finite). We show that

$$\mathbf{x}^{(k)} \to \mathbf{x}^* . \qquad (**)$$

Fix $n > N$; by (*) and continuity

$$\lim_{m\to\infty} |x_i^{(n+m)} - x_i^{(n)}| \leq \varepsilon$$

which gives $|x_i^* - x_i^{(n)}| \leq \varepsilon$ for all i. Hence

$$\sup_i |x_i^* - x_i^{(n)}| = ||\mathbf{x}^* - \mathbf{x}^{(n)}|| \leq \varepsilon \text{ for all } n > N$$

and we have established (**). Finally we must show that $\mathbf{x}^* \in c$ by showing that $\{x_i^*\}$ is a Cauchy sequence. Consider the difference

$$|x_n^* - x_m^*| \leq |x_n^* - x_n^{(k)}| + |x_n^{(k)} - x_m^{(k)}| + |x_m^{(k)} - x_m^*|$$

and use an $\varepsilon/3$ argument. Let $\varepsilon > 0$ be given. We can make the first and third terms on the right less than $\varepsilon/3$ for any n, m by fixing k sufficiently large. For this k, $\{x_j^{(k)}\}$ is a Cauchy sequence; hence we can make the second term on the right less than $\varepsilon/3$ by taking n and m sufficiently large. And so there is some M such that $|x_n^* - x_m^*| \leq \varepsilon$ whenever $n, m > M$, which means that $\{x_n^*\}$ is a Cauchy sequence. Therefore $\mathbf{x}^* \in c$.

1.13.2 Let $\mathbf{y} = (\eta_1, \eta_2, \ldots)$ be an accumulation point of S. There is a sequence $\{\mathbf{x}^{(j)}\} \subset S$ such that

$$\left\|\mathbf{y}-\mathbf{x}^{(j)}\right\|^2=\sum_{k=1}^{\infty}|\eta_k-\xi_k^{(j)}|^2\to 0 \text{ as } j\to\infty\,.$$

Hence for each $k = 1, 2, \ldots$, we have $|\eta_k - \xi_k^{(j)}| \to 0$ as $j \to \infty$. An application of the triangle inequality gives

$$|\eta_k|\le|\eta_k-\xi_k^{(j)}|+|\xi_k^{(j)}|\le|\eta_k-\xi_k^{(j)}|+a_k\,.$$

Letting $j \to \infty$, we see that $|\eta_k| \le a_k$ for each $k = 1, 2, \ldots$. Therefore $\mathbf{y} \in S$. Because S contains all of its accumulation points, it is closed.

1.14.1

$$\begin{aligned}\|x+y\|^2+\|x-y\|^2&=(x+y,x+y)+(x-y,x-y)\\&=(x,x+y)+(y,x+y)+(x,x-y)-(y,x-y)\\&=\overline{(x+y,x)}+\overline{(x+y,y)}+\overline{(x-y,x)}-\overline{(x-y,y)}\\&=\overline{(x,x)}+\overline{(y,x)}+\overline{(x,y)}+\overline{(y,y)}+\overline{(x,x)}-\overline{(y,x)}-\overline{(x,y)}+\overline{(y,y)}\\&=2\,(x,x)+2\,(y,y)=2\,\|x\|^2+2\,\|y\|^2\,.\end{aligned}$$

1.14.2

$$\left|\sum_{k=1}^{\infty}x_k\overline{y_k}\right|\le\left(\sum_{k=1}^{\infty}|x_k|^2\right)^{1/2}\left(\sum_{k=1}^{\infty}|y_k|^2\right)^{1/2},$$

$$\left|\int_\Omega f(\mathbf{x})\overline{g(\mathbf{x})}\,d\Omega\right|\le\left(\int_\Omega|f(\mathbf{x})|^2\,d\Omega\right)^{1/2}\left(\int_\Omega|g(\mathbf{x})|^2\,d\Omega\right)^{1/2}.$$

1.14.3 Apply the parallelogram equality to the points $\frac{1}{2}(z-x)$ and $\frac{1}{2}(z-y)$.

1.15.1 Let $\{x_n\}$ be a Cauchy sequence, and suppose its subsequence $\{x_{n_k}\}$ converges to x. Let $\varepsilon > 0$ be given. There exists N such that $d(x_m, x_n) < \varepsilon/2$ for $m, n \ge N$, and K such that $d(x, x_{n_k}) < \varepsilon/2$ for $k \ge K$. Take $M = \max\{N, K\}$. If $m \ge M$ (which also means that $n_m \ge M$), then

$$d(x,x_m)\le d(x,x_{n_m})+d(x_{n_m},x_m)<\varepsilon\,.$$

So $x_n \to x$.

1.16.1 Fix $n \in \mathbb{N}$ and let P_r^n denote the set of all polynomials of degree n having rational coefficients. Denote by $\mathbb{Q}$ the set of all rational numbers. The set P_r^n can be put into one-to-one correspondence with the countable set $\mathbb{Q}^{n+1} = \mathbb{Q}\times\mathbb{Q}\times\cdots\times\mathbb{Q}$ ($n + 1$ times). The set P_r of all polynomials having rational coefficients is given by

$$P_r=\bigcup_{n=0}^{\infty}P_r^n$$

and this is a countable union of countable sets.

1.16.2 In the spaces ℓ^p ($p \geq 1$) and c_0, the countable dense sets are the sets of sequences of rational numbers (in the complex case, the real and imaginary parts are rational) such that some finite number (the number is not fixed) of the terms is not zero and the rest are zero.

1.16.3 Suppose X is a separable metric space, containing a countable, dense subset S. The completion theorem places X into one-to-one correspondence with a set $\tilde{X}$ that is dense in the completion X^*. Let $\tilde{S}$ be the image of S under this correspondence. Since the correspondence is also an isometry, $\tilde{S}$ is dense in $\tilde{X}$. So we have $\tilde{S} \subseteq \tilde{X} \subseteq X^*$, where each set is dense in the next; therefore $\tilde{S}$ is dense in X^*. Since $\tilde{S}$ is evidently countable, the proof is complete.

1.17.1 Yes.

1.18.1 (a) Yes. M bounded in $C(\Omega)$ means there exists R such that for all $f \in M$,

$$\|f\|_{C(\Omega)} = \max_{\mathbf{x}\in\Omega} |f(\mathbf{x})| \leq R .$$

Based on condition (i) we can assert this with $R = c$. (b) Given ε, we can find the required δ_i for each $f_i \in M$ and then take $\delta = \min_i \delta_i$.

1.18.2 We say that $f(x)$ satisfies a Lipschitz condition with constant K on a segment $[a, b]$ if for any $x_1, x_2 \in [a, b]$ we have

$$|f(x_1) - f(x_2)| \leq K\,|x_1 - x_2| .$$

So let $\varepsilon > 0$ be given. Take $\delta = \varepsilon/K$. Then for $|x_1 - x_2| < \delta$ we have $|f(x_1) - f(x_2)| \leq K|x_1 - x_2| < K\delta = \varepsilon$ for any $f \in M$.

1.18.3 We have

$$\begin{aligned} \frac{df(s\mathbf{x} + (1-s)\mathbf{y})}{ds} &= f'(s\mathbf{x} + (1-s)\mathbf{y})\frac{d}{ds}(s\mathbf{x} + (1-s)\mathbf{y}) \\ &= f'(s\mathbf{x} + (1-s)\mathbf{y})(\mathbf{x} - \mathbf{y}) \end{aligned}$$

so

$$\begin{aligned} |f(\mathbf{x}) - f(\mathbf{y})| &= \left| \int_0^1 \frac{df(s\mathbf{x} + (1-s)\mathbf{y})}{ds}\,ds \right| = \left| \int_0^1 f'(s\mathbf{x} + (1-s)\mathbf{y})(\mathbf{x} - \mathbf{y})\,ds \right| \\ &\leq \int_0^1 |f'(s\mathbf{x} + (1-s)\mathbf{y})|\,|\mathbf{x} - \mathbf{y}|\,ds = \left(\int_0^1 |f'(s\mathbf{x} + (1-s)\mathbf{y})|\,ds \right) |\mathbf{x} - \mathbf{y}| . \end{aligned}$$

1.19.1 Consider the conditions for equality to hold in Minkowski's inequality. Alternatively, use the fact that ℓ^2 is an inner product space.

1.21.1 For any element x with $\|x\| \leq 1$, we have

$$\|Ax\| \leq \|A\|\,\|x\| \leq \|A\| .$$

Taking the sup of both sides for all such x, we get

$$\sup_{\|x\|\le 1} \|Ax\| \le \|A\| \; . \qquad (*)$$

Now we prove the reverse inequality. Let $\varepsilon > 0$ be given. According to property 2, there exists x_ε such that

$$\|Ax_\varepsilon\| > (\|A\| - \varepsilon) \|x_\varepsilon\| \; .$$

Setting $y_\varepsilon = x_\varepsilon / \|x_\varepsilon\|$, we get

$$\|Ay_\varepsilon\| = \frac{1}{\|x_\varepsilon\|} \|Ax_\varepsilon\| > \frac{1}{\|x_\varepsilon\|}(\|A\| - \varepsilon) \|x_\varepsilon\| = \|A\| - \varepsilon \; .$$

But $\|y_\varepsilon\| = 1$ and we have

$$\sup_{\|x\|\le 1} \|Ax\| \ge \|Ay_\varepsilon\|$$

(because the sup on the left-hand side is being taken over a set that includes all elements having unit norm) so that

$$\sup_{\|x\|\le 1} \|Ax\| > \|A\| - \varepsilon \; .$$

Since $\varepsilon > 0$ is arbitrary, we have

$$\sup_{\|x\|\le 1} \|Ax\| \ge \|A\| \; . \qquad (**)$$

Inequalities (*) and (**) show that $\|A\|$ is given by the indicated expression.

1.21.2 The differentiation operator d/dx is bounded from $C^{(1)}(-\infty, \infty)$ to $C(-\infty, \infty)$ because the inequality

$$\left\| \frac{df}{dx} \right\|_{C(-\infty,\infty)} \le \alpha \|f\|_{C^{(1)}(-\infty,\infty)} \; ,$$

i.e.,

$$\sup \left| \frac{df(x)}{dx} \right| \le \alpha \left(\sup |f(x)| + \sup \left| \frac{df(x)}{dx} \right| \right)$$

obviously holds with $\alpha = 1$.

Note, however, that d/dx is not bounded when acting in the space $C(-\infty, \infty)$.

Continuity of A from $C^{(m)}(\overline{\Omega})$ to $C(\overline{\Omega})$ is seen as follows:

$$\begin{aligned}
\|A(u)(\mathbf{x})\|_C &= \max_{\mathbf{x}\in\Omega}\left|\sum_{|\alpha|\le m} a_\alpha D^\alpha u(\mathbf{x})\right| \le \max_{\mathbf{x}\in\Omega}\sum_{|\alpha|\le m}|a_\alpha|\,|D^\alpha u(\mathbf{x})| \\
&\le \max_{|\alpha|\le m}|a_\alpha|\cdot\max_{\mathbf{x}\in\Omega}\sum_{|\alpha|\le m}|D^\alpha u(\mathbf{x})| \\
&\le \max_{|\alpha|\le m}|a_\alpha|\cdot\sum_{|\alpha|\le m}\max_{\mathbf{x}\in\Omega}|D^\alpha u(\mathbf{x})|\ ,
\end{aligned}$$

where the last factor is $\|u(\mathbf{x})\|_{C^{(m)}}$. Continuity of A from $W^{m,2}(\Omega)$ to $L^2(\Omega)$ is seen as follows:

$$\begin{aligned}
\|Au(\mathbf{x})\|_{L^2(\Omega)} &= \left[\int_\Omega\left|\sum_{|\alpha|\le m} a_\alpha D^\alpha u(\mathbf{x})\right|^2 d\Omega\right]^{1/2} \\
&\le \left[\int_\Omega\Big(\sum_{|\alpha|\le m}|a_\alpha|^2\Big)\Big(\sum_{|\alpha|\le m}|D^\alpha u(\mathbf{x})|^2\Big)d\Omega\right]^{1/2} \\
&= \Big(\sum_{|\alpha|\le m}|a_\alpha|^2\Big)^{1/2}\left[\int_\Omega\sum_{|\alpha|\le m}|D^\alpha u(\mathbf{x})|^2\,d\Omega\right]^{1/2} .
\end{aligned}$$

where the last two factors are a constant and $\|u(\mathbf{x})\|_{W^{m,2}(\Omega)}$, respectively. Continuity of B from $C(0,1)$ to $C(0,1)$ is seen as follows:

$$\begin{aligned}
\|B(u)(x)\|_{C(0,1)} &= \max_{x\in[0,1]}\left|\int_0^x u(s)\,ds\right| \le \max_{x\in[0,1]}\int_0^x |u(s)|\,ds \\
&\le \max_{s\in[0,1]}|u(s)|\cdot x \le \|u(x)\|_{C(0,1)}\ .
\end{aligned}$$

Continuity of B from $C(0,1)$ to $C^{(1)}(0,1)$ is seen as follows:

$$\begin{aligned}
\|B(u)(x)\|_{C^{(1)}(0,1)} &= \max_{x\in[0,1]}\left|\int_0^x u(s)\,ds\right| + \max_{x\in[0,1]}\left|\frac{d}{dx}\int_0^x u(s)\,ds\right| \\
&\le \max_{x\in[0,1]}\int_0^x |u(s)|\,ds + \max_{x\in[0,1]}|u(x)| \\
&\le 2\max_{x\in[0,1]}|u(x)| = 2\,\|u(x)\|_{C(0,1)}\ .
\end{aligned}$$

1.21.4 Use the notion of sequential continuity presented in Problem 1.21.3. Suppose $x_n \to x$ and $y_n \to y$. Write

$$\begin{aligned}
|(x_n, y_n) - (x, y)| &= |(x_n, y_n) - (x_n, y) + (x_n, y) - (x, y)| \\
&\le |(x_n, y_n) - (x_n, y)| + |(x_n, y) - (x, y)| \\
&\le \|x_n\|\,\|y_n - y\| + \|x_n - x\|\,\|y\|
\end{aligned}$$

Note that the convergent sequence $\{x_n\}$ is bounded, so there exists m such that $\|x_n\| \le m$ for all n.

1.21.5 Let $y \in Y$ and $\varepsilon > 0$ be given. Then $y = f(x)$ for some $x \in X$, and by continuity there exists $\delta > 0$ such that $f(B(x, \delta)) \subseteq B(y, \varepsilon)$. Since A is dense in X, there exists $x_* \in A$ such that $x_* \in B(x, \delta)$. Therefore $f(x_*) \in B(y, \varepsilon)$. So there is a point of $f(A)$ within distance ε of y.

1.21.6 Suppose $\{x_n\} \subset S$ is convergent to x_0. Since $x_n \in S$, we have $f(x_n) = g(x_n)$ for each n. Taking the limit of both sides of this equality as $n \to \infty$ and using the continuity of f and g, we obtain $f(x_0) = g(x_0)$. Therefore $x_0 \in S$. Because S contains the limits of all its convergent sequences, it is closed.

1.21.7 (a) Let S be a convex subset of $D(A)$. Take two points $y_1, y_2 \in A(S)$ and consider a point $y = ty_1 + (1-t)y_2$ where $0 \le t \le 1$. Since $y_1 = Ax_1$ and $y_2 = Ax_2$ for some elements $x_1, x_2 \in S$, we have

$$y = tAx_1 + (1-t)Ax_2 = A(tx_1 + (1-t)x_2) = Ax$$

for some $x \in S$. Therefore $y \in A(S)$. (b) Suppose $S \subset R(A)$ is a convex set. Take $x_1, x_2 \in A^{-1}(S)$; then there exist $y_1, y_2 \in S$ such that $y_1 = Ax_1$ and $y_2 = Ax_2$. If $z = tx_1 + (1-t)x_2$ for some $0 \le t \le 1$, then

$$Az = tAx_1 + (1-t)Ax_2 = ty_1 + (1-t)y_2 \in S$$

because S is convex. Since $Az \in S$, we have $z \in A^{-1}(S)$. This shows that $A^{-1}(S)$ is convex.

1.21.8 By linearity of A, the sum

$$S = c_1Ax_1 + \cdots + c_nAx_n = A(c_1x_1 + \cdots + c_nx_n)$$

is zero if and only if

$$c_1x_1 + \cdots + c_nx_n = 0\ .$$

If the x_i are linearly dependent, this last equation holds for some set of c_i that are not all zero. Therefore the sum S vanishes for some set of c_i that are not all zero.

1.21.9 (a) It is clear that the operator is linear. From the general relationship

$$\|Af\| \le \|A\|\,\|f\|$$

and the fact that

$$\begin{aligned}\|Af\| &= \max_{x\in[0,1]} \left| \int_0^x f(\xi)\,d\xi \right| \le \max_{x\in[0,1]} \int_0^x |f(\xi)|\,d\xi \\ &\le \max_{x\in[0,1]} |f(\xi)| \int_0^x d\xi \le \max_{x\in[0,1]} |f(\xi)| \int_0^1 d\xi = \|f\|\ ,\end{aligned}$$

we conclude that $\|A\| \le 1$. However, the ratio $\|Af\| / \|f\|$ attains the value 1 when f = constant, so $\|A\| = 1$. (b) Linearity of the operator is evident. The inequality

$$\|Af\| = \max_{x\in[0,1]} |f(x)| \le \max_{x\in[-1,1]} |f(x)| = \|f\|$$

shows that $\|A\| \le 1$. Since the ratio $\|Af\| / \|f\|$ attains the value 1 for $f(x) \equiv$ constant, we have $\|A\| = 1$. (c) Linearity of A is evident. We have

$$\|Af\| = \left\{\int_0^1 \left[x \int_0^1 f(x)\,dx\right]^2 dx\right\}^{1/2} = \left|\int_0^1 f(x)\,dx\right| \cdot \left\{\int_0^1 x^2\,dx\right\}^{1/2}$$
$$= \frac{1}{\sqrt{3}} \left|\int_0^1 f(x)\,dx\right| .$$

By the Schwarz inequality for integrals,

$$\|Af\| \le \frac{1}{\sqrt{3}}\left(\int_0^1 1^2\,dx\right)^{1/2}\left(\int_0^1 f^2(x)\,dx\right)^{1/2} = \frac{1}{\sqrt{3}}\|f\|$$

and equality holds if $f(x) \equiv$ constant on $[0, 1]$. Therefore $\|A\| = 1/\sqrt{3}$. (d) A is clearly linear. The inequality

$$\|Af\| = \max_{x\in[a,b]} |f(x)| \le \max_{x\in[a,b]} |f(x)| + \max_{x\in[a,b]} |f'(x)| = \|f\|$$

shows that $\|A\| \le 1$. Since the ratio $\|Af\| / \|f\|$ attains the value 1 for $f(x) \equiv$ constant on $[a, b]$, we have $\|A\| = 1$. (e) The inequality

$$\|Af\| = \max_{x\in[a,b]} |f'(x)| \le \max_{x\in[a,b]} |f(x)| + \max_{x\in[a,b]} |f'(x)| = \|f\|$$

shows that $\|A\| \le 1$. Consider $f_n(x) = \sin nx$. For sufficiently large n we have

$$\|Af_n\| = \max_{x\in[a,b]} |n \cos nx| = n\,, \qquad \|f_n\| = n + 1\,,$$

so

$$\|A\| \ge \|Af_n\| / \|f_n\| = n/(n + 1) \to 1 \ \text{ as } n \to \infty\,.$$

Together with the fact that $\|A\| \le 1$, this gives $\|A\| = 1$. (f) The inequality

$$\|Af\| = \left\{\int_0^1 \left|\int_0^x 1 \cdot f(\xi)\,d\xi\right|^2 dx\right\}^{1/2} \le \left\{\int_0^1 \left[\int_0^x 1^2\,d\xi \int_0^x |f(\xi)|^2\,d\xi\right] dx\right\}^{1/2}$$
$$\le \left\{\int_0^1 \left[x \int_0^1 |f(\xi)|^2\,d\xi\right] dx\right\}^{1/2} = \frac{1}{\sqrt{2}}\left\{\int_0^1 |f(\xi)|^2\,d\xi\right\}^{1/2} = \frac{1}{\sqrt{2}}\|f\|$$

shows that $\|A\| \le 1/\sqrt{2}$. (g) Linearity is evident. The inequality

$$\|Af\| = \max_{x\in[-1,1]} \left|\tfrac{1}{2}[f(x)+f(-x)]\right| \le \tfrac{1}{2}\left\{\max_{x\in[-1,1]}|f(x)| + \max_{x\in[-1,1]}|f(-x)|\right\}$$
$$= \max_{x\in[-1,1]}|f(x)| = \|f\|$$

shows that $\|A\| \le 1$. Equality is attained when $f =$ constant, so $\|A\| = 1$. (h) Linearity is evident. The inequality

$$\|\phi(f)\| = \left|\int_{-1}^{1} xf(x)\,dx\right| \le \left(\int_{-1}^{1} x^2\,dx\right)^{1/2}\left(\int_{-1}^{1}|f(x)|^2\,dx\right)^{1/2} = \sqrt{\tfrac{2}{3}}\,\|f\|_{L^2(-1,1)}$$

shows that ϕ is bounded with $\|\phi\| \le \sqrt{2/3}$. Since equality is attained for $f(x) = x$, we have $\|\phi\| = \sqrt{2/3}$. (i) Linearity is obvious. The inequality

$$\|\phi(f)\| = \left|\int_{-1}^{1} f(x)\,dx - f(0)\right| \le \int_{-1}^{1}|f(x)|\,dx + |f(0)|$$
$$\le \left(\int_{-1}^{1} dx + 1\right)\cdot \max_{x\in[-1,1]}|f(x)| = 3\,\|f\|_{C(-1,1)}$$

shows that ϕ is bounded with $\|\phi\| \le 3$. For each $n = 1, 2, \ldots$, the function

$$f_n(x) = \begin{cases} 2n|x| - 1\,, & |x| \le 1/n\,, \\ 1\,, & 1/n < |x| \le 1\,, \end{cases}$$

belongs to $C(-1,1)$. We have

$$\|f_n\|_{C(-1,1)} = \max_{x\in[-1,1]}|f_n(x)| = 1$$

and

$$\|\phi(f)\| = |2(1-1/n) - (-1)| = 3 - 2/n$$

so that

$$\frac{\|\phi(f_n)\|}{\|f_n\|_{C(-1,1)}} \le 3 - 2/n\,.$$

This shows that $\|\phi\| = 3$.

1.22.1 Take elements $u, v \in X$ and scalars α, β. Then

$$A(\alpha u + \beta v) = \lim_{n\to\infty} A_n(\alpha u + \beta v) = \lim_{n\to\infty}(\alpha A_n u + \beta A_n v)$$
$$= \alpha \lim_{n\to\infty} A_n u + \beta \lim_{n\to\infty} A_n v = \alpha A u + \beta A v\,.$$

Next we show that $A_n \to A$ in the operator norm. We have

$$\|(A - A_n)x\| = \left\|\left(\lim_{m\to\infty} A_m - A_n\right)x\right\| = \lim_{m\to\infty}\|(A_m - A_n)x\| \le \lim_{m\to\infty}\|A_m - A_n\|\,\|x\|\,,$$

so

$$\|A - A_n\| \le \lim_{m\to\infty} \|A_m - A_n\| .$$

Therefore

$$\lim_{n\to\infty} \|A - A_n\| \le \lim_{n\to\infty} \lim_{m\to\infty} \|A_m - A_n\| = 0 .$$

1.22.2 Since Y is a Banach space, so is $L(X, Y)$. But in any Banach space U, absolute convergence implies convergence. Indeed we have, for all n and $p \ge 1$,

$$\left\| \sum_{k=1}^{n+p} u_k - \sum_{k=1}^{n} u_k \right\| = \left\| \sum_{k=n+1}^{n+p} u_k \right\| \le \sum_{k=n+1}^{n+p} \|u_k\| = \sum_{k=1}^{n+p} \|u_k\| - \sum_{k=1}^{n} \|u_k\| ,$$

and therefore

$$\left\| \sum_{k=1}^{n+p} u_k - \sum_{k=1}^{n} u_k \right\| \le \left| \sum_{k=1}^{n+p} \|u_k\| - \sum_{k=1}^{n} \|u_k\| \right| .$$

By hypothesis the sequence $\sum_{k=1}^{n} \|u_k\|$ converges as $n \to \infty$ and is therefore a Cauchy sequence. By the inequality above, $\sum_{k=1}^{n} u_k$ is a Cauchy sequence and converges to an element of U by completeness.

1.22.3 We have

$$\begin{aligned} \|A_n B_n - AB\| &= \|A_n B_n - A_n B + A_n B - AB\| = \|A_n(B_n - B) + (A_n - A)B\| \\ &\le \|A_n(B_n - B)\| + \|(A_n - A)B\| \le \|A_n\| \cdot \|B_n - B\| + \|A_n - A\| \cdot \|B\| \end{aligned}$$

where $\|A_n\|$ is bounded since A_n is convergent.

1.22.4 Use orthogonality, orthonormality, and Bessel's inequality to write

$$\|P_n x\|^2 = \left\| \sum_{k=1}^{n} (x, g_k) g_k \right\|^2 = \sum_{k=1}^{n} \|(x, g_k) g_k\|^2 = \sum_{k=1}^{n} |(x, g_k)|^2 \le \|x\|^2.$$

Hence $\|P_n x\| \le \|x\|$ so that $\|P_n\| \le 1$. Since $\|P_n g_1\| = \|g_1\|$, it follows that $\|P_n\| = 1$.

1.23.1 First we recall that a is the limit inferior of a numerical sequence $\{a_n\}$, and we write

$$a = \liminf_{n\to\infty} a_n ,$$

if (1) for any $\varepsilon > 0$ only a finite number of the terms of $\{a_n\}$ are less than $a - \varepsilon$ and (2) for any integer M there is $a_k < a + \varepsilon$ with $k > M$. Thus if $\{a_n\}$ is bounded from below, then there exists $a = \liminf_{n\to\infty} a_n$. In this case, the definition implies that there is a subsequence $\{a_{n_k}\}$ of $\{a_n\}$ such that $a = \lim_{n_k\to\infty} a_{n_k}$. Note that if $\{a_n\}$ has a limit, then $\liminf_{n\to\infty} a_n = \lim_{n\to\infty} a_n$.

Now we prove (1.23.3). The numerical sequence $\{\|A_n\|\}$ is bounded from below by zero, and so there is a finite $\liminf_{n\to\infty} \|A_n\|$. By the above, from $\{A_n\}$ we can select a subsequence $\{A_{n_k}\}$ such that the sequence of its norms has a limit and

$$\liminf_{n\to\infty} \|A_n\| = \lim_{n_k\to\infty} \|A_{n_k}\| .$$

As for any x there exists $\lim_{n\to\infty} \|A_n x\|$, we see that

$$\lim_{n\to\infty} \|A_n x\| = \lim_{n_k\to\infty} \|A_{n_k} x\| .$$

Therefore

$$\|Ax\| = \lim_{n\to\infty} \|A_n x\| = \lim_{n_k\to\infty} \|A_{n_k} x\| \le \lim_{n_k\to\infty} \|A_{n_k}\| \, \|x\| = \liminf_{n\to\infty} \|A_n\| \, \|x\| .$$

1.24.1 Let $\{x_n'\}$ be another sequence in $D(A)$ such that $\|x_n' - x_0\| \to 0$. Then

$$\left\| \lim_{n\to\infty} Ax_n' - \lim_{n\to\infty} Ax_n \right\| = \lim_{n\to\infty} \|A(x_n' - x_n)\| \le \|A\| \lim_{n\to\infty} \|x_n' - x_n\|$$
$$\le \|A\| \lim_{n\to\infty} (\|x_n' - x_0\| + \|x_0 - x_n\|) = 0 .$$

1.25.1 Suppose A is a closed operator. Let (x, y) be a limit point of $G(A)$. Then there is a sequence $\{(x_n, Ax_n)\} \subset G(A)$ that converges to (x, y) in the norm of $X \times Y$. Evidently this implies that as $n \to \infty$ we have $x_n \to x$ in X and $Ax_n \to y$ in Y. Because A is closed, $x \in D(A)$ and $y = Ax$. Hence $(x, Ax) \in G(A)$ by definition of $G(A)$.

Conversely, suppose $G(A)$ is closed in $X \times Y$. Let $\{x_n\} \subset D(A)$ be such that, as $n \to \infty$, $x_n \to x$ in X and $Ax_n \to y$ in Y. The sequence $\{(x_n, Ax_n)\} \subset G(A)$ converges in the norm of $X \times Y$ to (x, y). Since $G(A)$ is closed, $(x, y) \in G(A)$. By definition of $G(A)$ this means that $x \in D(A)$ and $y = Ax$.

1.26.1 Suppose A is linear and A^{-1} exists. By definition, there is no more than one solution x to the equation $Ax = 0$. On the other hand, we see that $A0 = 0$ by putting $\alpha = 0$ in the equation $A(\alpha x) = \alpha Ax$. Conversely, suppose $x = 0$ is the only solution of $Ax = 0$. We want to show that there is no more than one solution to $Ax = y$ for given y. Suppose there were two solutions x_1, x_2: $Ax_1 = y$ and $Ax_2 = y$. Then $Ax_1 = Ax_2$, so $A(x_1 - x_2) = 0$, and we obtain $x_1 - x_2 = 0$ as desired.

Again suppose A is linear and A^{-1} exists. Let $y_1, y_2 \in R(A)$. Then

$$\alpha_1 y_1 + \alpha_2 y_2 = \alpha_1 Ax_1 + \alpha_2 Ax_2 = A(\alpha_1 x_1 + \alpha_2 x_2)$$

so that

$$A^{-1}(\alpha_1 y_1 + \alpha_2 y_2) = \alpha_1 x_1 + \alpha_2 x_2 = \alpha_1 A^{-1} y_1 + \alpha_2 A^{-1} y_2 .$$

This shows that A^{-1} is linear.

1.26.2 Let A map an element x from the space $(X, \|\cdot\|_2)$ into the same element x regarded as an element of the space $(X, \|\cdot\|_1)$. This operator is linear and, by hypothesis ($\|x\|_1 \le c_1 \|x\|_2$) it is bounded (continuous), hence it is closed. It is also one-to-one and onto. By Theorem 1.26.4, A^{-1} is continuous on $(X, \|\cdot\|_1)$; this gives the inequality $\|x\|_2 \le c_2 \|x\|_1$, as desired.

1.27.1 The representation $a(u,v) = (Au, v)$ yields

$$a(u_1, v) = (A(u_1), v) , \qquad a(u_2, v) = (A(u_2), v) , \tag{*}$$

for any $u_1, u_2, v \in H$ and

$$a(\alpha_1 u_1 + \alpha_2 u_2, v) = (A(\alpha_1 u_1 + \alpha_2 u_2), v) \tag{**}$$

for any scalars α_1, α_2. But by (*) and linearity of $a(u,v)$ in u,

$$\begin{aligned} a(\alpha_1 u_1 + \alpha_2 u_2, v) &= \alpha_1 a(u_1, v) + \alpha_2 a(u_2, v) = \alpha_1(A(u_1), v) + \alpha_2(A(u_2, v)) \\ &= (\alpha_1 A(u_1) + \alpha_2 A(u_2), v) . \end{aligned}$$

By (**) we get

$$(A(\alpha_1 u_1 + \alpha_2 u_2), v) = (\alpha_1 A(u_1) + \alpha_2 A(u_2), v) .$$

By arbitrariness of v we get

$$A(\alpha_1 u_1 + \alpha_2 u_2) = \alpha_1 A(u_1) + \alpha_2 A(u_2)$$

which means linearity of A.

1.32.1 Let $m \to \infty$ in the inequality $d(x_n, x_{n+m}) \le q^n d(x_0, x_m)$. The result is less useful than (1.32.4) because the right member involves the unknown quantity x_*.

1.32.2 We show that A^N is a contraction for some N if A acts in $C[0,T]$ and is given by

$$Ay(t) = \int_0^t g(t-\tau) y(\tau)\, d\tau .$$

Let M be the maximum value attained by $g(t)$ on $[0,T]$. For any $t \in [0,T]$ we have

$$\begin{aligned} |Ay_2(t) - Ay_1(t)| &\le \int_0^t |g(t-\tau)|\, |y_2(\tau) - y_1(\tau)|\, d\tau \\ &\le \max_{\tau\in[0,t]} |g(t-\tau)| \max_{\tau\in[0,t]} |y_2(\tau) - y_1(\tau)| \int_0^t d\tau \\ &\le \max_{\tau\in[0,T]} |g(t-\tau)| \max_{\tau\in[0,T]} |y_2(\tau) - y_1(\tau)|\, t = M\, t\, d(y_2, y_1) . \end{aligned}$$

Then

$$\begin{aligned} |A^2 y_2(t) - A^2 y_1(t)| &\le \int_0^t |g(t-\tau)|\, |Ay_2(\tau) - Ay_1(\tau)|\, d\tau \\ &\le M \cdot M\, d(y_2, y_1) \int_0^t \tau\, d\tau = M^2 \frac{t^2}{1\cdot 2} d(y_2, y_1) . \end{aligned}$$

Continuing in this manner we can show that

$$|A^k y_2(t) - A^k y_1(t)| \le M^k \frac{t^k}{k!} d(y_2, y_1)$$

for any positive integer k and any $t \in [0, T]$. Thus

$$\begin{aligned} d(A^k y_2 \,,\, A^k y_1) &= \max_{t\in[0,T]} |A^k y_2(t) - A^k y_1(t)| \\ &\le \max_{t\in[0,T]} M^k \frac{t^k}{k!} d(y_2, y_1) \\ &= \frac{(MT)^k}{k!} d(y_2, y_1)\,. \end{aligned}$$

Finally, we choose N so large that $(MT)^N/N! < 1$.

2.1.1 (a) Let $\{f_n\}$ be a representative sequence from $f \in W^{1,2}(a,b)$. Then $f_n \in C^{(1)}(a,b)$. Because $g \in C^{(1)}(a,b)$, we have

$$\max_{x\in[a,b]} |g(x)| + \max_{x\in[a,b]} |g'(x)| < \infty\,.$$

Hence each of the terms on the left-hand side is finite. Let m be the largest of these two terms. Then

$$\begin{aligned} \|f_n g\|_{W^{1,2}(a,b)} &= \left(\int_a^b [f_n(x)g(x)]^2\,dx + \int_a^b [f_n'(x)g'(x)]^2\,dx \right)^{1/2} \\ &\le m \left(\int_a^b [f_n(x)]^2\,dx + \int_a^b [f_n'(x)]^2\,dx \right)^{1/2} \\ &= m\,\|f_n\|_{W^{1,2}(a,b)}\,. \end{aligned}$$

Taking the limit as $n \to \infty$, we get

$$\|fg\|_{W^{1,2}(a,b)} \le m\,\|f\|_{W^{1,2}(a,b)}$$

which means that $fg \in W^{1,2}(a,b)$ because $f \in W^{1,2}(a,b)$. Note that this can be interpreted as the statement that the operator of multiplication by g, which is a linear operator in $W^{1,2}(a,b)$, is continuous if $g \in C^{(1)}(a,b)$. (b) The inequality

$$\|Af\| = \left(\int_0^1 f^2(x)\,dx \right)^{1/2} \le \left(\int_0^1 f^2(x)\,dx + \int_0^1 [f'(x)]^2\,dx \right)^{1/2} = \|f\|$$

shows that $\|A\| \le 1$. The ratio $\|Af\| \,/\, \|f\| = 1$ for $f =$ constant, so $\|A\| = 1$.

2.3.1 Let $\{\tilde{y}_n(x)\}$ be another representative of $y(x) \in E_B$ and suppose its limit is $\tilde{z}(x)$. Then for each $x \in [0, l]$ we have

$$\begin{aligned}|z(x)-\tilde{z}(x)| &= \left|\lim_{n\to\infty} y_n'(x) - \lim_{n\to\infty} \tilde{y}_n'(x)\right| = \lim_{n\to\infty} \left|y_n'(x) - \tilde{y}_n'(x)\right| \\ &= \lim_{n\to\infty} \left|\int_0^x y_n''(x)\,dx - \int_0^x \tilde{y}_n''(x)\,dx\right| \le \lim_{n\to\infty} \int_0^x |y_n''(x) - \tilde{y}_n''(x)|\,dx \\ &\le \lim_{n\to\infty} \int_0^l 1\cdot |y_n''(x) - \tilde{y}_n''(x)|\,dx \le l^{1/2} \lim_{n\to\infty} \left(\int_0^l [y_n''(x) - \tilde{y}_n''(x)]^2\,dx\right)^{1/2} \\ &\le c_1 \lim_{n\to\infty} \left(\frac{1}{2}\int_0^l B(x)[y_n''(x) - \tilde{y}_n''(x)]^2\,dx\right)^{1/2} \to 0 \text{ as } n\to\infty\,.\end{aligned}$$

2.3.2

$$\iint_\Omega F(x,y)\,dx\,dy + \sum_k F_k(x_k,y_k) + \int_\gamma F_l(x,y)\,ds = 0\,,$$

$$\iint_\Omega xF(x,y)\,dx\,dy + \sum_k x_k F_k(x_k,y_k) + \int_\gamma xF_l(x,y)\,ds = 0\,,$$

$$\iint_\Omega yF(x,y)\,dx\,dy + \sum_k y_k F_k(x_k,y_k) + \int_\gamma yF_l(x,y)\,ds = 0\,.$$

2.5.1 Assume $f(x,y) \in L^p(\Omega)$ for some $p > 1$, where $\Omega \subset \mathbb{R}^2$ is compact. The equilibrium problem for a membrane with clamped edge (i.e., the Dirichlet problem with homogeneous boundary condition) has a unique generalized solution in the sense of Definition 2.4.1.

2.7.1 Yes.

2.8.1 Countability is evident. To show denseness let $x \in X$ and $\varepsilon > 0$ be given. Write $x = \sum_{k=1}^\infty \alpha_k e_k$. Let $e = \sum_{k=1}^\infty r_k e_k$ where r_k is a rational number such that

$$|\alpha_k - r_k| < \frac{\varepsilon}{2^k\,\|e_k\|}\,.$$

Then

$$\|x - e\| = \left\|\sum_{k=1}^\infty (\alpha_k - r_k)e_k\right\| \le \sum_{k=1}^\infty |\alpha_k - r_k|\,\|e_k\| < \varepsilon \sum_{k=1}^\infty \frac{1}{2^k} = \varepsilon$$

as desired.

2.8.2 Any system that is complete in $C(\Omega)$ or $C^{(k)}(\Omega)$.

2.8.3 The set of finite linear combinations with rational coefficients is dense in the set of all linear combinations, and thus in the space.

2.8.4 Let x_k $(k = 1,\dots,n)$ be orthonormal. Then

$$\sum_{k=1}^n \alpha_k x_k = 0 \implies \left(\sum_{k=1}^n \alpha_k x_k, x_j\right) = \sum_{k=1}^n \alpha_k (x_k, x_j) = \alpha_j = 0 \qquad (j = 1,\dots,n)\,.$$

2.8.5 If $\{g_i\}$ is a complete orthonormal system in H, then any $f \in H$ can be written as

$$f = \sum_{k=1}^{\infty} (f, g_k) g_k$$

by Theorem 2.8.1. Since the assumption $(f, g_k) = 0$ for $k = 1, 2, 3, \ldots$ does entail $f = 0$, the system $\{g_i\}$ is closed by Definition 2.8.4

2.8.6 Let $\{g_i\}$ be a complete system. For each k we perform Gram–Schmidt on the finite set $\{g_1, \ldots, g_k\}$, discarding linearly dependent elements as needed; this gives a finite orthonormal set $\{e_1, \ldots, e_m\}$, $m \le k$, having the same span. It is clear that $\{e_j\}_{j=1}^{\infty}$ is a complete system (since, again, the finite spans are the same), hence it is closed in H by Theorem 2.8.2. Now suppose $f \in H$ is such that $(f, g_i) = 0$ for all i. Then $(f, e_j) = 0$ for all j because each e_j is a finite linear combination of the g_i. By closedness of $\{e_j\}_{j=1}^{\infty}$ we have $f = 0$. Hence $\{g_i\}_{i=1}^{\infty}$ is closed.

Conversely, let $\{g_i\}$ be a closed system. As above we perform Gram–Schmidt on the sets $\{g_1, \ldots, g_k\}$ to produce orthonormal sets $\{e_1, \ldots, e_m\}$. It is clear that $\{e_j\}_{j=1}^{\infty}$ is a closed system (since the e_j are finite linear combinations of the g_i), hence it is complete in H by Theorem 2.8.2. Completeness of $\{e_j\}_{j=1}^{\infty}$ in turn implies completeness of $\{g_i\}_{i=1}^{\infty}$ (since the spans of the finite sets of g_i are the same as those of the corresponding finite sets of e_j).

2.9.2 Suppose $\{x_n\}$ converges weakly to x_0 and $\|x_n\| \le M$. By the Schwarz inequality we have

$$|(x_n, x_0)| \le \|x_n\| \, \|x_0\| \le M \, \|x_0\| \, .$$

As $n \to \infty$ this inequality becomes $\|x_0\|^2 \le M \, \|x_0\|$ or $\|x_0\| \le M$ as desired.

2.9.3 Let S be weakly closed and $\{x_n\} \subset S$. If $x_n \to x_0$, then $x_n \rightharpoonup x_0$. The latter means that $x_0 \in S$, as S is weakly closed. So S contains the limits of all its convergent sequences and is therefore closed.

2.9.4 From (2.9.9) it follows there is $\varepsilon > 0$ and an increasing sequence $\{n_k\}$ such that $|(x_{n_k}, f) - (x^*, f)| > \varepsilon$. But $\{(x_{n_k}, f)\}$ is a bounded numerical sequence so it contains a Cauchy subsequence $\{(x_{n_{k_r}}, f)\}$ for which

$$\left| \lim_{r \to \infty} (x_{n_{k_r}}, f) - (x^*, f) \right| \ge \varepsilon \, .$$

2.12.1 Start with the integral representation

$$f(x) = f(y) + \int_y^x f'(s) \, ds$$

and integrate over y from 0 to 1:

$$\int_0^1 f(x) \, dy = \int_0^1 f(y) \, dy + \int_0^1 \int_y^x f'(s) \, ds \, dy \, .$$

We have

$$|f(x)| \le \int_0^1 |f(y)|\,dy + \int_0^1 \int_0^1 |f'(s)|\,ds\,dy = \int_0^1 |f(y)|\,dy + \int_0^1 |f'(s)|\,ds$$

$$= \int_0^1 1 \cdot (|f(y)| + |f'(y)|)\,dy$$

so that by the Schwarz inequality

$$|f(x)| \le \left[\int_0^1 1^2\,dy\right]^{1/2} \left[\int_0^1 \left(|f(y)|^2 + 2|f(y)f'(y)| + [f'(y)]^2\right) dy\right]^{1/2}$$

$$\le \left[\int_0^1 \left(|f(y)|^2 + \left\{f^2(y) + [f'(y)]^2\right\} + [f'(y)]^2\right) dy\right]^{1/2}$$

$$= \sqrt{2}\left[\int_0^1 \left(|f(y)|^2 + [f'(y)]^2\right) dy\right]^{1/2} = \sqrt{2}\,\|f\|_{W^{1,2}(0,1)} \le \sqrt{2}\,\|f\|_{W^{2,2}(0,1)}\ .$$

2.12.2 Take into account that $l(P_2(x_1, x_2)) = 0$ when $P_2(x_1, x_2)$ is a polynomial of second order.

3.2.1 Suppose A is continuous and $x_n \rightharpoonup x_0$ in X. We want to show that $\{Ax_n\}$ is weakly convergent. So we take any continuous linear functional F given on Y and apply it to $\{Ax_n\}$. But this amounts to applying FA, which is clearly a continuous linear functional on X, to $\{x_n\}$. We have $FAx_n \to FAx_0$ by definition of weak convergence of $\{x_n\}$.

3.2.3 We are considering a functional of the form

$$F(\varphi) = \int_\Omega \rho(\mathbf{x})\, u(\mathbf{x})\, \varphi(\mathbf{x})\, d\Omega \qquad (*)$$

for fixed ρ and u. It is obviously linear in φ. Let us show continuity. By the Schwarz inequality,

$$|F(\varphi)| \le \sup_{\mathbf{x}\in\Omega} |\rho(\mathbf{x})| \cdot \|u\|_{L^2(\Omega)} \cdot \|\varphi\|_{L^2(\Omega)}\ .$$

Each energy space E considered (membrane, plate, elastic body) imbeds continuously into $L^2(\Omega)$ so that

$$|F(\varphi)| \le m\,\|u\|_E\,\|\varphi\|_E\ . \qquad (**)$$

Because $F(\varphi)$ is a bounded linear functional on E with respect to $\varphi \in E$ for fixed $u \in E$, the Riesz representation theorem yields

$$F(\varphi) = (\varphi, h_u)_E = (h_u, \varphi)_E$$

where h_u is the Riesz representer of $F(\varphi)$. Denoting the element h_u by $K(u)$, we have the operator K of the problem statement along with the fact that

$$F(\varphi) = (K(u), \varphi)_E \, . \qquad (***)$$

To show that K is linear, take a scalar constant α and write

$$\begin{aligned}(K(\alpha u_1 + u_2), \varphi)_E &= \int_\Omega \rho(\mathbf{x})[\alpha u_1(\mathbf{x}) + u_2(\mathbf{x})]\varphi(\mathbf{x})\, d\Omega \\ &= \alpha \int_\Omega \rho(\mathbf{x})\, u_1(\mathbf{x})\, \varphi(\mathbf{x})\, d\Omega + \int_\Omega \rho(\mathbf{x})\, u_2(\mathbf{x})\, \varphi(\mathbf{x})\, d\Omega \\ &= \alpha(K(u_1), \varphi)_E + (K(u_2), \varphi)_E = (\alpha K(u_1) + K(u_2), \varphi)_E \, .\end{aligned}$$

Since $\varphi \in E$ is arbitrary, the linearity property follows. To show that K is continuous, substitute the representation (***) into (**) and put $\varphi = K(u)$:

$$|(K(u), K(u))_E| = \|K(u)\|_E^2 \le m\, \|u\|_E\, \|K(u)\|_E \, ,$$

hence $\|K(u)\|_E \le m\,\|u\|_E$ as required. Finally, since the right-hand side of (*) is symmetric in u and φ, we have

$$(K(u), \varphi)_E = \int_\Omega \rho(\mathbf{x})u(\mathbf{x})\varphi(\mathbf{x})\, d\Omega = \int_\Omega \rho(\mathbf{x})\varphi(\mathbf{x})u(\mathbf{x})\, d\Omega = (K(\varphi), u)_E = (u, K(\varphi))_E$$

for all $u, \varphi \in E$. This shows that K is self-adjoint.

3.3.1 Suppose A maps the closed unit ball of X to a precompact set in Y. Let $\{x_n\} \subset X$ be a bounded sequence with $\|x_n\| \le r$ for all n and some $r > 0$. Then the sequence $\{r^{-1}x_n\}$ belongs to the closed unit ball of X, and by hypothesis its image $\{A(r^{-1}x_n)\}$ has a Cauchy subsequence $\{A(r^{-1}x_{n_k})\}$. But

$$Ax_{n_k} = rA(r^{-1}x_{n_k}) \, ,$$

so $\{Ax_{n_k}\}$ is also a Cauchy sequence.

3.3.2 Take any bounded sequence $\{x_n\} \subset X$. Since A is a compact operator, the sequence $\{Ax_n\}$ contains a Cauchy subsequence $\{Ax_{n_k}\}$. Since B is a compact operator, the sequence $\{Bx_{n_k}\}$ contains a Cauchy subsequence $\{B_{n_{k_l}}\}$. Therefore the sequence $\{x_n\}$ contains a subsequence $\{x_{n_{k_l}}\}$ whose images $\{A_{n_{k_l}}\}$ and $\{B_{n_{k_l}}\}$ are both Cauchy sequences. This means that $\{\alpha A_{n_{k_l}} + B_{n_{k_l}}\}$, which is the image of $\{x_{n_{k_l}}\}$ under C, is a Cauchy sequence. So we have shown that $\{Cx_n\}$ has a Cauchy subsequence. Therefore C is a compact linear operator.

3.3.3 In order to invoke Arzelà's theorem we first show that B takes $L^2(0,1)$ into $C(0,1)$. We have

$$\begin{aligned}|(Bf)(t) - (Bf)(t_0)| &\le \int_0^1 |K(t,s) - K(t_0,s)|\, |f(s)|\, ds \\ &\le \max_{s\in[0,1]} |K(t,s) - K(t_0,s)| \int_0^1 |f(s)|\, ds \le \max_{s\in[0,1]} |K(t,s) - K(t_0,s)|\, \|f\|_{L^2(0,1)}\end{aligned}$$

by application of Schwarz's inequality in the form

$$\int_0^1 1 \cdot |f(s)|\, ds \le \left(\int_0^1 1^2\, ds \right)^{1/2} \left(\int_0^1 |f(s)|^2\, ds \right)^{1/2} = \|f\|_{L^2(0,1)} \,.$$

By continuity of K we can make $|(Bf)(t) - (Bf)(t_0)|$ as small as desired for sufficiently small $|t - t_0|$, uniformly with respect to s.

Following the argument in the text, we show that B takes the unit ball of $L^2(0,1)$ into a precompact subset S of $C(0,1)$. First S is uniformly bounded: by the inequality displayed above we have

$$\|(Bf)(t)\|_{C(0,1)} \le \max_{t\in[0,1]} \int_0^1 |K(t,s)|\,|f(s)|\, ds \le M\, \|f\|_{L^2(0,1)} \,,$$

where

$$M = \max_{t,s\in[0,1]} |K(t,s)| \,.$$

Equicontinuity follows from the inequality

$$|(Bf)(t+\delta) - (Bf)(t)| \le \max_{s\in[0,1]} |K(t+\delta,s) - K(t,s)|$$

on the unit ball in $L^2(0,1)$, and the uniform continuity of K.

Finally, we observe that a precompact set in $C(0,1)$ is precompact in $L^2(0,1)$. Indeed, if S is precompact in $C(0,1)$ then every sequence $\{f_j\} \subset S$ contains a Cauchy subsequence $\{f_{j_k}\}$: for every $\varepsilon > 0$ there exists N such that

$$\|f_{j_m} - f_{j_n}\|_{C(0,1)} = \max_{x\in[0,1]} |f_{j_m}(x) - f_{j_n}(x)| < \varepsilon \quad \text{for } m,n > N \,.$$

Then $\{f_{j_k}\}$ is also a Cauchy sequence in $L^2(0,1)$:

$$\|f_{j_m} - f_{j_n}\|_{L^2(0,1)} = \left(\int_0^1 |f_{j_m}(x) - f_{j_n}(x)|^2\, dx \right)^{1/2} \le \varepsilon \quad \text{for } m,n > N \,.$$

3.3.4 Clearly A is a linear operator; we recall that it is the imbedding operator from $C^{(1)}(a,b)$ to $C(a,b)$. A consequence of Arzelà's theorem is that a bounded set in $C^{(1)}(a,b)$ is precompact in $C(a,b)$. So A is compact.

3.3.5

$$\|A_n x\| \le \sum_{k=1}^{n} |\Phi_k(x)|\, \|y_k\| \le \left(\sum_{k=1}^{n} \|\Phi_k\|\, \|y_k\| \right) \|x\| \,.$$

3.3.6 Two preparatory steps are in order. First, consider the same integral operator A, but acting in the space $C(\Omega)$ and assuming that

$$K(\mathbf{x},\mathbf{y}) \in C(\Omega \times \Omega) \,. \qquad (*)$$

By a development analogous to that of the example on p. 185, show that it is compact. Then, as in Problem 3.3.3, show that the same integral operator is also compact in $L^2(\Omega)$, provided (*) still holds.

The rest of the solution makes use of Theorem 3.3.2 and is analogous to the example on p. 187. The estimate

$$||Af||_{L^2(\Omega)} \leq \left(\int_\Omega \int_\Omega |K(\mathbf{x},\mathbf{y})|^2 \, d\Omega_\mathbf{y} \, d\Omega_\mathbf{x} \right)^{1/2} ||f||_{L^2(\Omega)} , \qquad (**)$$

obtained using the Schwarz inequality, shows that A is continuous and that

$$||A|| \leq \left(\int_\Omega \int_\Omega |K(\mathbf{x},\mathbf{y})|^2 \, d\Omega_\mathbf{y} \, d\Omega_\mathbf{x} \right)^{1/2} .$$

To show that A is compact, we express it as a norm limit of a sequence $\{A_n\}$ of compact operators. Since $L^2(\Omega\times\Omega)$ is the completion of the set of $C(\Omega\times\Omega)$ functions in the L^2-norm, there is a sequence of kernels

$$K_n(\mathbf{x},\mathbf{y}) \in C(\Omega \times \Omega)$$

such that

$$\left(\int_\Omega \int_\Omega |K_n(\mathbf{x},\mathbf{y}) - K(\mathbf{x},\mathbf{y})|^2 \, d\Omega_\mathbf{y} \, d\Omega_\mathbf{x} \right)^{1/2} \to 0 \text{ as } n \to \infty . \qquad (***)$$

Define the compact operator A_n by

$$A_n f = \int_\Omega K_n(\mathbf{x},\mathbf{y}) f(\mathbf{y}) \, d\Omega .$$

Applying (**) to $A - A_n$, we have

$$||(A - A_n)f||_{L^2(\Omega)} \leq \left(\int_\Omega \int_\Omega |K(\mathbf{x},\mathbf{y}) - K_n(\mathbf{x},\mathbf{y})|^2 \, d\Omega_\mathbf{y} \, d\Omega_\mathbf{x} \right)^{1/2} ||f||_{L^2(\Omega)}$$

so that

$$||A - A_n|| \leq \left(\int_\Omega \int_\Omega |K(\mathbf{x},\mathbf{y}) - K_n(\mathbf{x},\mathbf{y})|^2 \, d\Omega_\mathbf{y} \, d\Omega_\mathbf{x} \right)^{1/2}$$

and $||A - A_n|| \to 0$ by (***).

3.6.1 For the coordinate operator Q, the operator $R(\lambda, Q)$ maps a function $f \in L^2(a,b)$ to a function $u \in L^2(a,b)$ such that

$$(\lambda - t)u(t) = f(t) .$$

First, $D(R(\lambda, Q)) \neq L^2(a,b)$. Indeed, $L^2(a,b)$ contains constant functions, but functions of the form $c/(\lambda - t)$ are not square integrable on $[a,b]$ when $\lambda \in (a,b)$. So constant functions are excluded from $D(R(\lambda, Q))$. However, $D(R(\lambda, Q))$ is dense in $L^2(a,b)$. To see this, we can take any $f \in L^2(a,b)$ and consider the sequence $\{f_n\} \subset L^2(a,b)$ given by

$$f_n(t) = \begin{cases} 0\,, & |t-\lambda| < 1/n\,, \\ f(t)\,, & \text{otherwise}\,. \end{cases}$$

It is easily verified that $f_n \in D(R(\lambda, Q))$ and $f_n \to f$. Hence the f_n can be used to approximate an arbitrary f in the norm of $L^2(a,b)$.

To show that $R(\lambda, Q)$ is not bounded, take a sequence $\{u_n\} \subset L^2(a,b)$ given by

$$u_n(t) = \begin{cases} \sqrt{n}\,, & |t-\lambda| < 1/2n\,, \\ 0\,, & \text{otherwise}\,. \end{cases}$$

We have

$$\|u_n\|_{L^2(a,b)} = 1 \text{ for all } n\,,$$

while the corresponding preimages

$$f_n(t) = (\lambda - t)u_n(t)$$

have norms

$$\|f_n\|_{L^2(a,b)} = \left(\int_{\lambda-1/2n}^{\lambda+1/2n} n(\lambda-t)^2\,dt \right)^{1/2} = \frac{1}{2\sqrt{3}n} \to 0 \text{ as } n \to \infty\,.$$

Therefore there is no constant m, independent of f, such that

$$\|u\|_{L^2(a,b)} \le m\,\|f\|_{L^2(a,b)} \text{ for all } f \in D(R(\lambda, Q))\,.$$

3.6.2 Recall that the null space of an invertible linear operator consists solely of the zero vector. Hence if $\lambda = 0$ in the equation $Ax = \lambda x$ where A^{-1} exists, then $x = 0$ and hence is not an eigenvector. So we can assume $\lambda \neq 0$. Let x be an eigenvector of A corresponding to λ. Operating on both sides of the equation $Ax = \lambda x$ with A^{-1}, we get

$$A^{-1}x = \lambda^{-1}x\,.$$

This shows that x is an eigenvector of A^{-1}, corresponding to the eigenvalue λ^{-1}.

3.6.3 The eigenvalue $\lambda = 1$ corresponds to the eigenspace of even functions, while $\lambda = -1$ corresponds to the eigenspace of odd functions.

3.7.1 We have

$$[(\lambda-\lambda_0)I+(\lambda_0 I-A)]\cdot R(\lambda_0,A)\left\{I+\sum_{n=1}^{\infty}(\lambda_0-\lambda)^n R^n(\lambda_0,A)\right\}$$
$$=[I-(\lambda_0-\lambda)R(\lambda_0,A)]\cdot\left\{I+\sum_{n=1}^{\infty}(\lambda_0-\lambda)^n R^n(\lambda_0,A)\right\}$$
$$=I-(\lambda_0-\lambda)R(\lambda_0,A)+\sum_{n=1}^{\infty}(\lambda_0-\lambda)^n R^n(\lambda_0,A)$$
$$-(\lambda_0-\lambda)R(\lambda_0,A)\sum_{n=1}^{\infty}(\lambda_0-\lambda)^n R^n(\lambda_0,A)$$
$$=I+\sum_{n=2}^{\infty}(\lambda_0-\lambda)^n R^n(\lambda_0,A)-\sum_{n=1}^{\infty}(\lambda_0-\lambda)^{n+1}R^{n+1}(\lambda_0,A)=I\,.$$

3.7.2

$$(\lambda I-B)\cdot\frac{1}{\lambda}\Big(I+\sum_{n=1}^{\infty}\lambda^{-n}B^n\Big)=(I-\lambda^{-1}B)\cdot\Big(I+\sum_{n=1}^{\infty}\lambda^{-n}B^n\Big)$$
$$=I-\lambda^{-1}B+\sum_{n=1}^{\infty}\lambda^{-n}B^n-\lambda^{-1}B\sum_{n=1}^{\infty}\lambda^{-n}B^n$$
$$=I+\sum_{n=2}^{\infty}\lambda^{-n}B^n-\sum_{n=1}^{\infty}\lambda^{-(n+1)}B^{n+1}=I\,.$$

3.8.1 By linearity, any finite linear combination of $x_1,\ldots,x_m\in H(\mu_0)$ also belongs to $H(\mu_0)$:

$$(I-\mu_0A)\Big(\sum_{i=1}^{m}\alpha_i x_i\Big)=\sum_{i=1}^{m}\alpha_i(I-\mu_0A)x_i=0\,.$$

By continuity of A, the limit $\bar{x}$ of any convergent sequence $\{x_n\}\subset H(\mu_0)$ lies in $H(\mu_0)$:

$$(I-\mu_0A)\bar{x}=(I-\mu_0A)\lim_{n\to\infty}x_n=\lim_{n\to\infty}(I-\mu_0A)x_n=0\,.$$

3.11.1 We have

$$\frac{d}{d\alpha}\left\{\frac{(Ax_0+\alpha Ay,x_0+\alpha y)}{(x_0+\alpha y,x_0+\alpha y)}\right\}\bigg|_{\alpha=0}$$
$$=\frac{d}{d\alpha}\left\{\frac{(Ax_0,x_0)+\alpha\operatorname{Re}(Ax_0,y)+\alpha^2(Ay,y)}{(x_0,x_0)+\alpha\operatorname{Re}(x_0,y)+\alpha^2(y,y)}\right\}\bigg|_{\alpha=0}$$
$$=\frac{(x_0,x_0)\operatorname{Re}(Ax_0,y)-(Ax_0,x_0)\operatorname{Re}(x_0,y)}{(x_0,x_0)^2}=0$$

giving

$$\frac{(x_0, x_0)\operatorname{Re}(Ax_0, y) - (Ax_0, x_0)\operatorname{Re}(x_0, y)}{(x_0, x_0)} = 0$$

hence

$$\operatorname{Re}(Ax_0, y) - \frac{(Ax_0, x_0)\operatorname{Re}(x_0, y)}{(x_0, x_0)} = 0 .$$

3.11.2 Let $\{g_k\}$ be an orthonormal system in a Hilbert space H. Suppose Parseval's equality

$$\sum_{k=1}^{\infty} |(f, g_k)|^2 = \|f\|^2$$

holds for all $f \in H$. Fix f and write

$$\left\| f - \sum_{k=1}^{\infty} (f, g_k) g_k \right\|^2 = \lim_{n\to\infty} \left\| f - \sum_{k=1}^{n} (f, g_k) g_k \right\|^2 = \lim_{n\to\infty} \left(\|f\|^2 - \sum_{k=1}^{n} |(f, g_k)|^2 \right) = 0 .$$

This shows that

$$f = \sum_{k=1}^{\infty} \alpha_k g_k \quad \text{where} \quad \alpha_k = (f, g_k) .$$

4.1.1 (a) Given $\mathbf{f}\colon \mathbb{R}^m \to \mathbb{R}^n$, we wish to examine the difference $\mathbf{f}(\mathbf{x}_0+\mathbf{h})-\mathbf{f}(\mathbf{x}_0)$. Let us introduce the standard orthonormal bases of $\mathbb{R}^m$ and $\mathbb{R}^n$, respectively: $\tilde{\mathbf{e}}_1, \ldots, \tilde{\mathbf{e}}_m$ and $\mathbf{e}_1, \ldots, \mathbf{e}_n$. Then

$$\mathbf{f}(\mathbf{x}) = \sum_{i=1}^{n} f_i(\mathbf{x})\, \mathbf{e}_i$$

and we have

$$\mathbf{f}(\mathbf{x}_0 + \mathbf{h}) - \mathbf{f}(\mathbf{x}_0) = \sum_{i=1}^{n} [f_i(\mathbf{x}_0 + \mathbf{h}) - f_i(\mathbf{x}_0)]\, \mathbf{e}_i .$$

But Taylor expansion to first order gives, for each i,

$$f_i(\mathbf{x}_0 + \mathbf{h}) - f_i(\mathbf{x}_0) = \sum_{j=1}^{m} \frac{\partial f_i(\mathbf{x}_0)}{\partial x_j} h_j + o(\|\mathbf{h}\|) ,$$

where the h_j are the components of $\mathbf{h}$:

$$\mathbf{h} = \sum_{j=1}^{m} h_j \tilde{\mathbf{e}}_j .$$

So we identify

$$\mathbf{f}'(\mathbf{x}_0)(\mathbf{h}) = \sum_{i=1}^{n} \mathbf{e}_i \sum_{j=1}^{m} \frac{\partial f_i(\mathbf{x}_0)}{\partial x_j} h_j ,$$

and observe that the right-hand side is represented in matrix-vector notation as

$$\begin{pmatrix} \frac{\partial f_1(\mathbf{x}_0)}{\partial x_1}h_1 + \cdots + \frac{\partial f_1(\mathbf{x}_0)}{\partial x_m}h_m \\ \vdots \\ \frac{\partial f_n(\mathbf{x}_0)}{\partial x_1}h_1 + \cdots + \frac{\partial f_n(\mathbf{x}_0)}{\partial x_m}h_m \end{pmatrix} = \begin{pmatrix} \frac{\partial f_1(\mathbf{x}_0)}{\partial x_1} & \cdots & \frac{\partial f_1(\mathbf{x}_0)}{\partial x_m} \\ \vdots & \ddots & \vdots \\ \frac{\partial f_n(\mathbf{x}_0)}{\partial x_1} & \cdots & \frac{\partial f_n(\mathbf{x}_0)}{\partial x_m} \end{pmatrix} \begin{pmatrix} h_1 \\ \vdots \\ h_m \end{pmatrix}.$$

(b) Suppose F is a bounded linear operator. Then

$$F(x_0 + h) - F(x_0) = F(x_0 + h - x_0) = Fh\,,$$

and comparison with Definition 4.1.1 shows that $F = F'(x_0)$.

4.1.3 The desired inequality follows from (4.1.2): replace both x and $K(x,\mu)$ by $x(\mu)$ to get

$$\|x(\mu) - x_0\| \le (q + \varepsilon)\,\|x(\mu) - x_0\| + \|A^{-1}\|\,\|F(x_0,\mu)\|\,.$$

4.1.4 Let us examine the behavior of the quantity

$$R = \frac{\|F'(x_0)\omega_1(z_0, z - z_0) + \omega(x_0, S(z) - S(z_0))\|_Y}{\|z - z_0\|_Z}$$

as $\|z - z_0\|_Z \to 0$. By the triangle inequality and the fact that $F'(x_0)$ is a bounded linear operator,

$$R \le \|F'(x_0)\|\frac{\|\omega_1(z_0, z - z_0)\|_X}{\|z - z_0\|_Z} + \frac{\|\omega(x_0, S(z) - S(z_0))\|_Y}{\|z - z_0\|_Z}\,.$$

We see that the first term on the right has limit zero. The second term can be rewritten as

$$\frac{\|S(z) - S(z_0)\|_X}{\|z - z_0\|_Z} \cdot \frac{\|\omega(x_0, S(z) - S(z_0))\|_Y}{\|S(z) - S(z_0)\|_X}$$

The second factor has limit zero, but we still need to show that the first factor is bounded. This fact follows from the equation

$$S(z) - S(z_0) = S'(z_0)(z - z_0) + \omega_1(z_0, z - z_0)$$

as follows:

$$\begin{aligned} \|S(z) - S(z_0)\|_X &= \|S'(z_0)(z - z_0) + \omega_1(z_0, z - z_0)\|_X \\ &\le \|S'(z_0)\|\,\|z - z_0\|_Z + \|\omega_1(z_0, z - z_0)\|_X\,, \end{aligned}$$

therefore

$$\frac{\|S(z) - S(z_0)\|_X}{\|z - z_0\|_Z} \le \|S'(z_0)\| + \frac{\|\omega_1(z_0, z - z_0)\|_X}{\|z - z_0\|_Z}\,.$$

4.3.1 For the functional $G(x) = \|x\|^2$, the term $\operatorname{grad} G(x_0)$ is calculated as follows:

$$
\begin{aligned}
\frac{d}{dt}\|x_0 + th\|^2\Big|_{t=0} &= \frac{d}{dt}(x_0 + th, x_0 + th)\Big|_{t=0} \\
&= \frac{d}{dt}[(x_0, x_0) + 2t(x_0, h) + t^2(h, h)]\Big|_{t=0} \\
&= [2(x_0, h) + 2t(h, h)]\Big|_{t=0} \\
&= (2x_0, h) \\
&= (\operatorname{grad} G(x_0), h) .
\end{aligned}
$$

References

1. Adams, R.A., and Fournier, J.H.F. *Sobolev Spaces*. 2nd ed. Elsevier B.V., Amsterdam, 2003.
2. Antman, S.S. The influence of elasticity on analysis: modern developments, Bull. Amer. Math. Soc. (New Series), 1983, 9, 267–291.
3. Antman, S.S. *Nonlinear Problems of Elasticity*. Springer-Verlag, New York, 1996.
4. Banach, S. *Théories des opérations linéaires*. Chelsea Publishing Company, New York, 1978 Translated as *Theory of linear operations*. North Holland, Amsterdam, 1987.
5. Bramble, J.H., and Hilbert, S.R. Bounds for a class of linear functionals with applications to Hermite interpolation. Numer. Math., 1971, 16, 362–369.
6. Brezis, H. *Functional Analysis, Sobolev Spaces and Partial Differential Equations*. Springer, NY, 2011.
7. Ciarlet, P.G. *The Finite Element Method for Elliptic Problems*. North Holland Publ. Company, 1978.
8. Ciarlet, P.G. *Mathematical Elasticity*, vol. 1–3. North Holland, 1988–2000.
9. Courant, R., and Hilbert, D. *Methods of Mathematical Physics*. Interscience Publishers, New York, 1953–62.
10. Destuynder, Ph.,and Salaun M. *Mathematical Analysis of Thin Plate Models*. Springer, Paris, 1996.
11. Fichera, G. Existence theorems in elasticity (XIII.15), and Boundary value problems of elasticity with unilateral constraints (YII.8, XIII.15, XIII.6), in Handbuch der Physik YIa/2, C. Truesdell, ed., Springer-Verlag, 1972.
12. Friedrichs, K.O. The identity of weak and strong extensions of differential operators. Trans. Amer. Math. Soc., 1944, vol. 55, pp. 132–151.
13. Gokhberg, I.Ts, and Krejn, M.G. *Theory of the Volterra Operators in Hilbert Space and Its Applications*. Nauka, Moscow, 1967.
14. Hardy, G.H., Littlewood, J.E., and Pólya, G. *Inequalities*. Cambridge University Press, 1952.
15. Hutson, V., Pym, J.S., and Cloud, M.J. *Applications of Functional Analysis and Operator Theory* (2nd ed). Elsevier, 2005.
16. Il'yushin, A.A. *Plasticity*. Gostekhizfat, Moscow, 1948 (in Russian).
17. Lax, P.D. *Functional Analysis*. Wiley –Interscience, 2002.
18. Lebedev, L.P., Vorovich, I.I., and Gladwell, G.M.L. *Functional Analysis: Applications in Mechanics and Inverse Problems*. Kluwer Academic Publishers, Dordrecht, 1996.
19. Lebedev, L.P., Cloud, M.J., and Eremeyev, V.A. *Advanced Engineering Analysis: The Calculus of Variations and Functional Analysis with Applications in Mechanics*. World Scientific, 2012.
20. Leray, J., and Schauder, J. Topologie et équations fonctionnelles. Ann. S.E.N., 1934, 51, 45–78.
21. Kantorovich, L.V., and Akilov, G.P. *Functional Analysis*. Pergamon, 1982.

22. Ladyzhenskya, O.A. *The Mathematical Theory of Viscous Incompressible Flow*. 2nd ed. Gordon and Breach, 1964.
23. Lax, P.D., and Milgram, A.N. "Parabolic equations" in *Contributions to the Theory of Partial Differential Equations*. Princeton, 1954.
24. Lebedev, L.P., and Cloud, M.J. *Introduction to Mathematical Elasticity*. World Scientific, 2009.
25. Lions, J.-L., and Magenes, E. Problèmes aux Limites Non Homogènes et Applications, Tome 1. Dunod, Paris, 1968.
26. Mikhlin, S.G. *The Problem of Minimum of a Quadratic Functional*. Holden–Day, San Francisco, 1965.
27. Mikhlin, S.G. *Variational Methods in Mathematical Physics*. Pergamon Press, Oxford, 1964.
28. Moore, R.E., and Cloud, M.J. *Computational Functional Analysis* (2nd ed). Ellis Horwood, 2007.
29. Neumann von, J. *Mathematical Foundations of Quantum Mechanics*. Princeton University Press, Princeton, 1996.
30. Riesz, F. and Sz.-Nagy, B. *Functional Analysis*. Dover Publications, Inc., NY, 1990.
31. Schechter, M. *Principles of Functional Analysis*, 2nd edition. American Mathematical Society, Providence, Rhode Island, 2002.
32. Schwartz, J.T. *Nonlinear Functional Analysis*. Gordon and Breach Sc. Publ. Inc., 1969.
33. Sobolev, S.L. *Some Applications of Functional Analysis to Mathematical Physics*. LGU, 1951 (Translated as *Some Applications of Functional Analysis to Mathematical Physics*. 3rd edition. American Mathematical Society, 1991.
34. Struwe, M. *Variational Methods*, 2nd ed. Springer-Verlag, Berlin, 1996.
35. Temam, R. *Navier–Stokes Equations: Theory and Numerical Analysis*. AMS Chelsea Publishing, 2001.
36. Timoshenko, S.P., and Goudier, J.N. *Theory of Elasticity*. McGraw-Hill Book Company, 1986.
37. Trenogin, V.A. *Functional Analysis*. Fizmatlit, 2002.
38. Trenogin, V.A., Pisarevskii, B.M., and Sobolev, T.C. *Problems and Solutions in Functional Analysis*. Fizmatlit, 2002.
39. Vitt, A., and Shubin, S. On tones of a membrane fixed in a finite number of points. Zhurn. Tekhn. Fiz., I (1931), no. 2–3, 163–175.
40. Vorovich, I.I., and Krasovskij, Yu.P. On the method of elastic solutions. Doklady Akad. Nauk SSSR, 126 (1959), no. 4, 740–743.
41. Vorovich, I.I. *Mathematical Problems of Nonlinear Theory of Shallow Shells*. Nauka, Moscow, 1989. (Translated as *Nonlinear Theory of Shallow Shells*. Springer-Verlag, New York, 1999.)
42. Vorovich, I.I., and Yudovich, V.I. Steady flow of viscous incompressible liquid. Mat. Sbornik, 1961, vol. 53 (95), no. 4, 361–428.
43. Vorovich, I.I. The problem of non-uniqueness and stability in the nonlinear mechanics of continuous media, Applied Mechanics. Proc. Thirteenth Intern. Congr. Theor. Appl. Mech., Springer-Verlag, 1973, 340–357.
44. Yosida, K. *Functional Analysis*. Springer-Verlag, New York, 1965.
45. Zeidler, E. *Nonlinear Functional Analysis and Its Applications, Parts 1–4*. Springer-Verlag, New York, 1985–1988.

In Memoriam: Iosif I. Vorovich

Iosif Izrailevich Vorovich was born on 21 June 1920 in Starodub, a small town within Russia's Briansk district. He entered Moscow State University and studied in the mathematics and mechanics department, hearing lectures by famous mathematicians A.N. Kolmogorov, I.G. Petrovskij, S.L. Sobolev, N.N. Bari, V.V. Golubev, B.N. Delone, A.G. Kurosh, and V.V. Stepanov, and by well-known mechanicists L.S. Lejbenzon, Yu.N. Rabotnov, and A.A. Iliushin. Future academician A.Yu. Ishlinskij was his thesis professor. As part of his military service during World War II, he attended the Zhukovski Flight Academy; there he studied under V.S. Pugachov, one of the founders of control theory. His doctoral work on the nonlinear theory of shells (1958) later appeared as one of his nine books. For his many contributions to mechanics, I.I. Vorovich received the high honor of membership in the Russian Academy of Sciences. During his later years, he served as director of the Research Institute for Mechanics and Applied Mathematics at Rostov State University. Professor Vorovich lectured and wrote prolifically until the end of his life. He is remembered as a helpful mentor and a warm friend. He passed away on 6 September 2001.

Index